Lecture Notes in Bioinformatics 7292
Edited by S. Istrail, P. Pevzner, and M. Waterman

Editorial Board: A. Apostolico S. Brunak M. Gelfand
T. Lengauer S. Miyano G. Myers M.-F. Sagot D. Sankoff
R. Shamir T. Speed M. Vingron W. Wong

Subseries of Lecture Notes in Computer Science

Leonidas Bleris   Ion Măndoiu
Russell Schwartz   Jianxin Wang (Eds.)

Bioinformatics Research and Applications

8th International Symposium, ISBRA 2012
Dallas, TX, USA, May 21-23, 2012
Proceedings

 Springer

Series Editors

Sorin Istrail, Brown University, Providence, RI, USA
Pavel Pevzner, University of California, San Diego, CA, USA
Michael Waterman, University of Southern California, Los Angeles, CA, USA

Volume Editors

Leonidas Bleris
University of Texas at Dallas
Richardson, TX 75080, USA
E-mail: bleris@utdallas.edu

Ion Măndoiu
University of Connecticut
Storrs, CT 06269, USA
E-mail: ion@engr.uconn.edu

Russell Schwartz
Carnegie Mellon University
Pittsburgh, PA 15213, USA
E-mail: russells@andrew.cmu.edu

Jianxin Wang
Central South University
Changsha 410083, China
E-mail: jxwang@mail.csu.edu.cn

ISSN 0302-9743　　　　　　　　　　e-ISSN 1611-3349
ISBN 978-3-642-30190-2　　　　　　e-ISBN 978-3-642-30191-9
DOI 10.1007/978-3-642-30191-9
Springer Heidelberg Dordrecht London New York

Library of Congress Control Number: 2012937075

CR Subject Classification (1998): J.3, H.2.8, H.3-4, F.1, I.5, J.3

LNCS Sublibrary: SL 8 – Bioinformatics

© Springer-Verlag Berlin Heidelberg 2012
This work is subject to copyright. All rights are reserved, whether the whole or part of the material is concerned, specifically the rights of translation, reprinting, re-use of illustrations, recitation, broadcasting, reproduction on microfilms or in any other way, and storage in data banks. Duplication of this publication or parts thereof is permitted only under the provisions of the German Copyright Law of September 9, 1965, in its current version, and permission for use must always be obtained from Springer. Violations are liable to prosecution under the German Copyright Law.
The use of general descriptive names, registered names, trademarks, etc. in this publication does not imply, even in the absence of a specific statement, that such names are exempt from the relevant protective laws and regulations and therefore free for general use.

Typesetting: Camera-ready by author, data conversion by Scientific Publishing Services, Chennai, India

Printed on acid-free paper

Springer is part of Springer Science+Business Media (www.springer.com)

Preface

The 8th International Symposium on Bioinformatics Research and Applications (ISBRA 2012) was held during May 21–23, 2012, in Dallas, Texas. The symposium provides a forum for the exchange of ideas and results among researchers, developers, and practitioners working on all aspects of bioinformatics and computational biology and their applications.

The technical program of the symposium included 26 extended abstracts, selected by the Program Committee from a number of 66 full submissions received in response to the call for papers. Additionally, the symposium included poster sessions and featured invited keynote talks by five distinguished speakers: Ambuj Singh from the University of California at Santa Barbara spoke on learning discriminative graph fragments, Bhaskar Dasgupta from the University of Illinois at Chicago spoke on models and algorithmic tools for computational processes in cellular biology, Cynthia Gibas from the University of North Carolina at Charlotte spoke on analytics approaches for the era of 10,000 genomes, Dong Xu from the University of Missouri spoke on protein structure prediction and clustering, and Michael Zhang from the University of Texas at Dallas and Tsinghua University spoke on computational modeling of mammalian promoters.

We would like to thank the Program Committee members and external reviewers for volunteering their time to review ISBRA submissions. We would like to extend a special thanks to the Steering and General Chairs of the symposium for their leadership, and the ISBRA 2012 Publicity Chair, webmaster, and local organizers for their hard work in making ISBRA 2012 a successful event. Last but not least we would like to thank all authors for presenting their work at the symposium.

May 2012

Leonidas Bleris
Ion Măndoiu
Russell Schwartz
Jianxin Wang

Organization

Steering Chairs

Dan Gusfield	University of California Davis, USA
Ion Măndoiu	University of Connecticut, USA
Yi Pan	Georgia State University, USA
Marie-France Sagot	INRIA, France
Alexander Zelikovsky	Georgia State University, USA

General Chairs

Ovidiu Daescu	University of Texas at Dallas, USA
Raj Sunderraman	Georgia State University, USA

Program Chairs

Leonidas Bleris	University of Texas at Dallas, USA
Ion Măndoiu	University of Connecticut, USA
Russell Schwartz	Carnegie Mellon University, USA
Jianxin Wang	Central South University, China

Publicity Chair

Sahar Al Seesi	University of Connecticut, USA

Webmaster

J. Steven Kirtzic	University of Texas at Dallas, USA

Program Committee

Srinivas Aluru	IIT Bombay, India and Iowa State University, USA
Danny Barash	Ben-Gurion University, Israel
Robert Beiko	Dalhousie University, Canada
Anne Bergeron	Université du Québec à Montréal, Canada
Daniel Berrar	Tokyo Institute of Technology, USA
Paola Bonizzoni	Università de Studi di Milano-Bicocca, Italy
Daniel Brown	University of Waterloo, Canada
Doina Caragea	Kansas State University, USA

Tien-Hao Chang	National Cheng Kung University, Taiwan
Chien-Yu Chen	National Taiwan University, Taiwan
Matteo Comin	Università degli Studi di Padova, Italy
Ovidiu Daescu	University of Texas at Dallas, USA
Bhaskar Dasgupta	University of Illinois at Chicago, USA
Douglas Densmore	Boston University, USA
Jorge Duitama	International Center for Tropical Agriculture, Colombia
Oliver Eulenstein	Iowa State University, USA
Guillaume Fertin	University of Nantes, France
Vladimir Filkov	University of California Davis, USA
Jean Gao	University of Texas at Arlington, USA
Katia Guimaraes	Universidade Federal de Pernambuco, Brazil
Jiong Guo	Universität des Saarlandes, Germany
Robert Harrison	Georgia State University, USA
Jieyue He	Southeast University, China
Steffen Heber	North Carolina State University, USA
Allen Holder	Rose-Hulman Institute of Technology, USA
Jinling Huang	East Carolina University, USA
Lars Kaderali	University of Technology Dresden, Germany
Iyad Kanj	DePaul University, USA
Ming-Yang Kao	Northwestern University, USA
Yury Khudyakov	Centers for Disease Control and Prevention, USA
Danny Krizanc	Wesleyan University, USA
Jing Li	Case Western Reserve University, USA
Yanchun Liang	Jilin University, China
Zhiyong Liu	Chinese Academy of Science, China
Fenglou Mao	University of Georgia, USA
Osamu Maruyama	Kyushu University, Japan
Li Min	Central South University, China
Ion Moraru	University of Connecticut Health Center, USA
Axel Mosig	CAS-MPG Partner Institute and Key Laboratory for Computational Biology, China and Ruhr University Bochum, Germany
Giri Narasimhan	Florida International University, USA
Yi Pan	Georgia State University, USA
Laxmi Parida	IBM T.J. Watson Research Center, USA
Bogdan Pasaniuc	Harvard School of Public Health, USA
Andrei Paun	Louisiana Tech University, USA
Itsik Pe'Er	Columbia University, USA
Weiqun Peng	George Washington University, USA
Nadia Pisanti	Università di Pisa, Italy

Maria Poptsova	University of Connecticut, USA
Teresa Przytycka	NCBI, USA
Sven Rahmann	University of Duisburg-Essen, Germany
Shoba Ranganathan	Macquarie University, Australia
S. Cenk Sahinalp	Simon Fraser University, Canada
David Sankoff	University of Ottawa, Canada
Daniel Schwartz	University of Connecticut, USA
Joao Setubal	Virginia Bioinformatics Institute, USA
Mona Singh	Princeton University, USA
Ileana Streinu	Smith College, USA
Raj Sunderraman	Georgia State University, USA
Wing-Kin Sung	National University of Singapore, Singapore
Sing-Hoi Sze	Texas A&M University, USA
Ilias Tagkopoulos	University of California Davis, USA
Marcel Turcotte	University of Ottawa, Canada
Gabriel Valiente	Technical University of Catalonia, Spain
Stéphane Vialette	Université Paris-Est Marne-la-Vallée, France
Panagiotis Vouzis	Carnegie Mellon University, USA
Jianxin Wang	Central South University, China
Li-San Wang	University of Pennsylvania, USA
Lusheng Wang	City University of Hong Kong, Hong Kong
Xiaowo Wang	Tsinghua University, China
Fangxiang Wu	University of Saskatchewan, Canada
Yufeng Wu	University of Connecticut, USA
Zhen Xie	Massachusetts Institute of Technology, USA
Jinbo Xu	Toyota Technical Institute at Chicago, USA
Zhenyu Xuan	University of Texas at Dallas, USA
Zuguo Yu	Xiangtan University, China and Queensland University of Technology, Australia
Alex Zelikovsky	Georgia State University, USA
Fa Zhang	Institute of Computing Technology, China
Yanqing Zhang	Georgia State University, USA
Leming Zhou	University of Pittsburgh, USA

Additional Reviewers

Bacciu, Davide
Badr, Ghada
Belcaid, Mahdi
Bernauer, Julie
Biswas, Ashis Kumer
Bulteau, Laurent
Caldas, José
Campo, David S.
Cliquet, Freddy

Della Vedova, Gianluca
Dondi, Riccardo
Du, Wei
Fang, Ming
Fox, Naomi
Guan, Renchu
Guillemot, Sylvain
Haiminen, Niina
Hayes, Matthew

Huang, Fangping
Huang, Yang
Hwang, Yih-Chii
Kang, Mingon
Kim, Yoo-Ah
Kopczynski, Dominik
Köster, Johannes
Labarre, Anthony
Lacroix, Vincent
Li, Fan
Li, Shuo
Liu, Bingqiang
Mendes, Nuno
Messina, Enza
Mina, Marco
Missirian, Victor
Mozhayskiy, Vadim
Nefedov, Alexey

Pyon, Yoon Soo
Rizzi, Raffaella
Ryvkin, Paul
Scornavacca, Celine
Skums, Pavel
Srichandan, Bismita
Thorne, Jeffrey
Tsirigos, Aristotelis
Utro, Filippo
Verzotto, Davide
Wang, Juexin
Wang, Lin
Wang, Yan
Wojtowicz, Damian
Yang, Sen
Zhang, Jin
Zhou, Chan
Zola, Jaroslaw

Table of Contents

Trie-based Apriori Motif Discovery Approach 1
   Isra Al-Turaiki, Ghada Badr, and Hassan Mathkour

Inapproximability of (1, 2)-Exemplar Distance 13
   Laurent Bulteau and Minghui Jiang

A Mixed Integer Programming Model for the Parsimonious Loss of
Heterozygosity Problem 24
   Daniele Catanzaro, Martine Labbé, and Bjarni V. Halldórsson

Reconstruction of Transcription Regulatory Networks by
Stability-Based Network Component Analysis 36
   Xi Chen, Chen Wang, Ayesha N. Shajahan, Rebecca B. Riggins, Robert Clarke, and Jianhua Xuan

A Transcript Perspective on Evolution 48
   Yann Christinat and Bernard M.E. Moret

A Fast Algorithm for Computing the Quartet Distance for Large Sets
of Evolutionary Trees 60
   Ralph W. Crosby and Tiffani L. Williams

Error Propagation in Sparse Linear Systems with Peptide-Protein
Incidence Matrices 72
   Peter Damaschke and Leonid Molokov

Models and Algorithmic Tools for Computational Processes in
Cellular Biology: Recent Developments and Future Directions
(Invited Keynote Talk) 84
   Bhaskar DasGupta

Identifying Rogue Taxa through Reduced Consensus: NP-Hardness and
Exact Algorithms 87
   Akshay Deepak, Jianrong Dong, and David Fernández-Baca

Analytics Approaches for the Era of 10,000 Genomes
(Invited Keynote Talk) 99
   Cynthia J. Gibas

GTP Supertrees from Unrooted Gene Trees: Linear Time Algorithms
for NNI Based Local Searches 102
   Paweł Górecki, J. Gordon Burleigh, and Oliver Eulenstein

A Robinson-Foulds Measure to Compare Unrooted Trees with Rooted
Trees 115
    Paweł Górecki and Oliver Eulenstein

P-Binder: A System for the Protein-Protein Binding Sites
Identification 127
    Fei Guo, Shuai Cheng Li, and Lusheng Wang

Non-identifiable Pedigrees and a Bayesian Solution 139
    Bonnie Kirkpatrick

Iterative Piecewise Linear Regression to Accurately Assess Statistical
Significance in Batch Confounded Differential Expression Analysis 153
    Juntao Li, Kwok Pui Choi, and R. Krishna Murthy Karuturi

Reconstruction of Network Evolutionary History from Extant Network
Topology and Duplication History 165
    Si Li, Kwok Pui Choi, Taoyang Wu, and Louxin Zhang

POPE: Pipeline of Parentally-Biased Expression 177
    Victor Missirian, Isabelle Henry, Luca Comai, and Vladimir Filkov

On Optimizing the Non-metric Similarity Search in Tandem Mass
Spectra by Clustering 189
    Jiří Novák, David Hoksza, Jakub Lokoč, and Tomáš Skopal

On the Comparison of Sets of Alternative Transcripts 201
    Aïda Ouangraoua, Krister M. Swenson, and Anne Bergeron

MURPAR: A Fast Heuristic for Inferring Parsimonious Phylogenetic
Networks from Multiple Gene Trees 213
    Hyun Jung Park and Luay Nakhleh

Large Scale Ranking and Repositioning of Drugs with Respect to
DrugBank Therapeutic Categories 225
    Matteo Re and Giorgio Valentini

Score Based Aggregation of microRNA Target Orderings 237
    Debarka Sengupta, Ujjwal Maulik, and Sanghamitra Bandyopadhyay

Modeling Complex Diseases Using Discriminative Network Fragments
(Invited Keynote Talk) 249
    Ambuj K. Singh

Novel Multi-sample Scheme for Inferring Phylogenetic Markers from
Whole Genome Tumor Profiles 250
    Ayshwarya Subramanian, Stanley Shackney, and Russell Schwartz

Algorithms for Knowledge-Enhanced Supertrees 263
    André Wehe, J. Gordon Burleigh, and Oliver Eulenstein

Improvement of BLASTp on the FPGA-Based High-Performance
Computer RIVYERA 275
   Lars Wienbrandt, Daniel Siebert, and Manfred Schimmler

A Polynomial Time Solution for Protein Chain Pair Simplification
under the Discrete Fréchet Distance 287
   Tim Wylie and Binhai Zhu

Designing RNA Secondary Structures in Coding Regions 299
   Rukhsana Yeasmin and Steven Skiena

Phylogenetic Tree Reconstruction with Protein Linkage 315
   *Junjie Yu, Henry Chi Ming Leung, Siu Ming Yiu, Yong Zhang,
Francis Y.L. Chin, Nathan Hobbs, and Amy Y.X. Wang*

Protein Structure Prediction and Clustering Using MUFOLD
(Invited Keynote Talk) 328
   Jingfen Zhang and Dong Xu

Computational Modeling of Mammalian Promoters
(Invited Keynote Talk) 330
   Michael Q. Zhang

Author Index 331

Trie-based Apriori Motif Discovery Approach

Isra Al-Turaiki, Ghada Badr, and Hassan Mathkour

King Saud University, College of Computer and Information Sciences,
Riyadh, Kingdom of Saudi Arabia
{ialturaiki,mathkour}@ksu.edu.sa, badrghada@hotmail.com

Abstract. One of the hardest and long-standing problems in Bioinformatics is the problem of motif discovery in biological sequences. It is the problem of finding recurring patterns in these sequences. Apriori is a well-known data mining algorithm. It is used to mine frequent patterns in large datasets. In this paper, we would like to apply Apriori to the common motifs discovery problem. We propose three modifications so that we can adapt the classic Apriori to our problem. First, the *Trie* data structure is used to store all biological sequences under examination. Second, both of the frequent pattern extraction and the candidate generation steps are done using the same data structure, the *Trie*. The Trie allows to simultaneously search all possible starting points in the sequence for any occurrence of the given pattern. Third, instead of using only the support as a measure to assess frequent patterns, a new measure, the *normalized information content* (normIC), is proposed which is able to distinguish motifs in real promoter sequences. Preliminary experiments are conducted on Tompa's benchmark to investigate the performance of our proposed algorithm, *the Trie-based Apriori Motif Discovery* (TrieAMD). Results show that our algorithm outperforms all of the tested tools on real datasets for average sensitivity.

Keywords: Bioinformatics, motif discovery, Apriori, DNA, Transcription factor binding site, data mining,Tries, Suffix trees.

1 Introduction

Finding recurring patterns, *motifs*, in biological sequences gives an indication of important functional or structural roles. Motif discovery is one of the hardest problems in Bioinformatics.

Motifs can be structural or sequential patterns. Motifs are structural when they represent patterns for RNA secondary structures [3,6]. Motifs are represented as sequences when they represent repeated patterns in sequences. It has been proven that the sequential motif discovery problem is an NP-Hard [9]. Sequential motifs can be either single or composite. Single motifs correspond to ordered residues conserved during evolution. Composite motifs are motifs composed of two or more parts separated by a constrained spacer [22]. There are two types of the sequential motif discovery problem: (1) *the repeated motifs problem* [7,16] and (2) *the common motifs problem* [15]. In the repeated motifs problem,

S1: GAAGATCC   S2:GAAAAGCT   S3:GAAGACCA   S4:CCCGATAT   S5:GAAGAAAG
Common motif: AAGA

Fig. 1. Example of five sequences along with the discovered motif for q=4

it is required to find at least a motif that occurs at least q times in a given sequence. A word is said to be *a motif occurrence* if the hamming distance between the motif and the word is less than or equal to a given error rate. In the common motifs problem, we are given a set of N sequences and we want to find at least a motif that occurs in a percentage of at least q of the given distinct sequences. In this research, we focus on discovering single sequential common motifs in the given biological sequences. Figure 1 shows an example for five sequences along with the corresponding discovered common motifs for $q = 80\%$.

An important application of a motif discovery is the identification of the regulatory elements such as *Transcription Factor Binding Sites* (TFBS) in the promoters of coagulated genes. TFBSs are short stretches of DNA where certain proteins called *Transcription Factors* (TF) bind to and cause genes to be expressed. Thus, the identification of such binding sites is an important step towards the understanding of the gene regulation. Although many TFs are known, binding sites have been fully characterized for only few of them in databases such as TRANSFAC [10] and JASPAR [17]. Furthermore, known TFBSs were derived experimentally using methods that are expensive and time consuming. Computational approaches have been developed over the past two decades for identifying novel binding sites. However, the process of motif discovery is not an easy task. This is because binding sites that are recognized by transcription factors are short DNA stretches (5-15 bp). Most motifs can occur by chance throughout the genome. Hence, distinguishing true from false binding sites in very difficult. Moreover, binding sites tolerate some degree of variability which is not well-understood. The choice of the motif model, the presence of bias in the genome composition, and the existence of noise in the experimentally generated sequences are some factors that contribute to the difficulty of the motif discovery problem.

1.1 Problem Definition

In this paper, we present an algorithm to solve the common motifs problem. The problem is defined as follows: Given a set of N sequences $S = s_1, s_2, ...s_N$ and two integers $e \geq 0$ and $2 \leq q \leq N$, it is required to find all motifs that are present in at least q sequences such that any motif occurs with at most e mismatches.

1.2 Contribution

A Trie-based Apriori Motif Discovery (TrieAMD) algorithm is proposed to solve the common motifs problem. The new approach is inspired by the classic Apriori algorithm, a well-known algorithm in data mining, with some modifications.

We use the Trie data structure for storing sequences as well as for extracting frequent patterns and for generating candidates. Combined with the classic support measure that is used for candidate selection in the Apriori algorithm, a new measure, the *normalized information content* (normIC), is proposed. The new measure is able to distinguish motifs in real promoter sequences. Preliminary experiments are conducted to test the sensitivity, specificity, and performance coefficient of TrieAMD as compared with other 13 well-known tools using the Tompa's benchmark [19]. The results show that our new approach outperforms most of the motif discovery tools when tested on synthetic data, and outperforms all of the tools when tested on real datasets.

2 Background

Over the past years, hundreds of algorithms have been developed to solve the motif discovery problem. In this section, we look at the different computational approaches that tackle the common motifs problem. We also shed light on the approaches that employ data structures such as suffix trees.

2.1 Computational Approaches in Common Motifs Discovery

Computational approaches for motif discovery can be divided into two main categories: pattern-driven (also enumerative) and alignment driven (also sequence-based or profile-driven) approaches [13]. In pattern driven approaches, such as Weeder [12], a motif is represented as a consensus sequence. An exhaustive search is made to find and report the most significant pattern of a given length l. Although pattern-driven approach are guaranteed to find optimal solutions in restricted search space, they become impractical when searching for long and complex patterns. In the alignment driven approaches it is required to find the location of the binding sites and the representative *Position Weight Matrix*. Motifs are the conserved regions in a multiple sequence alignment. Many heuristics have been proposed using this approach such as Gibbs sampling and Expectation Maximization. Unlike enumerative approaches, alignment-driven approaches employ local search. Thus, finding the optimal solution using these methods is not guaranteed. A very well-known algorithm in this category is MEME [4], which is an expectation maximization algorithm. In this paper, we propose a pattern-driven approach.

2.2 Suffix Trees in Common Motifs Discovery

A suffix tree [21] is a data structure that stores all suffixes of the given sequence(s). It is commonly used to accelerate word access and retrieval in a given sequence. A suffix tree can be constructed in linear time using Ukkonen's algorithm [20]. Figure 2 shows an example for a suffix tree that is built for the set of sequences that are shown in Figure 1. The use of suffix trees for motif discovery was first proposed by Sagot [15]. Sagot presented two pattern-driven algorithms

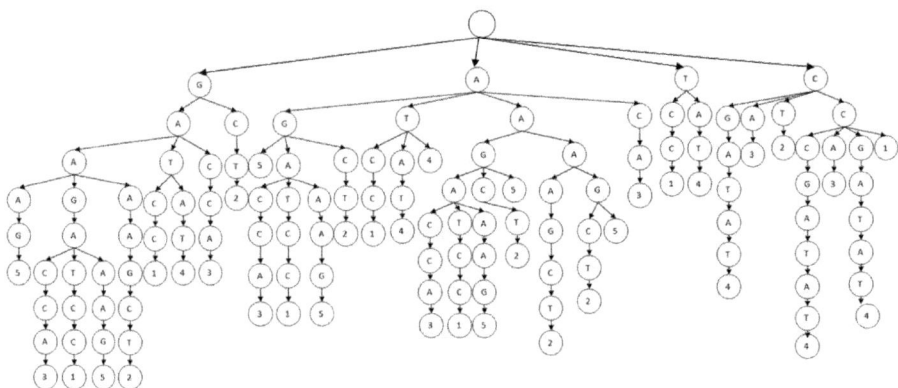

Fig. 2. A suffix tree for the sequences in Figure 1

one for extracting repeated motifs and the other for extracting common motifs. The algorithms worked by traversing a suffix tree in a recursive depth-first manner. Weeder algorithm [12] also used suffix trees and followed the idea introduced by Sagot in [15]. They used a dynamically chaining error threshold to weed out paths that were very unlikely to lead to valid occurrences. Jiang et al. [8] proposed a suffix-based algorithm to extract motifs that were modeled as degenerate motifs. Suffix trees can be used in combination with data mining algorithms in order to tackle the motif discovery problem [23]. We will discuss this in more details in the next section.

3 Apriori Approach in Motif Discovery

The progress in the data mining field has led to efficient and scalable algorithms in bioinformatics [5]. Data mining techniques can be applied in many bioinformatics tasks such as: mining of frequent sequential and structural pattern, clustering and classification of biological data, and modeling biological networks. The Apriori algorithm, a well-known data mining algorithm, was first introduced by Agrawal et al. [1]. It's goal is to discover association rules between frequent items in a large database of sales transactions. The algorithm employs an iterative approach where k-itemsets were used to find (k+1) itemsets. During candidate generation, candidates are pruned using the *apriori property*.

We propose an updated Apriori approach for motif discovery using a new technique for candidate generation along with a new measure of frequent itemsets.

Ozer and Ray [11] introduced a modification for the classic Apriori approach and evaluated motifs using information content. Yu et al. [23] presented an algorithm called Apriori-Motif to solve the Planted (l,d) Motif Problem [14]. The algorithm uses two data structures: a suffix tree and a consensus tree. Wijaya et al. [22] formulated the motif discovery problem as a frequent itemset mining problem.

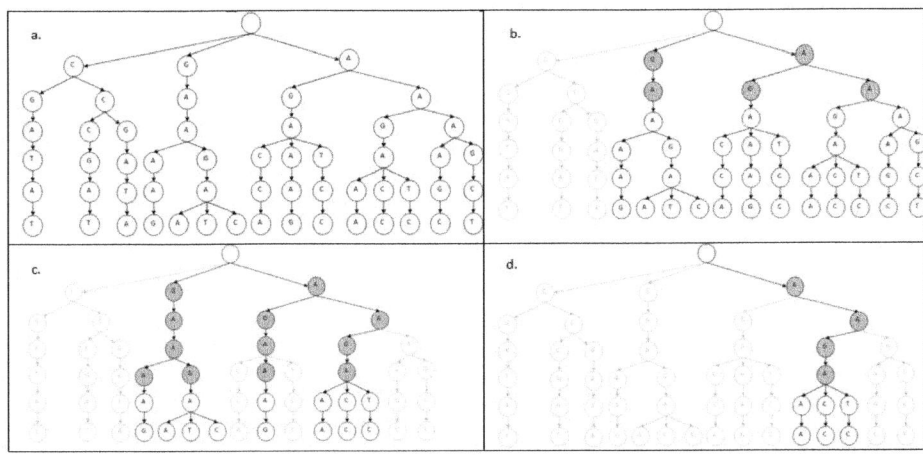

Fig. 3. An example of applying the TrieAMD approach on sequences that are shown in Figure 1. a) The Trie of depth 6, b) The frequent 2-mers, c) Candidate motifs, d) Reported motif when k=1. The Trie faded areas represent pruned paths

4 The Proposed Trie-based Apriori Approach

We present a *Trie-based Apriori motif discovery* (TrieAMD) approach that is inspired by the classic Apriori algorithm. Other than using a suffix tree, we use a window-based Trie for storing input sequences. Storing a sequence in the Trie structure allows to simultaneously search all possible starting points in the sequence for any occurrence of the given pattern. In [2], many exact and approximate matching algorithms were proposed in which the Tries were used to store the databases for strings. In TrieAMD, the Trie is also used for searching and storing candidate motifs with the help of a hash structure. The hash structure is built over the nodes of the Trie that represent the candidate motifs so that to facilitate combination. Initially we start from frequent 2-mers then generate 4-mer candidates. The resulting candidates are joined to produce 8-mers, and so on. This is unlike other incremental approaches for motif discovery that grow candidates one letter at a time [11,23]. Certain information is added to the nodes of the Trie to accelerate a candidate motif retrieval and support counting. Figure 3 shows an example of the complete approach applied on the sequences that are given in Figure 1. Each step is explained in the following subsections.

4.1 Window-Based Trie Construction

A Trie is constructed from the given input sequences. Each input sequence is divided into words with a user-defined window length. Words starting at each position in the sequence are then inserted into the Trie. The depth of the Trie is the window length. The length of the window corresponds to the maximum length of a motif. Each node in the Trie has a bit string of length that is equal to N. If $bit[i] = 1$ for a given node, then the word corresponding to the path ending at this node is contained in s_i of S. If the sum of the bits in a given node is greater than or equal q, then the motif that is represented by the node is a frequent n-mer, where n is the path length. Figure 3 (a) shows a Trie for a window length of 6, that is constructed for the sequences that are shown in Figure 1.

4.2 Trie-based 2-mer Extraction

After the Trie construction, the frequent 2-mers are extracted. This is done by looking only at the first two levels of the Trie. A frequent 2-mer corresponds to a path of length two that is present in more than q input sequences where q is equivalent to the value of *minimum support* and it is defined by the user. The number of sequences containing a path label is equivalent to the value of *support*. The support can be easily calculated using the bit string that is stored at the last node in the path as has been discussed before. Paths that do not represent frequent 2-mers are pruned from the search space. Figure 3 (b) shows frequent 2-mers of the constructed Trie, where faded areas represent the pruned paths. Dark nodes represent the generated frequent 2-mers. The frequent 2-mers are *GA, AG* and *AA*.

4.3 Trie-based Candidate Generation

Knowing the frequent 2-mers, the process of candidate generation can start. This step starts by combining every pair of frequent 2-mers in different orders. For each combination, we search for at least one exact occurrence in the Trie. If there is at least one occurrence, we look for the approximate occurrences. At the end, we have a set of motifs along with their exact and approximate occurrences. Frequent motifs are then determined using some measures (will be discussed in the next subsection). This step is repeated until no frequent motifs can be found. In each iteration, paths that do not lead to frequent motifs are pruned. Pruning is important to reduce the search space for motifs. The result of this step is shown in Figure 3 (c), where faded areas represent the pruned paths. Dark nodes represent candidates.

4.4 Measures for Frequent Motifs

To determine whether a candidate motif is frequent or not, we propose using an additional measure beside the minimum support. Like classic Apriori we use the minimum support to evaluate candidate. The support of a candidate motif is determined using exact and approximate occurrences. Number of exact occurrences is calculated using the bit string as described in Subsection??. Similarly, for the approximate occurrences, each node has a bit string, *Appbit*, for counting the approximate occurrences. If $Appbit[i] = 1$ then the path ending at the node has an approximate occurrence in s_i of S. The support is the sum of the two bit strings attached to the last node in the path representing the candidate motif. The sum is then compared to the minimum support, q.

Since the minimum support is defined by the user, it has a high impact on the resulting motifs. Failing to use the right minimum support can lead to a lot of trivial and meaningless motifs. It can also miss rare but important motifs. In [11], the authors observed this fact and used the information content measure [18] to evaluate motifs. However, with the information content threshold being also defined by the user, the problem remains.

We propose a new measure and combine it with the minimum support in order to evaluate candidate motifs. The measure is *the Normalized Information Content* (normIC). NormIC is defined as the ratio of motif's information content [11] to the motif's maximum possible information content. The proposed normIC measures how close the information content of the predicted motif to the maximum possible information content. A predicted motif with a value of normIC close to one could be a real motif. The new measure is able to distinguishes motifs in real promoter sequences.

In addition, we observed that candidates in early iterations have low normIC. Therefore, we use a dynamically changing normIC threshold. Initially, the value of normIC threshold is set to low. It is then incremented by the initial value in every iteration with the maximum value of normIC does not exceed one.

So for the frequent motif determination step, both the support and the normalized information content are used to evaluate candidate motifs as follows: if $support \geq q$ and $normIC \geq normICthreshold$, then the candidate motif is frequent. Motifs with k top normIC values are reported. In our example, Figure 3 (d) shows the reported motifs. With the minimum support set to 4, the algorithm finds three motifs, which are AAAA with normIC = 0.66, AAGA with normIC = 0.88 and GAAA with normIC = 0.79. The algorithm reports motif AAGA which has the highest normIC, when $k=1$.

4.5 Algorithm

Algorithm 1 presents the TrieAMD approach. Line 1 constructs a window-based Trie with a length defined by the user. In line 2, the frequent 2-mers are extracted. Lines 3-17 shows the modified candidate generation and the frequent motif determination. Initially, the frequent motifs are the frequent 2-mers. Then, inside the *while* loop, new candidate motifs are generated from current frequent motifs. The new candidates are evaluated and selected based on the calculated measures, then they are used to generate candidates for the next iteration. This continues until no more candidates are generated. Frequent motif determination, using the *support* and normIC as measures, is shown in line 11. The algorithm reports the best k motifs.

Algorithm 1. TrieAMD

1: Construct Trie
2: $Candmotifs \leftarrow Freq2mers$
3: **while** $Candmotifs \neq \emptyset$ **do**
4:     $Prevmotifs \leftarrow Candmotifs$
5:     $Candmotifs \leftarrow \emptyset$
6:     **for all** pairs of motifs $m1$ and $m2$ in $Prevmotifs$ **do**
7:         $newMotif \leftarrow combination(m1,m2)$
8:         **if** $newMotif$ has at least one occurrence in the $Trie$ **then**
9:             Find approximate matches for the $newMotif$
10:             **if** approximate matches exist **then**
11:                 **if** (support of $newMotif \geq minSup$)AND($normIC \geq normICThreshold$) **then**
12:                     Add $newMotif$ to $Candmotifs$
13:                 **end if**
14:             **end if**
15:         **end if**
16:     **end for**
17: **end while**
18: Report top k motifs from $Prevmotifs$

5 Empirical Results

5.1 Experimental Setup

We tested the proposed algorithm using datasets selected from Tompa's benchmark [19]. We compare the results of our proposed approach, TrieAMD, with the published results of the other available tools using the benchmark developed by Tompa *et al.* in [19]. The benchmark is made of 56 datasets from human, mouse, fly and yeast. Each dataset is constructed from one of the three different types of background sequences, where known binding sites are implanted at their original positions. The three types of background sequence are: real promoters, randomly chosen promoters from the same genome, and sequences generated by a Markov chain of order 3. The benchmark was used by Tompa to assess the accuracy of the available 13 motif discovery tools.

Our preliminary results are based on 45 datasets. The average number of sequences in our dataset is 5.7 sequences. Each sequence has an average length of 977.77. The algorithm reports the best k motifs in terms of normIC. The minimum support values are between 0.6 and 0.7. Initially, normIC is set to 0.16. We evaluated the performance of different tools using three measures [19]: *sensitivity* (nSn), *specificity* (nSp) and, *necleoside level performance coefficient* (nPC).

5.2 Results and Discussion

We now show and discuss the results obtained by the TrieAMD as compared with 13 other tools using the three measures as discussed before. These results

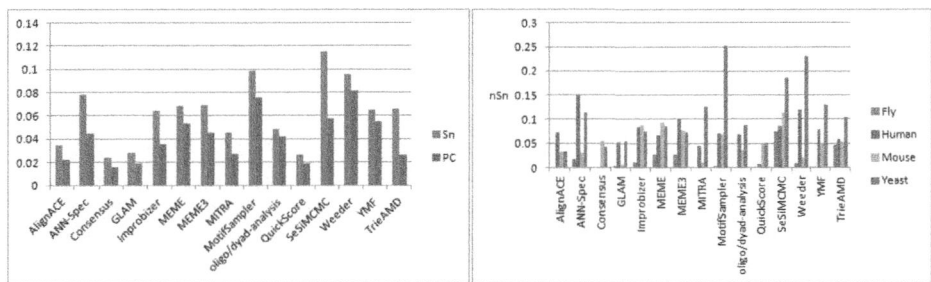

Fig. 4. Left: Average sensitivity (nSn) and performance coefficient (nPC) over all datasets for all species. Right:Average sensitivity over all datasets for each species.

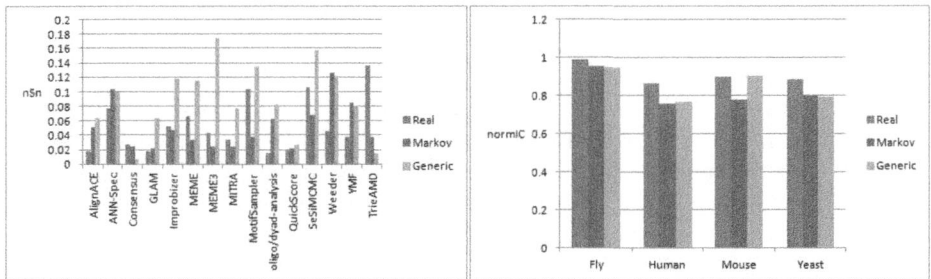

Fig. 5. Left: Average sensitivity over all species for real, Markovian and, generic datasets. Right: Aveage normalized information content (normIC) overall datasets.

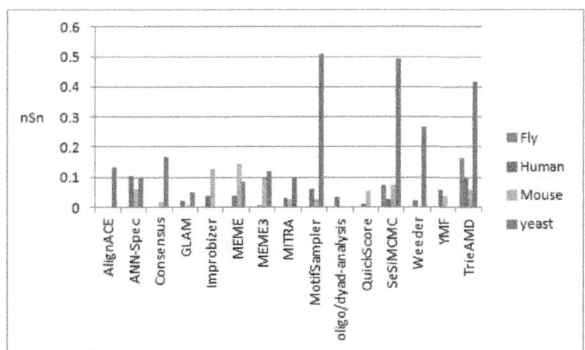

Fig. 6. Average sensitivity of TrieAMD as compared with other tools using real datasets for each species

are preliminary and are obtained for only 45 datasets from Tompa's benchmark because of machine limitations. The values obtained are very promising and show the benefits of our approach on real datasets. We present the results of applying TrieAMD on all datasets and then show them in more details for the real datasets.

Figure 4 (left) shows the average sensitivity (nSn) and performance coefficient (nPC) over all species for all tools. As for sensitivity, the results show that TrieAMD performs better than some tools. For example, it is better than AlignACE by 89%, Consensus by 173%, GLAM by 130 %, MITRA by 43%, QuickScore by 152%, Oligo/dyad by 35%, and YMF by 1%. In terms of nPC, TrieAMD also outperforms the same tools except for MITRA and Oligo/dyad. For specificity nSp, in general it performs as good as the other tools. It is slightly better than SeSiMCMC. The results for nSp are not shown for clarity of the figure, where they are much pretty the same.

We also evaluate the performance of TrieAMD species-wise. Figure 4 (right) shows the sensitivity measure for all tools per species. On the fly datasets, TrieAMD outperforms all tools except for SeSiMCMC. For the real fly dataset, dm02r, all tools achieve sensitivity of zero, whereas TrieAMD archives sensitivity of 0.1837. On human datasets, TrieAMD has a better performance than Consensus, GLAM, MITRA, and QuickScore. For mouse and yeast datasets TrieAMD outperforms some of the 13 tools.

We also evaluate the performance of TrieAMD based on the different background types. Figure 5 (left) shows the results for the sensitivity measure averaged over each background type. TrieAMD outperforms all of the 13 tools on real datasets, which is the most difficult type of background. The results of nSp are not shown because they are much pretty the same as the other 13 tools.

Figure 5 (right) shows the normIC for the three different background types over all species. This figure gives explanation on why TrieAMD performs better in real data. It also shows the influence of using normIC in our approach. Results show that the real background has the highest normIC value overall species.. The normIC value for the fly datasets is the highest, reaching a value of 0.99. We observe that the Markovian and generic backgrounds have lower *normIC*. This is because the background of those types may contain large number of approximate matches for a motif, which lowers the information content of the motif. This proves that this new measure is able to distinguish motifs in real promoter sequences.

To show the advantages of our proposed approach over other available tools we also present the results for only the real datasets. Figure 6 shows the average sensitivity over all real datasets for each species. TrieAMD always outperform the other tools in real fly datasets. For human, yeast and mouse, TrieAMD outperforms most of the tools.

6 Conclusion and Future Work

In this paper, we proposed a Trie-based algorithm, TrieAMD, to solve the common motifs problem. Our approach is inspired by the classic Apriori algorithm for mining frequent patterns. We used one data structure, namely the Trie, to store and simultaniously process multiple input sequences. We modified the candidate generation step in Apriori by combining frequent mers on the same structure, the Trie, to produce motifs. We also introduced a new measure to evaluate motifs and combined it with support measure. Instead of using only the information content as a measure, we used normalized information content, which is the ratio of actual information content to the maximum possible information content. We compared TrieAMD with 13 available motif discovery tools using 45 datasets. Preliminary results showed that TrieAMD outperforms all of them for real datasets for average senstivity. An explanation was given showing the effect of using the normalized information content as a measure for different types of data. It proves that the new measure is able to distinguish motifs in real promoter sequences.

As a future work, more experimental results will be conducted for the remaining large datasets. In addition, we would like to modify the candidate generation step in order to allow a motif to be generated even if it does not have any exact occurrence. The proposed approach will be also investigated to discover more complex motifs.

References

1. Agrawal, R., Srikant, R.: Fast algorithms for mining association rules in large databases. In: Proceedings of the 20th International Conference on Very Large Data Bases, San Francisco, CA, USA, pp. 487–499 (1994)
2. Badr, G.: Tries in information retrieval and syntactic pattern recognition. Ph.D. Thesis, School of Computer Science, Carleton University (June 2006)
3. Badr, G., Turcotte, M.: Component-Based Matching for Multiple Interacting RNA Sequences. In: Chen, J., Wang, J., Zelikovsky, A. (eds.) ISBRA 2011. LNCS, vol. 6674, pp. 73–86. Springer, Heidelberg (2011)
4. Bailey, T.L., Elkan, C.: Fitting a mixture model by expectation maximization to discover motifs in biopolymers. In: Proceedings of International Conference on Intelligent Systems for Molecular Biology, ISMB, vol. 2, pp. 28–36 (1994)
5. Bajcsy, P., Han, J., Liu, L., Yang, J.: Survey of biodata analysis from a data mining perspective. In: Wu, X., Jain, L., Wang, J.T., Zaki, M.J., Toivonen, H.T., Shasha, D. (eds.) Data Mining in Bioinformatics, pp. 9–39. Springer, London (2005)
6. Carvalho, A.M., Freitas, A.T., Oliveira, A.L., Sagot, M.: An efficient algorithm for the identification of structured motifs in DNA promoter sequences. IEEE/ACM Trans. Comput. Biol. Bioinformatics 3(2), 126–140 (2006)
7. Goos, G., Hartmanis, J., Leeuwen, J., Crochemore, M., Iliopoulos, C.S., Mohamed, M., Sagot, M.: Longest Repeats with a Block of Don't Cares. In: Farach-Colton, M. (ed.) LATIN 2004. LNCS, vol. 2976, pp. 271–278. Springer, Heidelberg (2004)
8. Jiang, H., Zhao, Y., Chen, W., Zheng, W.: Searching maximal degenerate motifs guided by a compact suffix tree. Advances in Experimental Medicine and Biology 680, 19–26 (2010)
9. Li, M., Ma, B., Wang, L.: Finding similar regions in many sequences. Journal of Computer and System Sciences 65, 73–96 (2002)
10. Matys, V., Kel-Margoulis, O.V., Fricke, E., Liebich, I., Land, S., Barre-Dirrie, A., Reuter, I., Chekmenev, D., Krull, M., Hornischer, K., Voss, N., Stegmaier, P., Lewicki-Potapov, B., Saxel, H., Kel, A.E., Wingender, E.: TRANSFAC and its module TRANSCompel: transcriptional gene regulation in eukaryotes. Nucleic Acids Research 34(Database issue), D108–D110 (2006)

11. Ozer, H.G., Ray, W.C.: Informative Motifs in Protein Family Alignments. In: Giancarlo, R., Hannenhalli, S. (eds.) WABI 2007. LNCS (LNBI), vol. 4645, pp. 161–170. Springer, Heidelberg (2007)
12. Pavesi, G., Mauri, G., Pesole, G.: An algorithm for finding signals of unknown length in DNA sequences. Bioinformatics 17(suppl. 1), S207–S214 (2001)
13. Pavesi, G., Mauri, G., Pesole, G.: In silico representation and discovery of transcription factor binding sites. Briefings in Bioinformatics 5(3), 217–236 (2004)
14. Pevzner, P.A., Sze, S.H.: Combinatorial approaches to finding subtle signals in DNA sequences. In: Proceedings of the Eighth International Conference on Intelligent Systems for Molecular Biology, pp. 269–278. AAAI Press (2000)
15. Sagot, M.-F.: Spelling Approximate Repeated or Common Motifs Using a Suffix Tree. In: Lucchesi, C.L., Moura, A.V. (eds.) LATIN 1998. LNCS, vol. 1380, pp. 374–390. Springer, Heidelberg (1998)
16. Sagot, M.F., Escalier, V., Viari, A., Soldano, H.: Searching for repeated words in a text allowing for mismatches and gaps. In: Baeza-Yates, R., Manber, U. (eds.) Second South American Workshop on String Processing, pp. 87–100. University of Chile (1995)
17. Sandelin, A., Alkema, W., Engstrm, P., Wasserman, W.W., Lenhard, B.: JASPAR: an open-access database for eukaryotic transcription factor binding profiles. Nucleic Acids Research 32(Database issue), D91–D94 (2004)
18. Schneider, T.D., Stormo, G.D., Gold, L.: Information content of binding sites on nucleotide sequences. Journal of Molecular Biology 188(3), 415–431 (1986)
19. Tompa, M., Li, N., Bailey, T.L., Church, G.M., De Moor, B., Eskin, E., Favorov, A.V., Frith, M.C., Fu, Y., Kent, W.J., Makeev, V.J., Mironov, A.A., Noble, W.S., Pavesi, G., Pesole, G., Rgnier, M., Simonis, N., Sinha, S., Thijs, G., van Helden, J., Vandenbogaert, M., Weng, Z., Workman, C., Ye, C., Zhu, Z.: Assessing computational tools for the discovery of transcription factor binding sites. Nature Biotechnology 23(1), 137–144 (2005)
20. Ukkonen, E.: On-Line construction of suffix trees. Algorithmica 14, 249–260 (1995)
21. Weiner, P.: Linear pattern matching algorithms. In: IEEE Conference Record of 14th Annual Symposium on Switching and Automata Theory, SWAT 2008, pp. 1–11. IEEE (October 1973)
22. Wijaya, E., Rajaraman, K., Yiu, S., Sung, W.: Detection of generic spaced motifs using submotif pattern mining. Bioinformatics 23(12), 1476–1485 (2007)
23. Jia, C., Lu, R., Chen, L.: A Frequent Pattern Mining Method for Finding Planted (l, d)-motifs of Unknown Length. In: Yu, J., Greco, S., Lingras, P., Wang, G., Skowron, A. (eds.) RSKT 2010. LNCS, vol. 6401, pp. 240–248. Springer, Heidelberg (2010)

Inapproximability of $(1,2)$-Exemplar Distance

Laurent Bulteau[1] and Minghui Jiang[2]

[1] Laboratoire d'Informatique de Nantes-Atlantique (LINA), UMR CNRS 6241
Université de Nantes, 2 rue de la Houssinière, 44322 Nantes Cedex 3, France
Laurent.Bulteau@univ-nantes.fr
[2] Department of Computer Science, Utah State University
Logan, UT 84322-4205, USA
mjiang@cc.usu.edu

Abstract. Given two genomes possibly with duplicate genes, the exemplar distance problem is that of removing all but one copy of each gene in each genome, so as to minimize the distance between the two reduced genomes according to some measure. Let (s,t)-EXEMPLAR DISTANCE denote the exemplar distance problem on two genomes G_1 and G_2 where each gene occurs at most s times in G_1 and at most t times in G_2. We show that the simplest non-trivial variant of the exemplar distance problem, $(1,2)$-EXEMPLAR DISTANCE, is already hard to approximate for a wide variety of distance measures, including popular genome rearrangement measures such as adjacency disruptions and signed reversals, and classic string edit distance measures such as Levenshtein and Hamming distances.

Keywords: Comparative genomics, hardness of approximation, adjacency disruption, sorting by reversals, edit distance, Levenshtein distance, Hamming distance.

1 Introduction

In the study of genome rearrangement, a *gene* is usually represented by a signed integer: the absolute value of the integer (the unsigned integer) denotes the gene family to which the gene belongs; the sign of the integer denotes the orientation of the gene in its chromosome. Then a *chromosome* is a sequence of signed integers, and a *genome* is a collection of chromosomes. Given two genomes possibly with duplicate genes, the *exemplar distance* problem [14] is that of removing all but one copy of each gene in each genome, so as to minimize the distance between the two reduced genomes according to some measure. The reduced genomes are said to be *exemplar subsequences* of the original genomes. This approach amounts to considering that, in the evolution history, duplications have taken place after the speciation of the genomes (or more generally, that we are able to distinguish genes that have been duplicated before the speciation). Hence, in each genome, only one copy of each gene may be matched to an ortholog gene in the other genome.

For example, the following two monochromosomal genomes

$$G_1: \quad -4\ +1\ +2\ +3\ -5\ +1\ +2\ +3\ -6$$
$$G_2: \quad -1\ -4\ +1\ +2\ -5\ +3\ -2\ -6\ +3$$

can both be reduced to the same genome

$$G': \quad -4\ +1\ +2\ -5\ +3\ -6$$

by removing duplicates, thus they have exemplar distance zero for any reasonable distance measure. In general, unless we are to decide simply whether two genomes can be reduced to the same genome by removing duplicates, the exemplar distance problem is not a single problem but a group of related problems because the choice of the distance measure is not unique.

We denote by (s,t)-EXEMPLAR DISTANCE the exemplar distance problem on two genomes G_1 and G_2 where each gene occurs at most s times in G_1 and at most t times in G_2. It is known [5,12] that for any reasonable distance measure, $(2,2)$-EXEMPLAR DISTANCE does not admit any approximation. This is because to decide simply whether two genomes with maximum occurrence 2 can be reduced to the same genome by removing duplicates is already NP-hard. In this paper, we focus on the simplest non-trivial variant of the exemplar distance problem: $(1,2)$-EXEMPLAR DISTANCE.

The problem $(1,t)$-EXEMPLAR DISTANCE has been studied for several distance measures commonly used in genome rearrangement. Angibaud et al. [2] showed that $(1,2)$-EXEMPLAR BREAKPOINT DISTANCE, $(1,2)$-EXEMPLAR COMMON INTERVAL DISTANCE, and $(1,2)$-EXEMPLAR CONSERVED INTERVAL DISTANCE are all APX-hard. Blin et al. [4] showed that $(1,9)$-EXEMPLAR MAD DISTANCE is NP-hard to approximate within $2-\epsilon$ for any $\epsilon > 0$, and that $(1,\infty)$-EXEMPLAR SAD DISTANCE is NP-hard to approximate within $c\log n$ for some constant $c > 0$, where n is the number of genes in G_1. See also [8,6,7] for related results.

The two distance measures we first consider, MAD and SAD, were introduced by Sankoff and Haque [15]. For two permutations $\pi' = \pi'_1 \ldots \pi'_n$ and $\pi'' = \pi''_1 \ldots \pi''_n$ of n distinct elements, define $\tau'(i)$ as the index j such that $\pi'_j = \pi''_i$, and $\tau''(i)$ as the index j such that $\pi''_j = \pi'_i$. Then the maximum adjacency disruption (MAD) and the summed adjacency disruption (SAD) between π' and π'' are

$$\mathrm{MAD}(\pi',\pi'') = \max_{1 \le i \le n-1} \left\{ |\tau'(i) - \tau'(i+1)|, |\tau''(i) - \tau''(i+1)| \right\},$$
$$\mathrm{SAD}(\pi',\pi'') = \sum_{1 \le i \le n-1} \left(|\tau'(i) - \tau'(i+1)| + |\tau''(i) - \tau''(i+1)| \right).$$

Our first two theorems sharpen the previous results on the inapproximability on $(1,t)$-EXEMPLAR DISTANCE for both MAD and SAD measures:

Theorem 1. $(1,2)$-EXEMPLAR MAD DISTANCE *is NP-hard to approximate within $2-\epsilon$ for any $\epsilon > 0$.*

Theorem 2. $(1,2)$-EXEMPLAR SAD DISTANCE *is NP-hard to approximate within* $10\sqrt{5} - 21 - \epsilon = 1.3606\ldots - \epsilon$, *and is NP-hard to approximate within* $2 - \epsilon$ *if the unique games conjecture is true, for any* $\epsilon > 0$.

For an unsigned permutation $\pi = \pi_1 \ldots \pi_n$, an *unsigned reversal* (i,j) with $1 \leq i \leq j \leq n$ turns it into $\pi_1 \ldots \pi_{i-1} \pi_j \ldots \pi_i \pi_{j+1} \ldots \pi_n$, where the substring $\pi_i \ldots \pi_j$ is reversed. For a signed permutation $\sigma = \sigma_1 \ldots \sigma_n$, a *signed reversal* (i,j) with $1 \leq i \leq j \leq n$ turns it into $\sigma_1 \ldots \sigma_{i-1} -\sigma_j \ldots -\sigma_i \sigma_{j+1} \ldots \sigma_n$, where the substring $\sigma_i \ldots \sigma_j$ is reversed and negated. The *unsigned reversal distance* (resp. *signed reversal distance*) between two unsigned (resp. signed) permutations is the minimum number of unsigned (resp. signed) reversals required to transform one to the other. Computing the unsigned reversal distance is APX-hard [3], although the signed reversal distance can be computed in polynomial time [11].

Our next theorem answers an open question of Blin et al. [4] on the inapproximability of the exemplar reversal distance problem:

Theorem 3. $(1,2)$-EXEMPLAR SIGNED REVERSAL DISTANCE *is NP-hard to approximate within* $1237/1236 - \epsilon$ *for any* $\epsilon > 0$.

In the last theorem of this paper, we present the first inapproximability result on the exemplar distance problem using the classic string edit distance measure:

Theorem 4. $(1,2)$-EXEMPLAR EDIT DISTANCE *is APX-hard to compute when the cost of a substitution is* 1 *and the cost of an insertion or a deletion is at least* 1.

Note that both Levenshtein distance and Hamming distance are special cases of the string edit distance: for Levenshtein distance, the cost of every operation (substitution, insertion, or deletion) is 1; for Hamming distance, the cost of a substitution is 1 and the cost of an insertion or a deletion is $+\infty$. Thus we have the following corollaries:

Corollary 1. $(1,2)$-EXEMPLAR LEVENSHTEIN DISTANCE *is APX-hard.*

Corollary 2. $(1,2)$-EXEMPLAR HAMMING DISTANCE *is APX-hard.*

2 MAD Distance

In this section we prove Theorem 1. We prove that EXEMPLAR MAD DISTANCE is NP-hard by a reduction from the well-known NP-hard problem 3SAT [10]. Let (V, C) be a 3SAT instance, where $V = \{v_1, \ldots, v_n\}$ is a set of n boolean variables, $C = \{c_1, \ldots, c_m\}$ is a conjunctive boolean formula of m clauses, and each clause in C is a disjunction of exactly three literals of the variables in V. The problem 3SAT is that of deciding whether (V, C) is satisfiable, i.e., whether there is a truth assignment for the variables in V that satisfies all clauses in C.

Let M be a large number to be specified. We will construct two sequences (genomes) G_1 and G_2 over $L = 3m + (n+1) + (2n+1) + (m+1) + (2M+2) = 2M + 3n + 4m + 5$ distinct markers (genes):

- 3 *literal markers* r_j, s_j, t_j for the 3 literals of each clause c_j, $1 \le j \le m$;
- $n+1$ *variable markers* x_i, $0 \le i \le n$;
- $2n+1$ *separator markers* y_i, $0 \le i \le 2n$;
- $m+1$ *clause markers* z_j, $0 \le j \le m$;
- $2M+2$ *dummy markers* ϕ_k and ψ_k, $0 \le k \le M$.

For each clause c_j, let $O_j = r_j s_j t_j$ be the concatenation of the three literal markers of c_j. For each variable v_i, let $P_i = p_{i,1} \ldots p_{i,k_i}$ be the concatenation of the k_i literal markers of the positive literals of v_i, and let $Q_i = q_{i,1}, \ldots, q_{i,l_i}$ be the concatenation of the l_i literal markers of the negative literals of v_i. Without loss of generality, assume that $\min\{k_i, l_i\} \ge 1$. Note that the two concatenated sequences $O_1 \ldots O_m$ and $P_1 Q_1 \ldots P_n Q_n$ are both permutations of the $3m$ literal markers.

The two sequences G_1 and G_2 are represented schematically as follows. G_1 contains exactly one copy of each marker, and has length L; G_2 contains exactly two copies of each literal marker and exactly one copy of each non-literal marker, and has length $L + 3m$.

$G_1:\quad \ldots z_3 z_1 \quad \phi_0 \quad \ldots x_2 x_0 \quad \phi_M \ldots \phi_1 \quad y_0 P_1 y_1 Q_1 y_2 \ldots P_n y_{2n-1} Q_n y_{2n} \quad \psi_1 \ldots \psi_M \quad z_0 z_2 \ldots \quad \psi_0 \quad x_1 x_3 \ldots$
$G_2:\quad x_n P_n Q_n \ldots x_1 P_1 Q_1 x_0 \quad \phi_M \ldots \phi_1 \phi_0 \quad y_0 y_1 y_2 \ldots y_{2n-1} y_{2n} \quad \psi_0 \psi_1 \ldots \psi_M \quad z_0 O_1 z_1 \ldots O_m z_m$

Lemma 1. *If (V, C) is satisfiable, then G_2 has an exemplar subsequence G_2' that satisfies* $\mathrm{MAD}(G_1, G_2') \le M + 3n + 4m + 5$.

Proof. Let f be a truth assignment for the variables in V that satisfies all clauses in C. For each variable v_i, compose a subsequence V_i of $P_i Q_i$ such that $V_i = Q_i$ if $f(v_i)$ is true and $V_i = P_i$ if $f(v_i)$ is false. For each clause c_j, compose a subsequence C_j of O_j containing only the literal markers of the literals that are true under the assignment f. Then $V_1 \ldots V_n C_1 \ldots C_m$ is a permutation of the $3m$ literal markers. It is straightforward to verify that the exemplar subsequence G_2' of G_2 in the following satisfies $\mathrm{MAD}(G_1, G_2') \le L - M = M + 3n + 4m + 5$:

$G_1:\quad \ldots z_3 z_1 \quad \phi_0 \quad \ldots x_2 x_0 \quad \phi_M \ldots \phi_1 \quad y_0 P_1 y_1 Q_1 y_2 \ldots P_n y_{2n-1} Q_n y_{2n} \quad \psi_1 \ldots \psi_M \quad z_0 z_2 \ldots \quad \psi_0 \quad x_1 x_3 \ldots$
$G_2:\quad x_n P_n Q_n \ldots x_1 P_1 Q_1 x_0 \quad \phi_M \ldots \phi_1 \phi_0 \quad y_0 y_1 y_2 \ldots y_{2n-1} y_{2n} \quad \psi_0 \psi_1 \ldots \psi_M \quad z_0 O_1 z_1 \ldots O_m z_m$
$G_2':\quad x_n V_n \ldots x_1 V_1 x_0 \quad \phi_M \ldots \phi_1 \phi_0 \quad y_0 y_1 y_2 \ldots y_{2n-1} y_{2n} \quad \psi_0 \psi_1 \ldots \psi_M \quad z_0 C_1 z_1 \ldots C_m z_m$

□

Lemma 2. *If (V, C) is not satisfiable, then every exemplar subsequence G_2' of G_2 satisfies* $\mathrm{MAD}(G_1, G_2') > 2M$.

Proof. We prove the contrapositive. Suppose G_2 has an exemplar subsequence G_2' that satisfies $\mathrm{MAD}(G_1, G_2') \le 2M$. We will find a truth assignment f for the variables in V that satisfies all clauses in C.

First, we claim that for each variable v_i, the literal markers of the positive literals of v_i must appear in G_2' either all before ϕ_M or all after ψ_M. Suppose the contrary. Then there would be two literal markers of v_i, one before ϕ_M and one after ψ_M in G_2', that are adjacent in the substring P_i in G_1, incurring a

MAD distance larger than $2M$. Similarly, we claim that the literal markers of the negative literals of each variable v_i must appear in G_2' either all before ϕ_M or all after ψ_M.

Next, we claim that for each variable v_i, the literal markers of either all positive literals of v_i or all negative literals of v_i must appear in G_2' before ϕ_M, between x_i and x_{i-1}. Suppose the contrary that all literal markers of both the positive and the negative literals of v_i appear in G_2' after ψ_M. Then the two variable markers x_i and x_{i-1}, one before ϕ_M and one after ψ_M in G_1, would become adjacent in G_2', incurring a MAD distance larger than $2M$.

Finally, we claim that for each clause c_j, at least one of the three literal markers r_j, s_j, t_j must appear in G_2' after ψ_M, between z_{j-1} and z_j. Suppose the contrary. Then the two clause markers z_{j-1} and z_j, one before ϕ_M and one after ψ_M in G_1, would become adjacent in G_2', again incurring a MAD distance larger than $2M$.

Now compose a truth assignment f for the variables in V such that $f(v_i)$ is true if the literal markers for the negative literals of v_i appear before ϕ_M, and is false otherwise. Then f satisfies all clauses in C. □

For any constant ϵ, $0 < \epsilon < 2$, we can get a gap of $2M/(M+3n+4m+5) = 2-\epsilon$ by setting $M = (\frac{2}{\epsilon}-1)(3n+4m+5)$. Thus the NP-hardness of 3SAT and the two preceding lemmas together imply that EXEMPLAR MAD DISTANCE is NP-hard to approximate within $2 - \epsilon$ for any $\epsilon > 0$.

3 SAD Distance

In this section we prove Theorem 2. We show that EXEMPLAR SAD DISTANCE is NP-hard to approximate by a reduction from another well-known NP-hard problem MINIMUM VERTEX COVER [10]. Let (V, E) be a graph, where $V = \{v_1, \ldots, v_n\}$ is a set of n vertices, and $E = \{e_1, \ldots, e_m\}$ is a set of m edges. The problem MINIMUM VERTEX COVER is that of finding a subset $C \subseteq V$ of the minimum cardinality such that each edge in E is incident to at least one vertex in C. Dinur and Safra [9] showed that MINIMUM VERTEX COVER is NP-hard to approximate within any constant less than $10\sqrt{5} - 21 = 1.3606\ldots$. Khot and Regev [13] showed that MINIMUM VERTEX COVER is NP-hard to approximate within any constant less than 2 if the unique games conjecture is true.

Let $M = 2(n+m)^2$. We will construct two sequences (genomes) G_1 and G_2 over $L = n + m + M + 1$ distinct markers (genes):

- n vertex markers v_i, $1 \le i \le n$;
- m edge markers e_j, $1 \le j \le m$;
- M dummy markers ϕ_k, $0 \le k \le M$.

For each vertex v_i, let $E_i = e_{i,1} \ldots e_{i,k_i}$ be the concatenation of the edge markers of all edges incident to v_i, where k_i is the degree of v_i. The two sequences G_1 and G_2 are represented schematically as follows. G_1 contains exactly one copy of each marker, and has length L; G_2 contains exactly two copies of each edge marker and exactly one copy of each non-edge marker, and has length $L + m$.

$$G_1: \quad e_1 \ldots e_m \quad \phi_0\phi_1\ldots\phi_M \quad v_1\ldots v_n$$
$$G_2: \quad \phi_0\phi_1\ldots\phi_M \quad E_1v_1\ldots E_nv_n$$

Lemma 3. *G has a vertex cover of size at most k if and only if G_2 has an exemplar subsequence G'_2 that satisfies $\mathrm{SAD}(G_1, G'_2) \leq (2k+4)M$.*

Proof. We first prove the direct implication. Let C be a vertex cover of size at most k in G. Extract a subsequence E'_i of E_i for each vertex v_i in C such that the concatenated sequence $E'_1 \ldots E'_n$ contains each edge marker e_j exactly once. From G_2, remove E_i for each vertex v_i not in C, and replace E_i by E'_i for each vertex v_i in C. Then we obtain an exemplar subsequence G'_2 of G_2.

The two sequences G_1 and G'_2 have the same length $L = n + m + M + 1$ and together have $2n + 2m + 2M$ adjacencies. The contributions of these adjacencies to $\mathrm{SAD}(G_1, G'_2)$ are as follows:

1. The shared adjacencies $\phi_i\phi_{i+1}$ in G_1 and G'_2, $0 \leq i \leq M-1$, contribute a total value of exactly $2M$.
2. The adjacency $e_m\phi_0$ in G_1 contributes a value of at least M and at most $M + n + m$.
3. Each adjacency between an edge marker and a non-edge marker in G'_2 contributes a value of at least M and at most $M + n + m$.
4. Each remaining adjacency contributes a value of at least 1 and at most $n+m$.

The number of adjacencies between an edge marker and a non-edge marker in G'_2 is exactly twice the size of the vertex cover C. Thus we have

$$\begin{aligned}\mathrm{SAD}(G_1, G'_2) &\leq 2M + (2k+1)(M + n + m) \\ &\quad + (2n + 2m + 2M - 2M - 2k - 1)(n+m) \\ &= (2k+3)M + 2(n+m)^2 = (2k+4)M.\end{aligned}$$

We next prove the reverse implication. Let G'_2 be an exemplar subsequence of G_2 such that $\mathrm{SAD}(G_1, G'_2) \leq (2k+4)M$. Refer back to the list of contributions to $\mathrm{SAD}(G_1, G'_2)$. Let l be the number of adjacencies between an edge marker and a non-edge marker in G'_2. Then we have the following inequality:

$$\mathrm{SAD}(G_1, G'_2) \geq 2M + (l+1)M = (l+3)M.$$

Since $\mathrm{SAD}(G_1, G'_2) \leq (2k+4)M$, we have $l + 3 \leq 2k+4$ and hence $l \leq 2k+1$. Note that l must be an even number: for each adjacency between an edge marker in E_i and a non-edge marker to its left, there must be another adjacency between an edge marker in E_i and a non-edge marker (indeed a vertex marker) to its right, and vice versa. It follows that there are at most k vertex markers v_i that are adjacent to an edge marker to its left. The corresponding at most k vertices v_i form a vertex cover of G. □

The inapproximability of MINIMUM VERTEX COVER and the preceding lemma together imply that EXEMPLAR SAD DISTANCE is NP-hard to approximate within $10\sqrt{5} - 21 - \epsilon$, and is NP-hard to approximate within $2 - \epsilon$ if the unique games conjecture is true, for any $\epsilon > 0$.

4 Signed Reversal Distance

In this section we prove Theorem 3. We show that $(1,2)$-EXEMPLAR SIGNED REVERSAL DISTANCE is APX-hard by a reduction from the problem MIN-SBR [3], which asks for the minimum number of unsigned reversals to sort a given unsigned permutation into the identity permutation.

Let $\pi = \pi_1 \ldots \pi_n$ be an unsigned permutation of $1 \ldots n$. We construct two sequences $G_1 = 1 \ldots n$ and $G_2 = \pi_1 - \pi_1 \ldots \pi_n - \pi_n$.

Lemma 4. *π can be sorted into the identity permutation $1 \ldots n$ by at most k unsigned reversals if and only if G_2 has an exemplar subsequence G_2' with signed reversal distance at most k from G_1.*

We leave the proof of Lemma 4 to the reader as an easy exercise. Since MIN-SBR is NP-hard to approximate within $1237/1236 - \epsilon$ for any $\epsilon > 0$ [3], $(1,2)$-EXEMPLAR SIGNED REVERSAL DISTANCE is NP-hard to approximate within $1237/1236 - \epsilon$ for any $\epsilon > 0$ too.

5 Edit Distance

In this section we prove Theorem 4. For any edit distance where the cost of a substitution is 1 and the cost of an insertion or a deletion is at least 1 (possibly $+\infty$), we show that the problem $(1,2)$-EXEMPLAR EDIT DISTANCE is APX-hard by a reduction from the problem MINIMUM VERTEX COVER IN CUBIC GRAPHS.

Let $G = (V, E)$ be a cubic graph of n vertices and m edges, where $3n = 2m$. We will construct two sequences (genomes) G_1 and G_2 over an alphabet of

$$3m + 4n + 2(m + 7n) + 2(m - 1) + (n - 1)$$

distinct markers (genes). For each edge $e = \{u, v\} \in E$, we have three edge markers e, e_u, and e_v. For each vertex $v \in V$, we have a vertex marker v and 3 dummy markers v_1', v_2', v_3'. In addition, we have $2(m + 7n) + 2(m - 1) + (n - 1)$ markers for separators.

The two sequences G_1 and G_2 are composed from $m + n + 1$ gadgets: an edge gadget for each edge, a vertex gadget for each vertex, and a tail gadget. The $m + n + 1$ gadgets are separated by $m + n$ separators of total length $2(m + 7n) + 2(m - 1) + (n - 1)$:

- two long separators, each of length $m + 7n$: one between the last edge gadget and the first vertex gadget, one between the last vertex gadget and the tail gadget;

– $m+n-2$ short separators: a length-2 separator between any two consecutive edge gadgets, and a length-1 separator between any two consecutive vertex gadgets.

For each edge $e = \{u,v\}$, the edge gadget for e is

$$G_1 \langle e \rangle = e$$
$$G_2 \langle e \rangle = e_u e_v$$

For each vertex v incident to edges e, f, g, the vertex gadget for v is

$$G_1 \langle v \rangle = v\, v_1'\, v_2'\, v_3'$$
$$G_2 \langle v \rangle = e_v f_v g_v\, v\, e\, f\, g$$

Let V' be the $3n$ markers v_1', v_2', v_3' for $v \in V$. Let E' be the $2m = 3n$ markers e_u and e_v for $e = \{u,v\} \in E$. The tail gadget is

$$G_1 \langle tail \rangle = E'$$
$$G_2 \langle tail \rangle = V'$$

This completes the construction.

Lemma 5. *G has a vertex cover of size at most k if and only if G_2 has an exemplar subsequence G_2' with edit distance at most $m + 6n + k$ from G_1.*

Proof. We first prove the direct implication. Let X be a vertex cover of G with $|X| \le k$. Create G_2' as follows. For each edge $e = \{u,v\}$, at least one vertex, say u, is in X. Remove e_u and retain e_v in the edge gadget $G_2 \langle e \rangle$, and correspondingly retain e_u in the vertex gadget $G_2 \langle u \rangle$ and remove e_v in the vertex gadget $G_2 \langle v \rangle$, then remove e in $G_2 \langle u \rangle$ and retain e in $G_2 \langle v \rangle$. We claim that the edit distance from G_1 to G_2' is at most $m + 6n + k$.

It suffices to show that the Hamming distance of G_1 and G_2' is at most $m + 6n + k$ since, for the edit distance that we consider, the cost of a substitution is 1. Observe that in both G_1 and G_2', each edge gadget has length 1, and each vertex gadget has length 4. Thus all gadgets are aligned and all separators are matched. The Hamming distance for each edge gadget is 1, so the total Hamming distance over all edge gadgets is m. The Hamming distance for each vertex gadget is at most 4. Moreover, for each vertex $v \notin X$ (v incident to edges e, f, g), since the markers e_v, f_v, g_v are removed (and the markers e, f, g are retained) in the vertex gadget, the marker v is matched, which reduces the Hamming distance by 1. Thus the total Hamming distance over all vertex gadgets is at most $4n - (n - |X|) = 3n + |X|$. Finally, since the Hamming distance for the tail gadget is $3n$, the overall Hamming distance of G_1 and G_2' is at most $m + 6n + |X| \le m + 6n + k$.

We next prove the reverse implication. Let G_2' be an exemplar subsequence of G_2 with edit distance at most $m + 6n + k$ from G_1. Compute an alignment of G_1 and G_2' corresponding to the edit distance, then obtain the following three sets $X_E(G_2')$, $X_V(G_2')$, and $X(G_2')$:

- The set $X_E(G_2') \subseteq E$ contains every edge $e = \{u,v\}$ such that either $G_2'\langle e\rangle$ contains both e_u and e_v, or $G_1'\langle e\rangle$ has an adjacent separator marker which is unmatched.
- The set $X_V(G_2') \subseteq V$ contains every vertex v (v incident to edges e, f, g) such that either $G_2'\langle v\rangle$ contains one of $\{e_v, f_v, g_v\}$, or $G_1'\langle v\rangle$ has an adjacent separator marker (to its left) which is unmatched.
- The set $X(G_2') \subseteq V$ is the union of $X_V(G_2')$ and a set composed by arbitrarily choosing one vertex from each edge in $X_E(G_2')$ (thus $|X(G_2')| \leq |X_V(G_2')| + |X_E(G_2')|$).

We first show that the edit distance from G_1 to G_2' is at least $m + 6n + |X(G_2')|$. If a long separator (with $m + 7n$ markers) is completely unmatched, then the edit distance is at least $m + 7n \geq m + 6n + |X(G_2')|$. Hence we can assume that there is at least one matched marker in each long separator. Consequently, the markers e, e_u, e_v for all $e \in E$ and v_1', v_2', v_3' for all $v \in V$ are unmatched.

Consider an edge $e = \{u,v\} \in E$. If $e \notin X_E(G_2')$, then the edit distance for $G_1\langle e\rangle$ is at least 1 since the marker e is unmatched. If $e \in X_E(G_2')$, then consider the substring of $G_1\langle e\rangle$ containing the marker e and the at most two separator markers adjacent to it (for the first edge gadget, there is only one separator marker adjacent to e, to its right). The edit distance for this substring is at least 2: the marker e is unmatched, and moreover either an adjacent separator marker is unmatched or an insertion is required. The total edit distance over all edge gadgets is at least $m + |X_E(G_2')|$.

Consider a vertex $v \in V$ incident to three edges e, f, g. If $v \notin X_V(G_2')$, then the edit distance for $G_1\langle v\rangle$ is at least 3 since the markers v_1', v_2', v_3' are unmatched. If $v \in X_V(G_2')$, then consider the substring of G_1 containing $G_1\langle v\rangle$ and the separator to its left. The edit distance for this substring is at least 4: the markers v_1', v_2', v_3' are unmatched, and moreover at least one insertion is required unless either the marker v or the separator marker to its left is unmatched. The total edit distance over the vertex gadgets is at least $3n + |X_V(G_2')|$.

Finally, the edit distance over the tail gadget is at least the length of $G_1\langle tail\rangle$, which is $3n$. Hence the overall edit distance is at least

$$m + |X_E(G_2')| + 3n + |X_V(G_2')| + 3n \geq m + 6n + |X(G_2')|.$$

Since the edit distance from G_1 to G_2' is at most $m + 6n + k$, it follows that $|X(G_2')| \leq k$.

To complete the proof, we show that $X(G_2')$ is a vertex cover of G. Consider any edge $e = \{u,v\}$. If $e \in X_E(G_2')$, then, by our choice of $X(G_2')$, either $u \in X(G_2')$ or $v \in X(G_2')$. Otherwise, if $e \notin X_E(G_2')$, then in the edge gadget $G_2\langle e\rangle = e_u e_v$, at least one marker is removed to obtain $G_2'\langle e\rangle$. Assume that e_u is removed: then the second copy, in $G_2\langle u\rangle$, is retained, and $u \in X_V(G_2') \subseteq X(G_2')$. Likewise if e_v is removed, then $v \in X(G_2')$. In summary, $X(G_2')$ contains a vertex from every edge in E, hence it is a vertex cover of G. □

The problem MINIMUM VERTEX COVER IN CUBIC GRAPHS is APX-hard; see e.g. [1]. For a cubic graph G of n vertices and m edges, where $3n = 2m$, the minimum size k^* of a vertex cover is $\Theta(m+n)$. By Lemma 5, the exemplar edit distance of the two sequences G_1 and G_2 in the reduced instance is also $\Theta(m+n)$. Thus by the standard technique of L-reduction, it follows that (1, 2)-EXEMPLAR EDIT DISTANCE, when the cost of a substitution is 1 and the cost of an insertion or a deletion is at least 1, is APX-hard too. Then the APX-hardness of (1, 2)-EXEMPLAR LEVENSHTEIN DISTANCE and the APX-hardness of (1, 2)-EXEMPLAR HAMMING DISTANCE follow as special cases. Moreover, since the lengths of the two sequences G_1 and G_2 in the reduced instance are both $\Theta(m+n)$ as well, it follows that the complementary maximization problem (1, 2)-EXEMPLAR HAMMING SIMILARITY is also APX-hard, if we define the *Hamming similarity* of two sequences of the same length ℓ as ℓ minus their Hamming distance.

6 Concluding Remarks

We find it most intriguing that although the problem (1, 2)-EXEMPLAR DISTANCE has been shown to be APX-hard for a wide variety of distance measures, including breakpoints, conserved intervals, common intervals, MAD, SAD, signed reversals, Levenshtein distance, Hamming distance..., no constant approximation is known for any one of these measures, while on the other hand, it seems difficult to improve the constant lower bound in any one of these APX-hardness results into a lower bound that grows with the input size similar to the logarithmic lower bound for MINIMUM SET COVER.

References

1. Alimonti, P., Kann, V.: Some APX-completeness results for cubic graphs. Theoretical Computer Science 237, 123–134 (2000)
2. Angibaud, S., Fertin, G., Rusu, I., Thévenin, A., Vialette, S.: On the approximability of comparing genomes with duplicates. Journal of Graph Algorithms and Applications 13, 19–53 (2009)
3. Berman, P., Karpinski, M.: On Some Tighter Inapproximability Results. In: Wiedermann, J., van Emde Boas, P., Nielsen, M. (eds.) ICALP 1999. LNCS, vol. 1644, pp. 200–209. Springer, Heidelberg (1999)
4. Blin, G., Chauve, C., Fertin, G., Rizzi, R., Vialette, S.: Comparing genomes with duplications: a computational complexity point of view. IEEE/ACM Transactions on Computational Biology and Bioinformatics 4, 523–534 (2007)
5. Blin, G., Fertin, G., Sikora, F., Vialette, S.: The EXEMPLAR BREAKPOINT DISTANCE for Non-trivial Genomes Cannot Be Approximated. In: Das, S., Uehara, R. (eds.) WALCOM 2009. LNCS, vol. 5431, pp. 357–368. Springer, Heidelberg (2009)
6. Bonizzoni, P., Della Vedova, G., Dondi, R., Fertin, G., Rizzi, R., Vialette, S.: Exemplar longest common subsequence. IEEE/ACM Transactions on Computational Biology and Bioinformatics 4, 535–543 (2007)
7. Chen, Z., Fowler, R.H., Fu, B., Zhu, B.: On the inapproximability of the exemplar conserved interval distance problem of genomes. Journal of Combinatorial Optimization 15, 201–221 (2008); A preliminary version appeared in Proceedings of the 12th Annual International Conference on Computing and Combinatorics, COCOON 2006, pp. 245–254 (2006)

8. Chen, Z., Fu, B., Zhu, B.: The Approximability of the Exemplar Breakpoint Distance Problem. In: Cheng, S.-W., Poon, C.K. (eds.) AAIM 2006. LNCS, vol. 4041, pp. 291–302. Springer, Heidelberg (2006)
9. Dinur, I., Safra, S.: On the hardness of approximating minimum vertex cover. Annals of Mathematics 162, 439–485 (2005)
10. Garey, M.R., Johnson, D.S.: Computers and Intractability: A Guide to the Theory of NP-Completeness. W.H. Freeman and Company (1979)
11. Hannenhalli, S., Pevzner, P.: Transforming cabbage into turnip: polynomial algorithm for sorting signed permutations by reversals. Journal of the ACM 46, 1–27 (1999)
12. Jiang, M.: The zero exemplar distance problem. Journal of Computational Biology 18, 1077–1086 (2011)
13. Khot, S., Regev, O.: Vertex cover might be hard to approximate to within $2 - \epsilon$. Journal of Computer and System Sciences 74, 335–349 (2008)
14. Sankoff, D.: Genome rearrangement with gene families. Bioinformatics 15, 909–917 (1999)
15. Sankoff, D., Haque, L.: Power Boosts for Cluster Tests. In: McLysaght, A., Huson, D.H. (eds.) RECOMB 2005. LNCS (LNBI), vol. 3678, pp. 121–130. Springer, Heidelberg (2005)

A Mixed Integer Programming Model for the Parsimonious Loss of Heterozygosity Problem

Daniele Catanzaro[1], Martine Labbé[1], and Bjarni V. Halldórsson[2]

[1] Graphs and Mathematical Optimization Unit, Computer Science Department, Université Libre de Bruxelles (ULB)
[2] School of Science and Engineering, Reykjavík University

Abstract. We investigate the *Parsimonious Loss of Heterozygosity Problem (PLOHP)*, i.e., the problem of partitioning suspected polymorphisms of a set of individuals into the minimum number of deletion areas. We generalize the work of Halldórsson *et al.*' by showing how one can incorporate prior knowledge about the location of deletion; we prove the general $\mathcal{NP}$-hardness of the problem and we provide a state-of-the-art mixed integer programming formulation and a number of possible strengthening valid inequalities able to exactly solve practical instances of the PLHOP containing up to 9.000 individuals and 3000 SNPs within 12 hours compute time.

Keywords: Clique partitioning, submodular functions, polymatroid rank functions, rule-constrained interval graphs, loss of heterozygosity.

1 Introduction

The recent completion of the Hap Map project [24] has shown that any two copies of the human genome differ from one another by approximately 0.1% of nucleotide sites, i.e., one variant per 1,000 nucleotides on average [2,15,18,25]. The most common variants, called *Single Nucleotide Polymorphisms* (SNPs), together with the recombination process, constitute the predominant form of human variation as well as determinants of human disease [3,23].

A large number of other types of variations exist, including insertions, inversions, translocations. One type of variation being *deletions*, which occur when a subsequence of the human genome is present in a reference genome but is not in the genome of an individual being analyzed. When the genotypes of a child and its two parents are known a deletion polymorphism may be observed as a *Loss of Heterozygosity* (LOH) event on the child chromosome. Specifically, the laws of Mendelian inheritance dictate that each individual inherits one copy of a chromosome from the father and one from the mother. Hence, for a fixed SNP of a chromosomal region, an individual can be either *homozygous*, i.e., the nucleotides of the parental DNA strands are equal, or *heterozygous*, i.e., the nucleotides of the parental DNA strands are different.

A Mixed Integer Programming Model for the Parsimonious Loss

Individuals	Molecular sequence of a specific chromosome region
Individual 1 (father - 1st DNA strand)	... A C A **A** G C C A T T C G **A** G G T C A G T C **C** A C C G ...
Individual 1 (father - 2nd DNA strand)	... A C A **A** G C C A T T C G **A** G G T C A G T C **C** A C C G ...
Individual 2 (mother - 1st DNA strand)	... A C A **G** G C C A T T C G **A** G G T C A G T C A A C C G ...
Individual 2 (mother - 2nd DNA strand)	... A C A **G** G C C A T T C G **C** G G T C A G T C A A C C G ...
Individual 3 (child - 1st DNA strand)	... A C A **A** G C C A T T C G **A** G G T C A G T C **C** A C C G ...
Individual 3 (child - 2nd DNA strand)	... A C A **G** G C C A T T C G **A** G G T C A G T C **C** A C C G ...

Trio	SNPs			
Individual 1 (father - 1st DNA strand)	...	A	A	C	...
Individual 1 (father - 2nd DNA strand)	...	A	A	C	...
Individual 2 (mother - 1st DNA strand)	...	G	A	A	...
Individual 2 (mother - 2nd DNA strand)	...	G	C	A	...
Individual 3 (child - 1st DNA strand)	...	A	A	C	...
Individual 3 (child - 2nd DNA strand)	...	G	A	C	...

Fig. 1. An example of a *trio* and their genotypes; A set of SNPs in the genomes of two parents and their offspring. The first highlighted column in the molecular sequence of the trio represents a SNP inconsistent with a loss of heterozygosity; the second highlighted column represents a SNP consistent with a loss of heterozygosity; the third highlighted column represents a SNP showing an evidence of a loss of heterozygosity.

When a deletion polymorphism occurs, an individual carries only a single copy of the chromosomal segment while the other is missing. If the deletion is *de novo*, the lack of information concerns only the individual and not the respective parents. Otherwise, the deletion is said to be *inherited* i.e., passed from one of the two parents to the child. Deletions may be deleterious for an individual and give rise to several human diseases. For example, recent studies have shown examples of deletions in schizophrenia [22], multiple sclerosis [14] and many other diseases.

As the location of deletions is not known, a number of methods have been suggested in the literature to restrict the areas of the genomes to be analyzed, including tiling arrays [6] and high throughput sequencing [4,8]. In this paper we focus on detecting deletions from genotype data (similar to those described in [5,9,14] and [19]). These data may be derived from SNP arrays, which have been used for genome-wide association at a number of laboratories, or other genotyping platforms.

Current SNP genotyping technology is not able to discern easily the difference between a homozygous site and a deletion, hence the output will always be a homozygous SNP even if the true genotype of the individual may carry only a single copy of the genotype. Moreover, even if a deletion polymorphisms were observed in molecular data, such events could be due either to the presence of real deletions or to *genotyping errors*, i.e., misreadings caused by the genotyping technology.

Trios	SNPs									
Trio 1	X	1	0	0	X	0	X	0	1	0
Trio 2	0	X	0	1	0	0	0	X	0	0
Trio 3	0	0	X	0	0	0	0	0	X	0

Fig. 2. An example of three trios having 10 SNPs each

2 Notation and Problem Formulation

Consider a *trio* t, i.e., a set of two parents and an offspring, and let s denote a given SNP genotyped in t. Then, one of the following three situations may occur:

1. The SNP s can be *Inconsistent with a Loss of Heterozygosity* (ILOH), a situation that occurs when the child is heterozygous. In this case the alleles must have been inherited from each parent (cf. first column of Figure 1).
2. The SNP s can be *Consistent with a Loss of Heterozygosity* (CLOH), a situation that occurs when a deletion may (but needs not) be introduced to explain the trio's inheritance pattern (cf. second column of Figure 1).
3. The SNP s can show *Evidence of a Loss of Heterozygosity* (ELOH), a situation that occurs when a deletion or a genotyping error are the only possible explanation for the trio inheritance pattern. (cf. third column of Figure 1)

Using the definitions described above, a trio genotyped at m SNPs can be encoded as a string of length m over an alphabet $\Sigma = \{1, 0, x\}$, where '1' codes for a SNP inconsistent with having a loss of heterozygosity; '0' codes for a SNP consistent with a loss of heterozygosity; and 'x' codes for a SNP showing evidence of loss of heterozygosity [14]. For example, the string $t = \langle x100x0x010 \rangle$ in Figure 2 represents a trio genotyped at 10 SNPs, thereof 5 consistent with having a loss of heterozygosity, 3 showing evidence of a loss of heterozygosity, and 2 inconsistent with having a loss of heterozygosity.

Denote $\mathcal{SNP}$ as the set of m SNPs to analyze and $\mathcal{T} = \{t^p\}$ as a set of n trios genotyped at the m SNPs in $\mathcal{SNP}$. Moreover, given a trio $t^p \in \mathcal{T}$, let t_s^p denote both the value and the position of s-th SNP in t^p. Fixed a trio $t^p \in \mathcal{T}$ and an ELOH x in t^p, denote $I = ([l,r], x)$ as the *pointed interval* induced by x, i.e., as the contiguous subset of SNPs in t^p delimited by the ILOHs l and r, including x, and containing no other ILOH in $[l, r]$. We set $l = 1$ if there is no ILOH SNP on the left of x and $r = m$ if there is no ILOH SNP on the right of x, respectively. For example, by considering the ELOH at position t_5^1 in Figure 2 we have that $l_5^1 = 3$ and $r_5^1 = 8$. Similarly, by considering the ELOH at position t_3^3 we have that $l_3^3 = 1$ and $r_3^3 = 10$. Finally, by considering the ELOH at position t_1^1 we have that $l_1^1 = r_1^1 = 1$.

Denoted $\mathcal{T}_x = \{t_s^p \in \mathcal{T} \times \mathcal{SNP} : t_s^p = \text{'x'}, t^p \in \mathcal{T}, s \in \mathcal{SNP}\}$, consider two ELOHs x_{k_1} and x_{k_2} in $\mathcal{T}_x$ and their corresponding induced pointed intervals

$I_{k_1} = ([l_{k_1}, r_{k_1}], x_{k_1})$ and $I_{k_2} = ([l_{k_2}, r_{k_2}], x_{k_2})$, respectively. We say that x_{k_1} and x_{k_2} are *mutually compatible with a deletion* if both $x_{k_1} \in [l_{k_2}, r_{k_2}]$ and $x_{k_2} \in [l_{k_1}, r_{k_1}]$. For example, the ELOHs at position t_3^3 and t_5^1 in Figure 2 are mutually compatible with a deletion; in fact the position of the ELOH t_3^3 falls inside $[l_5^1, r_5^1]$ and vice-versa the position of the ELOH t_1^5 falls inside $[l_3^3, r_3^3]$. In contrast, the ELOH t_1^1 is not compatible with any other ELOH in $\mathcal{T}_x$ as its position does not fall inside any other induced pointed interval.

The relationship of mutual compatibility between ELOHs plays a central role in detecting deletions in the human genome. By generalizing the definition of mutual compatibility, we define a *Region Compatible with a Deletion* (RCD) in $\mathcal{T}$ as a subset mutually compatible ELOHs in $\mathcal{T}_x$. For example, the ELOHs t_5^1, t_7^1 and t_8^2 in Figure 2 form a RCD, as their corresponding pointed intervals are mutually compatible. Note that a RCD provides more information that the simple mutual compatibility, as it can be used also to discern between true deletions and genotyping errors. In fact, since genotyping errors are usually sporadic during the genotyping phase, a high concentration of ELOHs in a specific area of $\mathcal{T}$ (i.e., a presence of a RCD) is likely to indicate the presence of true underlying deletions in that area. However, also note that the definition of a RCD alone cannot be used to classify all of the ELOHs in $\mathcal{T}_x$ as it does not provide a criterion to select RCDs from among plausible alternatives. In this context, Halldórsson et al. [14] observed that as deletion events are rare in nature, the number of RCDs can be expected to be small. Hence, the criterion of minimizing the overall number of RCDs in $\mathcal{T}$ can be considered a plausible approach to detecting deletions in the human genome. Denoted $\mathcal{R}$ and $\mathcal{E}$ as the set of RCDs and the set of genotyping errors in $\mathcal{T}$, respectively, and $h(\rho)$ and $g(\eta)$ as the costs of detecting a RCD $\rho \in \mathcal{R}$ and a genotyping error $\eta \in \mathcal{E}$, respectively, the authors proposed therefore to solve the following optimization problem to accomplish the task:

Problem. *The Parsimonious Loss of Heterozygosity Problem (PLOHP)*. Given a set $\mathcal{T}$ of n trios having m SNP each, minimize the overall cost

$$\chi = \sum_{\rho \in \mathcal{R}} h(\rho) + \sum_{\eta \in \mathcal{E}} g(\eta)$$

such that each entry in $\mathcal{T}_x$ is either compatible with a deletion or is classified as a genotyping error.

Halldórsson et al. [14] assumed that functions $h(\rho)$ and $g(\eta)$ always assign the same cost to each $\rho \in \mathcal{R}$ and $\eta \in \mathcal{E}$, respectively. In this article we relax this constraint and generalize the PLOHP to the case where we can have different costs depending on the SNP and deletion being considered. In fact, genotyping technologies are usually characterized by a high variability in the quality of the SNP genotypes produced [7]. Similarly, different regions in the genome may have different propensity for carrying deletions [8]. This fact justifies the need to weigh the different SNPs based on their probability of being a genotyping error. Hence, in what follows we shall assume that functions h and g are generic functions.

Halldórsson et al. [14] conjectured the general $\mathcal{NP}$-hardness of the PLOHP but did not investigate the issue any further. In the next sections we shall address this major issue and provide an exact algorithm able to exactly solve practical-use instances of the problem.

2.1 The LOH Graph

We revisit the graphs defined in [14], which we shall term *LOH Graphs* and turn out useful in transforming the PLOHP into a particular version of the Minimum Clique Partition Problem (MCPP) [12].

In order to characterize such a class of graphs, consider a set of trios $\mathcal{T}$ and denote $I_k = ([l_k, r_k], x_k)$ as the pointed interval induced by the k-th trio/SNP pair in $\mathcal{T}_x$, $\mathcal{I}$ as the set of pointed intervals induced by $\mathcal{T}$, and set $\nu = |\mathcal{I}|$. Consider a graph G_π having a vertex for each interval I_k, $k = 1, \ldots, |\mathcal{T}_x|$, and an edge between two vertices if a given intersection rule π is satisfied. If π concerns just the presence/absence of an intersection between two distinct pointed intervals then G_π is a classical *interval graph* (see [11]). If π concerns the presence/absence of mutual compatibility between two distinct pointed intervals then G_π is called a *LOH Graph* (LOHG).

Given an instance of the PLOHP and its corresponding LOHG G, we observe that by definition, the RCDs in $\mathcal{T}$ correspond to *cliques* in G, i.e., to complete subgraphs of G. We also recall that a *maximal clique* of a graph is a clique that is not a subset of a larger clique and a *maximum clique* is a clique of maximum size. Then, the following result holds for LOHGs:

Proposition 1. *Let G be a LOHG. Then G contains at most $\nu(\nu-1)/2$ maximal cliques.*

Proof. Take any set of vertices that forms a maximal clique C in G and consider the corresponding set of pointed intervals in $\mathcal{I}$. Let v_l and v_r be the nodes whose corresponding 'x' values x_{v_l} and x_{v_r} are the furthest to the left and to the right, respectively, in the set of pointed intervals induced by C. Each vertex v in the clique is connected to these two vertices and its corresponding pointed interval is such that $x_v \in [x_{v_l}, x_{v_r}] \subseteq [l_v, r_v]$. Hence, a clique can be defined by the leftmost and rightmost vertices, respectively, and as consequence G contains at most $\nu(\nu-1)/2$ maximal cliques.

We note that Proposition 1 implies that the maximum clique problem can be solved in polynomial time by enumerating all cliques. This approach can be performed in polynomial time by choosing all possible distinct pairs of values x_{v_l} and x_{v_r} in $\mathcal{I}$ and, for each of them, by listing the pointed intervals such that $x_v \in [x_{v_l}, x_{v_r}] \subseteq [l_v, r_v]$.

In the Section 4 we shall exploit Proposition 1 to develop an exact approach for solving the PLOHP based on mixed integer programming.

3 The Complexity of the PLOHP

Given an instance of the PLOHP and its corresponding LOHG G, we note that in any optimal solution to the instance, a genotyping error will always be a SNP that does not belong to any RCD selected. Hence, in any optimal solution to the instance, a genotyping error will correspond to a clique of G having cardinality 1. This insight allows us to consider the PLOHP as an instance of the MCPP in a particular class of graphs [12]. The MCPP is known to be $\mathcal{NP}$-hard in general[12]; in this section we will show that the MCPP (i) remains hard even when restricted to the class of the LOHGs and (ii) can be solved in polynomial time if the LOHG and the cost functions h and g satisfy some specific properties. Before proceeding, we shall introduce some notation that will prove useful throughout the section.

We say that a set function f is *zero-cardinal* if $f(\emptyset) = 0$; *non-negative* if f assumes only non-negative values; and *non-decreasing* if $f(T) \leq f(S)$ for any $T \subseteq S \subseteq V$. We say that f is *submodular* if it satisfies the following property [21]:

$$f(S \cup \{u\}) + f(T) \leq f(S) + f(T \cup \{u\}) \quad \forall\, T \subseteq S \subseteq V,\, u \in V \setminus S.$$

We say that f is a *polymatroid rank function* if it is zero-cardinal, non-decreasing, non-negative, and submodular. Moreover, similarly to [10], we define a *value-polymatroid* set function f as a zero-cardinal, non-decreasing, non-negative set function that satisfies the following property:

$$f(S \cup \{u\}) + f(T) \leq f(S) + f(T \cup \{u\})$$

for any $S, T \subseteq V : f(S) \geq f(T)$, $u \in V \setminus (S \cup T)$. Note that a value-polymatroid set function is also polymatroidal, but the converse is generally not true [10]. Finally, a set function f is *size-defined submodular* if there exists a function $\psi : [0 \dots |V|] \to \mathbb{R}_0^+$ such that $f(S) = \psi(|S|)$, for any $S \subseteq V$. As shown in [10], a size-defined submodular set function f is both value-polymatroidal and polymatroidal.

The previous definitions turn out to be useful to investigate the complexity of the PLOHP. Specifically, denote $\mathcal{C}(G)$ the set of cliques of G and set

$$f(C) = \begin{cases} 0 & \text{if } C = \emptyset \\ g(C) & \text{if } |C| = 1 \\ h(C) & \text{if } |C| \geq 2. \end{cases} \quad \forall\, C \in \mathcal{C}(G)$$

Then, the following proposition holds:

Proposition 2. *The decision version of the PLOHP is $\mathcal{NP}$-complete even when the cost function f is restricted to polymatroidal set functions.*

Proof. The statement follows by observing that the class of the LOHGs strictly contains the class of interval graphs and that the minimum clique partition problem on an interval graph is $\mathcal{NP}$-complete when the cost function is polymatroidal [10].

In general, it is easy to verify that the decision version of the PLOHP is $\mathcal{NP}$-complete for any cost function $f(C) = \psi(|C|)\sigma(C)$ such that $\psi : [0\ldots|V|] \to \mathbb{R}_0^+$ and $\sigma(C)$ is a generic function on C. In fact, such a case also includes the rooted-TSP cost function on a tree (see [10]) which is trivially polymatroidal.

Although Proposition 2 states that the decision version of the PLOHP is in general $\mathcal{NP}$-complete, in some cases the problem can be still solved in polynomial time. For example, the following proposition holds:

Proposition 3. Let $G = (V, E)$ be a LOHG and $f : \mathcal{C}(G) \to \mathbb{R}_0^+$ a value-polymatroidal cost function. If G is also an interval graph then it is possible to compute a minimum cost partition into cliques of G in polynomial time.

Proof. The statement follows by observing that the class of the LOHGs strictly contains the class of interval graphs and that the minimum clique partition problem on an interval graph can be solved in polynomial time when the cost function is a value-polymatroidal set function [10].

In their article, Halldórsson et al. [14] introduced the following cost function for PLOHP:

$$f_\alpha(C) = \begin{cases} 0 & \text{if } C = \emptyset \\ c_1 & \text{if } |C| = 1 \\ c_2 & \text{if } |C| \geq 2, \end{cases} \quad \forall \, C \in \mathcal{C}(G) \qquad (1)$$

where c_1 and c_2 are two constants such that $0 < c_1 \leq \alpha c_1 < c_2 \leq (\alpha+1)c_1$, and α is a positive integer such that $2 \leq \alpha \leq |V|-1$. Interestingly, Proposition 3 proves useful to show that such a version of the PLOHP can be solved in polynomial-time if G is an interval graph. In fact, in such a case it is easy to see that the set function $f_\alpha(C)$ is size-defined submodular, hence value-polymatroidal, thus if G is an interval graph by Proposition 3 the PLOHP can be solved in polynomial time.

4 A Mixed Integer Programming Model for the PLOHP

The $\mathcal{NP}$-hardness of the PLOHP justifies the development of exact and approximate solution approaches similar to those described in [14]. In this section we present a mixed integer programming model for the PLOHP. The algorithm is guarenteed to return an optimal solution and its time performance is significantly faster than the current state-of-the-art exact algorithm described in [14].

To this end, given a vertex $v \in V$, we denote $\mathcal{C}_v = \{C \in \mathcal{C}(G) : v \in C\}$. Moreover, we denote y_C as a decision variable equal to 1 if the clique $C \in \mathcal{C}(G)$ is selected in the optimal solution to the problem and 0 otherwise. Then, a valid formulation for the PLOHP is the following:

Formulation 1

$$\min \sum_{C \in \mathcal{C}(G)} f(C) y_C \tag{2a}$$

$$\text{s.t.} \sum_{C \in \mathcal{C}_v} y_C = 1 \quad \forall\, v \in V \tag{2b}$$

$$y_C \in \{0,1\} \quad \forall\, C \in \mathcal{C}(G). \tag{2c}$$

Constraints (2b) impose that each vertex $v \in V$ belongs to the clique $C \in \mathcal{C}(G)$ and constraints (2c) impose the integrality on variables y_C.

Formulation 1 is characterized by an exponential number of variables and constraints and its linear relaxation can be exactly solved by using column generation techniques. Specifically, observe that a variable with negative reduced cost in the linear relaxation of Formulation 1 corresponds to a dual constraint violated by the current dual solution. Denoted μ_v as the dual variables associated with constraints (2b), the dual of the linear relaxation of Formulation 1, denoted by LP1, is characterized by the following constraints:

$$\sum_{v \in V : v \in C} \mu_v \leq f(C) \quad \forall\, C \in \mathcal{C}(G). \tag{3}$$

Constraints (3) are violated if there exists a clique $\hat{C} \in \mathcal{C}(G)$ such that

$$\sum_{v \in V : v \in C} \mu_v > f(C)$$

. The existence of such a clique can be checked in polynomial time by using Proposition 1 and this in turn implies that the linear relaxation of LP1 can be solved in polynomial time.

Interestingly, if the cost function f is defined as in (1) then Formulation 1 can be rewritten as follows. Denote x_v as a decision variable equal to 1 if vertex $v \in V$ forms a clique of cardinality 1 in the optimal solution to the problem and 0 otherwise. Moreover, denote $\hat{\mathcal{C}}(G)$ as the set of all maximal cliques in G having cardinality greater or equal to 2 and $\hat{\mathcal{C}}_v(G) = \{C \in \mathcal{C}(G) : v \in C, |C| \geq 2\}$. Then a valid formulation for the PLOHP is the following:

Formulation 2

$$\min \sum_{C \in \hat{\mathcal{C}}(G)} c_2 y_C + \sum_{v \in V} c_1 x_v \tag{4a}$$

$$\text{s.t.} \sum_{C \in \hat{\mathcal{C}}_v(G)} y_C + x_v \geq 1 \quad \forall\, v \in V \tag{4b}$$

$$x_v \in \{0,1\} \quad \forall\, v \in V \tag{4c}$$

$$y_C \in \{0,1\} \quad \forall\, C \in \hat{\mathcal{C}}(G). \tag{4d}$$

Formulation 2 has the benefit of being polynomial-sized, due to Proposition 1, hence in principle its relaxation does not require the use of column generation techniques to be solved. Moreover, note that both Formulations 1 and 2 can be strengthened by adapting appropriately the inequalities described in [1,13,20].

5 Experiments

Halldórsson et al. [14] first proposed an exact approach to solution of the PLOHP based on the explicit enumeration of all the possible clique partitions of the LOHG G. The algorithm was tested on a set of 5 artificial instances of the PLOHP, generated by using Hudson Simulator [16] and containing 3000 trios having 3575 SNP each. Halldórsson et al. observed that their algorithm was unable to solve these instances for this reason the authors developed a possible greedy algorithm to approximate the optimal solution to the problem.

In order to measure the performances of our model we considered the same instances of Halldórsson et al. and assumed 12 hours as limit runtime for our model. As shown in the next section, Halldórsson et al.'s instances proved to be quite easy to solve by our model, hence conservatively we also generated harder instances of the PLOHP, characterized by a smaller number of trios and SNPs and a higher density of ELOHs.

We implemented our model in ANSI C++, using FICO Xpress 64-bit BCL libraries to encode the model and FICO Xpress Optimizer v20.00.11 to exactly solve the instances of the problem. The model ran on a Pentium 4, 3.2 GHz, equipped with 2 GByte RAM and release 7 of the Gentoo operating system (linux kernel 2.6.17). In our experiments we deactivated FICO Xpress pre-solving strategy and proprietary cuts and activated FICO Xpress primal heuristic to generate the first upper bound to the problem. This particular combination was chosen by observing, in preliminary experiments, that the computational overhead introduced by the pre-solving strategy and cut generation did not lead to a substantial improvement of the solution times.

As regards to the objective function, in order to compare the performances of our model versus Halldórsson et al.' exact algorithm, we used function (1) to deal with the random instances of the PLOHP and used the same coefficients described in [14] to set the constants c_1 and c_2. Moreover, in order to simulate the high variability in the quality of the SNP genotypes produced by genotyping technologies and the different propensity of the regions in the genome to carry deletions, we also considered an alternative objective function to analyze Halldórsson et al.' instances. Specifically, we used the first 3500 recombination rates between sexes, populations, and individuals [17] (appropriately rescaled in the interval $[0, 1]$) relative to chromosome 1 in order to associate a weight to each SNP of the considered real instances. Then, we computed the objective function as

$$f(C) = \begin{cases} 0 & \text{if } C = \emptyset \\ b * \alpha_r & \text{if } |C| = 1 \\ |C|\gamma_C & \text{if } |C| \geq 2, \end{cases} \quad \forall\, C \in \mathcal{C}(G) \qquad (5)$$

where b is a random number uniformly distributed in $[0, 1]$, α_r equal to the average rate in the considered chromosomic region, and γ_C is the average of the rates associated to the SNPs involved in the clique C. Codes and datasets can be downloaded upon request.

Table 1. Performances on Halldórsson et al.' instances for $c_1 = 1$ and $c_2 = 2$

	$c_1 = 1$ and $c_2 = 2$			$c_1 = 2$ and $c_2 = 11$		
Instance	Gap (%)	Nodes	Time (sec.)	Gap (%)	Nodes	Time (sec.)
gens1	0	1	14741.9	0	1	15049.5
gens2	0	1	17479.6	0	1	15975.8
gens3	0	1	17205.7	0	1	15234.1
gens4	0	1	13914.5	0	1	13490.8
gens5	0	1	8318.1	0	1	15337.4

Table 2. Performance on Halldórsson et al.' instances (for 1000 Trios and 1500 SNPs) when assuming objective function (1)

	$c_1 = 1$ and $c_2 = 2$			$c_1 = 2$ and $c_2 = 11$		
Instance	Gap (%)	Nodes	Time (sec.)	Gap (%)	Nodes	Time (sec.)
gens1	0	1	404.76	0	1	982.49
gens2	0	1	401.23	0	1	900.74
gens3	0	1	442.85	0	1	886.69
gens4	0	1	399.51	0	1	920.76
gens5	0	1	444.49	0	1	985.95

5.1 Numerical Results

Tables 1-2 show the results of the model when tackling Halldórsson et al.' instances of the PLOHP under different objective functions. Specifically, the columns of each table evidence the Gap expressed in percentage (i.e., the difference between the optimal value found and the value of linear relaxation at the root node of the search tree, divided by the optimal value), the number of branches performed, and the solution time taken to exactly solve each instance.

As general trend, we observed that our model constitutes a very tight formulation for the PLOHP, being characterized by Gap is in all cases 0% and by an average number of branches are in all cases 1. The results showed that there is not a big variation in terms of runtime performance when considering different value for c_1 and c_2 (see e.g., Table 1). However, there is a notable variation when considering function (5) with respect to function (1).

We observed a generalized increment of the runtimes independently from the considered objective function. Specifically, Table 1 and show that, when using the same objective function, the solution times took a minimum of 2.5 hours (see, e.g., instance gens5) up to almost 5 hours (see, e.g., instance gens2) to exactly analyze a specific instance of the PLOHP. This fact is mainly justified by the larger size of Halldórsson et al.' instances with respect to the random instances previously described, which in turn implies a heavier computational overhead necessary to enumerate all possible maximal cliques in the LOH graph.

Table 3. Performances on Halldórsson et al.' instances (for 1000 Trios and 1500 SNPs) when assuming objective function (5)

Instance	1000 × 1500			2000 × 2000		
	Gap (%)	Nodes	Time (sec.)	Gap (%)	Nodes	Time (sec.)
gens1	0	1	458.36	0	1	2823.77
gens2	0	1	457.89	0	1	2883.34
gens3	0	1	442.85	0	1	2852.16
gens4	0	1	352.61	0	1	2926.84
gens5	0	1	999.07	0	1	2878.87

Similarly, we observed longer runtimes also when considering the objective function (5). In order to provide a better evidence of this phenomenon, we analyzed the leading principal submatrices 1000x1500 and 2000x2000 of each instance in Halldórsson et al.' dataset, both in absence and in presence of rates (i.e., both when considering the objective functions (1) and (5), respectively; see Tables 2 and 3). As general trend, the results showed that, when considering the rates, the solution times increased from 105% up to 317%, by keeping a similar trend in terms of Gaps and nodes. As for the random instances, possibly the use of column generation techniques and divide and conquer strategies (e.g., graph decomposition methods), out of the scope of the present article, could improve the solution time of the model and make it more prone to tackle Halldórsson et al.' instances in presence of recombination rates.

Acknowledgments. The first author acknowledges support from the Belgian National Fund for Scientific Research (F.N.R.S.), of which he is "Chargé de Recherches". Both the first and the second authors also acknowledge support from Communauté Française de Belgique — Actions de Recherche Concertées (ARC) and "Ministerio de Ciencia e Innovación" through the research project MTM2009-14039-C06. Part of this work has been developed when Dr. Catanzaro was visiting Reykjavik University.

References

1. Bandelt, H.J., Oosten, M., Rutten, J.H.G.C., Spieksma, F.C.R.: Lifting theorems and facet characterization for a class of clique partitioning inequalities. Operations Research Letters 24(5), 235–243 (1999)
2. Cargill, M., et al.: Characterization of single-nucleotide polymorphisms in coding regions of human genes. Nature Genetics 22, 231–238 (1999)
3. Catanzaro, D., Andrien, M., Labbé, M., Toungouz-Nevessignsky, M.: Computer-aided human leukocyte antigen association studies: A case study for psoriasis and severe alopecia areata. Human Immunology 71(8), 783–788 (2010)
4. Chen, K., et al.: BreakDancer: an algorithm for high resolution mapping of genomic structural variation. Nature Methods 6, 677–681 (2009)

5. Conrad, D.F., Andrews, T.D., Carter, N.P., Hurles, M.E., Pritchard, J.K.: A high-resolution survey of deletion polymorphism in the human genome. Nature Genetics 38, 75–81 (2006)
6. Conrad, D.F., et al.: Origins and functional impact of copy number variation in the human genome. Nature (464), 704–712 (2009)
7. The International HapMap Consortium, The international hapmap project. Nature 426(18), 789–796 (2003)
8. Corbel, J.A., et al.: PEMer: a computational framework with simulation-based error models: for inferring genome structaral variants from massive paired-end sequencing data. Genome Biology 38(10), R23 (2009)
9. Corona, E., Raphael, B., Eskin, E.: Identification of Deletion Polymorphisms from Haplotypes. In: Speed, T., Huang, H. (eds.) RECOMB 2007. LNCS (LNBI), vol. 4453, pp. 354–365. Springer, Heidelberg (2007)
10. Dion, G., Jost, V., Queyranne, M.: Clique partitioning of interval graphs with submodular costs on the cliques. RAIRO Operations Research 41, 275–287 (2007)
11. Fishburn, P.C.: Interval orders and interval graphs: Study of partially ordered sets. John Wiley and Sons Inc., NY (1985)
12. Garey, M.R., Johnson, D.S.: Computers and intractability: A guide to the theory of NP-Completeness. Freeman, NY (2003)
13. Grötschel, M., Wakabayashi, Y.: Facets of the clique partitioning polytope. Mathematical Programming 47, 367–387 (1990)
14. Halldórsson, B., Aguiar, D., Tarpine, R., Istrail, S.: The Clark phase-able sample size problem: Long-range phasing and loss of heterozygosity in GWAS. Journal of Computational Biology 18(3), 323–333 (2011)
15. Halushka, M.K., Fan, J.B., Bentley, K., Hsie, L., Shen, N., Weder, A., Cooper, R., Lipshutz, R., Chakravarti, A.: Patterns of single nucleotide polymorphisms in candidate genes of blood pressure homeostasis. Nature Genetics 22, 239–247 (1999)
16. Hudson, R.R.: Generating samples under a Wright-Fisher neutral model of genetic variation. Bioinformatics 18, 337–338 (2002)
17. Kong, A., et al.: Fine-scale recombination rate differences between sexes, populations and individuals. Nature 467, 1099–1103 (2010)
18. Li, W.H., Sadler, L.A.: Low nucleotide diversity in man. Genetics 129, 513–523 (1991)
19. McCarroll, S.A., et al.: Integrated detection and population-genetic analysis of snps and copy number variation. Nature Genetics 40, 1166–1174 (2008)
20. Oosten, M., Rutten, J.H.G.C., Spieksma, F.C.R.: The clique partitioning problem: Facets and patching facets. Networks 38(4), 209–226 (2001)
21. Schrijver, A.: Combinatorial optimization: Polyhedra and efficiency. Springer, NY (2003)
22. Stefansson, H., et al.: Large recurrent microdeletions associated with schizophrenia. Nature 455, 232–236 (2008)
23. Terwilliger, J., Weiss, K.: Linkage disequilibrium mapping of complex disease: Fantasy and reality? Current Opinions in Biotechnology 9, 579–594 (1998)
24. The International HapMap Consortium, A second generation human haplotype map of over 3.1 million SNPs. Nature 449(18), 851–861 (2007)
25. Wang, D.G., et al.: Large-scale identification, mapping, and genotyping of single-nucleotide polymorphisms in the human genome. Science 280(5366), 1077–1082 (1998)

Reconstruction of Transcription Regulatory Networks by Stability-Based Network Component Analysis

Xi Chen[1], Chen Wang[1], Ayesha N. Shajahan[2], Rebecca B. Riggins[2], Robert Clarke[2], and Jianhua Xuan[1]

[1] Department of Electrical & Computer Engineering, Virginia Polytechnic Institute and State University, Arlington, VA 22203
{xichen86,topsoil,xuan}@vt.edu
[2] Lombardi Comprehensive Cancer Center and Department of Oncology, Georgetown University, Washington, DC 20057
{ans33,rbr7,clarke}@georgetown.edu

Abstract. Reliable inference of transcription regulatory networks is still a challenging task in the field of computational biology. Network component analysis (NCA) has become a powerful scheme to uncover the networks behind complex biological processes, especially when gene expression data is integrated with binding motif information. However, the performance of NCA is impaired by the high rate of false connections in binding motif information and the high level of noise in gene expression data. Moreover, in real applications such as cancer research, the performance of NCA in simultaneously analyzing multiple candidate transcription factors (TFs) is further limited by the small sample number of gene expression data. In this paper, we propose a novel scheme, stability-based NCA, to overcome the above-mentioned problems by addressing the inconsistency between gene expression data and motif binding information (i.e., prior network knowledge). This method introduces small perturbations on prior network knowledge and utilizes the variation of estimated TF activities to reflect the stability of TF activities. Such a scheme is less limited by the sample size and especially capable to identify condition-specific TFs and their target genes. Experiment results on both simulation data and real breast cancer data demonstrate the efficiency and robustness of the proposed method.

Keywords: transcription regulatory network, network component analysis, stability analysis, transcription factor activity, target genes identification.

1 Introduction

Transcription factors (TFs) are special proteins that control and affect the rate of downstream genes' mRNA expression. TFs can either activate or inhibit gene expression, usually by binding to short and highly conserved DNA sequences in promoter (or upstream) regions. Such transcriptional regulatory relationships can be explicitly described as transcriptional regulatory networks [1], consisting of TFs as regulators and downstream genes as targets.

It has been known that a TF may play different regulatory roles to its downstream target genes or even has different downstream targets under different conditions [2]. Motivated by this understanding, many computational algorithms [3, 4] are proposed to discover condition-specific regulatory networks. By incorporating protein-DNA interaction (PDI) information into a linear latent variable model, several methods have also been proposed to solve the inference problem using regression or constraint regression like Network Component Analysis (NCA) [1] and Least Square (LS) regression [5]. Among them, NCA is a prominent method, which has mathematically derived identifiable conditions [6] and resulted in successful applications for several real biological problems [7, 8].

However, most of the existing methods including NCA do not consider the false positives in given network connections. Integrating gene expression data with inconsistent biological knowledge is a true challenge for the identification of TFs and their target genes. Another issue has to be considered is that NCA requires that the number of samples must be larger than the number of TFs [1]. However, in real practice, the number of gene expression data samples is usually smaller than the number of candidate TFs. Mainly because we don't know whether a certain TF is active or not in the condition where the gene expression data is sampled. Many candidate TFs should be tested in order to find the 'true' regulators. This limits the application of NCA to real biological studies, such as cancer research, for active TF identification.

In this paper, we propose a novel stability-based NCA (sNCA) approach to tackle the problem of regulatory network inference. By assuming that the consistency relationship between expression data and biological knowledge remains stable after small perturbations introduced to the prior knowledge of network topology, we can assign the knowledge-derived estimation (i.e., transcription factor activity (TFA), regulatory strength (RS), etc.) an instability score (ISS), which represents the average distance of multiple estimations upon different perturbations. To solve the NCA limitation about the small sample size (relative to the number of TFs) in sNCA, at each time, we randomly sample a small number of TFs (<number of samples), calculate and store their ISS value. Then, after multiple times of randomly sampling, the averaged ISS value of each TF is calculated and used to prioritize all the TFs under investigation. Another scheme adopts local rank method. Here, we just change a small part of the selected TFs and only record the relatively stable TFs. After multiple times, the TFs counted for the most times are the global stable ones. Having the estimated activities of the TFs prioritized by sNCA, we further propose to use multivariate regression to rank the condition-specific target genes of active TFs.

To assess the performance of sNCA for TF identification, we have first applied it to simulation data in comparison with NCA and LS regression methods. The comparison results clearly demonstrate that sNCA outperforms the other two methods, especially when the network topology is incomplete or has false connections. Then, we apply sNCA to one breast cancer microarray dataset [9] to identify condition-specific TFs, aiming to provide insight into the underlying regulatory mechanism related to the development and progression of breast cancer. The experimental results show that our method can identify biologically meaningful TFs and target genes associating with estrogen signaling and action in breast cancer.

2 Motif-directed NCA and Stability Analysis

The sNCA approach is aimed to prioritize significant condition-specific TFs and their target genes, and uncover regulatory networks utilizing both gene expression data and binding motif information. Fig. 1 shows a flowchart of the proposed sNCA approach consisting of the following major components: (1) motif-direct NCA (m-NCA) and (2) stability analysis (SA). In the following subsections, we will give a description of each component in details.

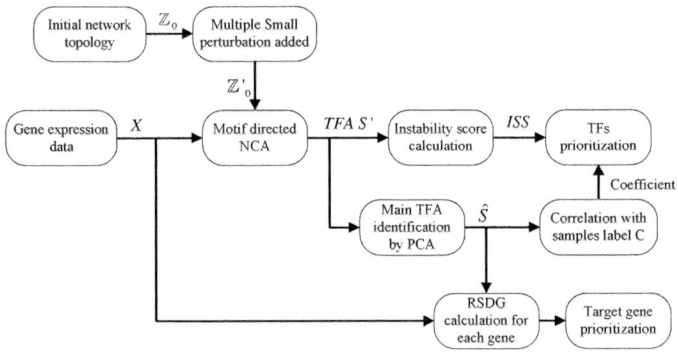

Fig. 1. Flowchart of the proposed sNCA approach

2.1 Motif-directed NCA

m-NCA scheme was proposed by us in [10] to infer regulatory networks, which utilizes sequence motif information to construct initial network connections, and later integrates with gene expression data to estimate the TFA and their downstream targets with the NCA algorithm. To construct the initial network, $Match^{TM}$ [11] can be used to search the transcription factor binding sites (TFBS) for their position-weighted matrices (PWMs) that can be extracted from the TRANSFAC 11.1 professional Database [12]. According to the PWMs, a motif score can be calculated for each TF-gene pair. These motif scores provide the initial connection information for m-NCA. Based on motif scores, we encode regulation relationship from L TFs to N target genes as a network connectivity pattern $B \in (0,1)^{N \times L}$, which is a binary matrix with element $b_{n,l} = 1$ indicating potential regulatory relationship from l-th TF to n-th gene.

The gene expression data could be modeled as follows:

$$x_n = \sum_{l=1}^{L} a_{nl} s_l + \gamma_n \qquad (1)$$

where $x_n = [x_{n1}, \cdots, x_{nM}]$ and $\gamma_n = [\gamma_{n1}, \cdots, \gamma_{nM}]$ are gene expression profile and noise vector of n-th gene, respectively. M is the number of gene expression samples. $a_n = [a_{n1}, \cdots, a_{nL}]$ represents the regulatory strength for each TF on the n-th gene and $s_l = [s_{l1}, \cdots, s_{lM}]$ is the hidden activity vector of l-th TF.

We further define regulatory strength (RS) matrix $A = [a_1, \cdots, a_N]^T$, and TFA matrix $S = [s_1, \cdots, s_L]^T$ light of their biological implication. Finally, the gene expression data $X = [x_1, \cdots, x_N]^T$ can be represented in a matrix-vector form as follows:

$$X = AS + \Gamma \quad (2)$$

where $\Gamma = [\gamma_1, \cdots, \gamma_N]^T$ is the noise matrix, subject to zero-mean Gaussian distribution.

To solve Eq. (2), based on available biological knowledge B and gene expression data, the NCA algorithm is designed to estimate A and S by minimizing the following fitting errors:

$$(\hat{A}, \hat{S}) = \arg\min_{(A,S)} \|X - AS\|_2^2, \quad s.t.\ A \in \mathbb{Z}_0 \quad (3)$$

where $\mathbb{Z}_0$ is a regulatory matrix set, deriving from $B \in (0,1)^{N \times L}$:

$$\mathbb{Z}_0 \triangleq \left\{ A \in \mathbb{R}^{N \times L} \mid a_{nl} = 0 \text{ for } b_{nl} = 0 \right\} \quad (4)$$

2.2 Stability Analysis

Stability analysis (SA) is used to identify true positive connection in the TF-gene networks. Actually, as TF binding motif is a relatively short sequence pattern, the topology obtained from motif information is quite noisy and contains many false positive connections. Since the initial topology information is often unreliable for specific TF-gene pairs, we are going to address the inconsistency between biological knowledge derived from DNA motif information and expression data through our proposed sNCA scheme. By adding small perturbations to the given topological connections, sNCA can be used to prioritize condition-specific TFs and their target genes.

We aim to study the regulatory role of the l-th TF through its TFA estimation s_l, but the quality of biological knowledge, network information, is our concern. We can keep the expression unchanged and generated a perturbed $\mathbb{Z}'_0$ by intentionally altering a small amount of connections in given $\mathbb{Z}_0$. We would expect a deviation (denoted as d) between estimated s'_l (based on the perturbed $\mathbb{Z}'_0$) and original estimation s_l (based on $\mathbb{Z}_0$). The rationale behind this is that the TFA estimation of a condition-specific TF should be more robust to a small amount of perturbation than any non-specific TF that generally lacks of support in data-network consistency.

Multiple perturbations are added to network topology for SA. With each individual perturbation, a different estimated TFA can be generated. P independent perturbations on original $\mathbb{Z}_0$ obtain P versions of $\mathbb{Z}_p, 1 \leq p \leq P$ and the degree of perturbation should be very small (FP=FN=$\alpha \ll 1$). Mathematically, we denote the perturbation function as follows:

$$\mathbb{Z}_p = Perturb_\alpha(\mathbb{Z}_0, p), 1 \le p \le P \tag{5}$$

The estimated TFA $\hat{s}_{l,p}$ of the l-th TF and associated RS $\hat{a}_{l,p}$ (with respect to expression data X) for the p-th perturbed regulatory matrix $\mathbb{Z}_p$ can be decomposed by NCA algorithm as

$$(\hat{a}_{l,p}, \hat{s}_{l,p}) = NCA_{l-thTFA}(X, \mathbb{Z}_p) \tag{6}$$

where $NCA_{l-thTFA}$ represents the NCA estimation process for the l-th TF.

After P times of above procedure, a stability measure, namely instability score (ISS) of TFA, is introduced here as follows:

$$ISS_l = \frac{1}{P(P-1)} \sum_{p1=1}^{P} \sum_{p2=1, p2 \ne p1}^{P} d(\hat{s}_{l,p1}, \hat{s}_{l,p2}) \tag{7}$$

where the distance function $d(v_1, v_2)$ is defined as

$$d(v_1, v_2) = \log[d_0(v_1, v_2) / (1 - d_0(v_1, v_2))] \tag{8}$$

$$d_0(v_1, v_2) = \min(0.9, \cos(v_1, v_2), \cos(v_1, -v_2)) \tag{9}$$

where, $\cos(\cdot, \cdot)$ is the cosine distance. To prevent d from approaching infinity when d_0 is close to 1, here, we limit the maximum value of d_0 to 0.9.

If the candidate TF number is larger than the sample number, to address an NCA limitation, here we randomly sample a small number of TFs (< sample number) at each time, then calculate and store their ISS values. After multiple times of randomly sampling, the averaged ISS value of each TF is calculated, which is further used to prioritize all the TFs. We call this process as 'average ISS based method'.

Considering another selection scheme, first, we randomly sample a small number L of TFs. Then another L' TFs are selected from the rest ones, and totally $L+L'$ TFs (< sample number) are tested as Eq. (6). Based on their ISS value, the top L TFs are retained and the count for each is added by 1. In each round, we randomly select a new set of L' TFs and do stability analysis. After multiple times of sampling, the count of each TF is used to prioritize all the TFs. This is named as 'rank based' sampling method, compared to the 'average ISS based' sampling scheme.

Based on identified TFs, we further utilize multivariable regression scheme to identify target genes controlled by them. For the n-th gene, we use the activities of its potential regulators, TFA, to regress its expression profile x_n and calculate the resulted p-value of regression coefficient. p-value reflects how well the gene expression data can be fitted by corresponding TFAs. Here, principle component analysis (PCA) [13] is adopted to estimate the main TFA $\hat{s}_{n,l}$ among P times of m-NCA results.

With the estimated TFA matrix $\hat{S}$, considering formula (2), by using the least square errors regression method, the associated RS matrix $\hat{A}$ is obtained as:

$$\hat{A}^T = (\hat{S}\hat{S}^T)^{-1}\hat{S}X^T \tag{10}$$

We simply definite the relevance score of downstream gene (RSDG) based on its average significance level according to TFA regression analysis.

$$RSDG_n = \frac{-\sum_{l=1}^{L} \log(p - value(\hat{a}_{n,l}))}{count(\hat{a}_{n,l} \neq 0)} \tag{11}$$

where function $count(\cdot)$ is to count the number of regulatory strength $\hat{a}_{n,l}$ which does not equal to zero. The higher the RSDG is, the more likely this gene is truly regulated by l-th TF under this condition.

To summarize, for inference of transcription regulatory networks using the proposed stability-based approach, we propose two consecutive steps: 1) identify condition-specific TFs by sNCA; 2) identify condition-specific downstream target genes based on stable TFA regression.

3 Simulation and Experiment Results

3.1 Simulation Data

To test the performance of the sNCA approach, we generate 100 different simulated regulatory networks consisting of 30 TFs and 500 target genes. On average, each gene is regulated by 3 TFs. The gene expression data is formed by using the model defined in formula (1), where the RS and TFA both follow Gaussian distribution. The purpose of this study is to first identify true regulators and then their significant target genes.

Here we design two kinds of simulation data. For 'none-overlap' case, 50 gene expression data samples are generated as regulated by 15 TFs, and another 15 TFs have no impact on the gene expression data. So just some unrelated network information is

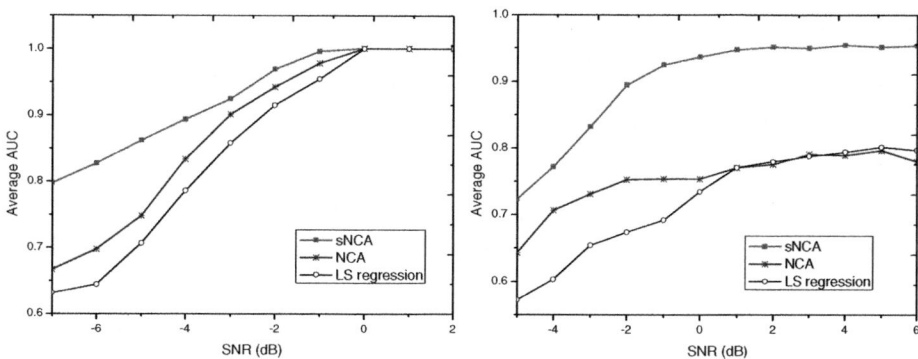

Fig. 2. Comparison of TF identification performances by ROC studies: (a) AUC of the 'none – overlap' case; (b) AUC of the 'overlap' case

added to the 'true' network, which has 'none overlap' with the regulatory mechanism of the gene expression data. Comparably, for the 'overlap' case, the gene expression samples are regulated by all 30 TFs, within which 15 TFs have false positive/false negative connections of 5% and the rest is contaminated with false connections of 30%. Here, apparently, the network with high false connections has 'overlap' with the regulatory mechanism of the gene expression data. Actually, sometimes we don't know whether a certain TF has any relationship with condition specific gene expression data. At this time, in order to not miss important TFs, we have to enlarge the range of candidate TFs. If the added TFs have no relationship with gene expression data, like the 'none-overlap' case, the true regulatory mechanism is not affected. But if the added TFs have certain relationship with the data but with high noise in their network topology, the false positive of the network information is increased, as the 'overlap' case. Here, we compare sNCA TF identification performance with NCA and LS regression under different SNR conditions. The performance is comprehensively measured by receiver operating characteristic studies; in particular, the area-under-ROC curve (AUC) is used, as shown in Fig. 2 (a) and (b).

In Fig.2, we can see that the LS regression always has the worst performance compared to the others. It is reasonable because there is no refining estimation about the regulation strength between TFs and their target genes. In Fig.2 (a), 'none-overlap' case, NCA has a comparable performance as sNCA, mainly due to the correct network topology information assumed in the 'none-overlap' case. However, if the signal to noise ratio (SNR) is low, the consistency between gene expression data and network topology is affected severely by the noise in expression data, which makes NCA cannot estimate the TFA with high accuracy. So its performance comes down close to the LS regression. This proves that NCA algorithm is quite sensitive to inconsistency between the network topology and expression data. In Fig.2 (b), the NCA performs much worse than sNCA for the 'overlap' case. Apparently, the false connections in network topology severely affect NCA algorithm accuracy. Comparing the performance of sNCA in Fig.2 (a) and (b), we can find that if the network is more complicated (regulated by 15 vs. 30 TFs), the sNCA scheme is more robust, although false connections exist. Therefore, sNCA improves the robustness of the TF identification compared to the NCA scheme, and could potentially be more effective when applied to the inference of complex networks.

Further, sNCA is used to identify the target genes of stable TFs, where neither NCA nor LS regression could work. The performance of target gene identification by sNCA in the above two simulation data sets is summarized in Table 1.

Considering the 5% false connection in the 'none-overlap' case, the AUC value of target gene identification comes close to 0.95 when the SNR is high. In the 'overlap' case, the TF - gene regulation mechanism is affected more by the high level false

Table 1. Average AUC of target gene identification by sNCA based on RSDG value

SNR (dB)	-6	-4	-2	0	2	4	6
none-overlap	0.705	0.769	0.817	0.855	0.890	0.917	0.938
overlap	0.699	0.731	0.774	0.800	0.836	0.854	0.885

connections. But the AUC value for target gene identification does not degrade significantly. This point proves that sNCA could the estimate TFA with high robustness against the inconsistency between gene expression data and network topology. And also, the RSDG is a proper measure to prioritize the target gene.

To investigate the small sample number vs. large TF number problem, here, we increase the number of TFs to 80, and keep sample number still 50. For the 'none-overlap' case, gene expression data is regulated by 40 TFs, and the rest TFs have no impact on the gene expression data. For the 'overlap' case, gene expression data is regulated by 80 TFs, within which 40 TFs are contaminated with 30% of false connections. Because NCA scheme requires that the sample number should be larger than the TF number, so it cannot be used in this case due to its theoretical limitation. As mentioned in section 2.2, sNCA could solve this problem by sampling a subset of TFs and calculating the mean ISS value after multiple randomly sampling, or counting the relatively stable TFs within each round of sampling. Thus, we test the TF identification performance of sNCA with both sampling methods, which are shown in Table 2.

Table 2. Average AUC for the identification of TFs by sNCA

SNR (dB)	-6	-4	-2	0	2	4	6
none overlap	0.654	0.805	0.874	0.934	0.935	0.949	0.960
overlap: average ISS based	0.538	0.613	0.660	0.710	0.720	0.734	0.755
overlap: rank based	0.560	0.652	0.701	0.748	0.775	0.792	0.836

Compared to Fig. 2 and Table 1, in Table 2, for the 'none-overlap' case, although the performance of TFs identification by sNCA degrades slightly due to the incomplete network knowledge, it could still keep a relatively high accuracy to identify true regulatory modules. Here the 'average ISS based' method and rank based method (not listed here) have similar performance. For the 'overlap' case, if we use the average ISS value after multiple times of sampling to identify stable TFs, the AUC performance degrades more compared to the rank based method. So 'rank based' method could identify global optimal TFs from local prioritization selection. Although the overall performance degrades when the sample number is smaller than the number of TFs, to certain extent, the sNCA approach can overcome the limitation of NCA and provide us a possible solution to analyze multiple candidate TFs simultaneously with limited samples, which is especially useful in cancer research.

3.2 Breast Cancer Expression Data

The sNCA scheme is further applied to a breast cancer expression data set [9] to identify condition specific TFs and significant target genes. According to the survival time, we group the gene expression data samples into two groups, 49 samples in 'early' group (< 3 years) and another 51 samples in 'late' group (> 8 years). T-test is used to select top ~4,000 genes that have different patterns between the early and late groups. Then, these genes are further divided into two clusters. Genes in 'Cluster 1' have higher expression values in the early group, and comparatively, genes in 'Cluster 2' express higher in the late group. From literature and biological databases, we have nearly 240 TFs, which are likely associated with breast cancer and estrogen signaling. Finally, sNCA is applied to cluster 1 and 2 respectively, to test each candidate TF.

Actually, we are more focusing on the TFs whose activities have different pattern between the two sample groups. In this case, their target genes are more probably associated with the survival time of breast cancer patients. Here, defining a phenotype vector $C = [c_1, \cdots, c_M]$ for M samples (100 in our case), with -1 for each sample in the early group and 1 for each sample in the late group. Among all the TFs selected by sNCA, the coefficient (Coef.) of the l-th TF's activity s_l and vector C is calculated as

$$Coef_l = \sum_{m=1}^{M} s_{l,m} c_m \bigg/ \sqrt{\operatorname{var}(s_l) \operatorname{var}(C)} \qquad (12)$$

The absolute value of *Coef.* is used to further refine our selected TFs pool. Based on the ISS value from sNCA and the *Coef.*, the top stable TFs are listed in Table 3.

Table 3. Top ranked TFs based on their ISS and absolute value of Coef

TF name	ISS	Coef.	TF name	ISS	Coef.
OCT1_03	0.0119	0.3949	CEBPB_01	0.0337	-0.3097
ELK1_02	0.0142	0.4068	ETS_Q6	0.0342	0.3567
E2F_Q4	0.017	-0.3499	CREBP1CJUN_01	0.0373	-0.376
T3R_Q6	0.0172	0.4531	CETS1P54_01	0.0385	0.3809
NFY_C	0.0202	0.3283	PBX1_01	0.0416	0.4512
FREAC4_01	0.0245	0.4144	WT1_Q6	0.0459	0.4831
OCT1_02	0.0253	-0.3441	NFY_Q6_01	0.0482	-0.3186
HMGIY_Q3	0.0259	0.4285	PPARA_02	0.0493	-0.3475

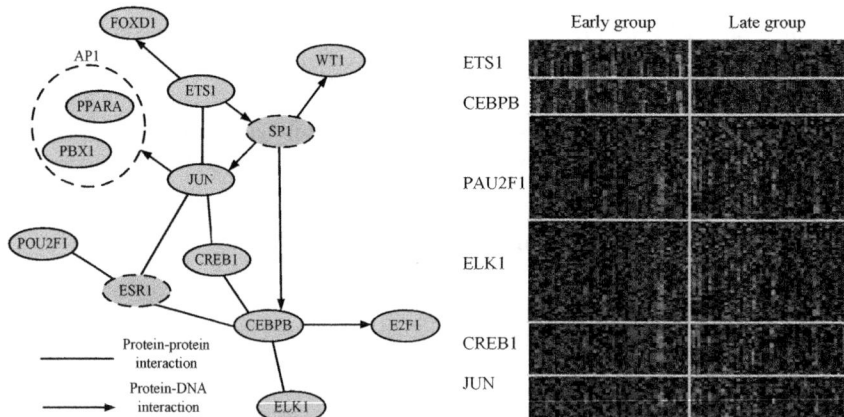

Fig. 3. (a) Identified TFs network and (b) associated target genes pattern

As shown in Fig. 3 (a), these tumor and breast cancer related TFs have direct relationships through protein-protein interaction (PPI) and protein-DNA interaction (PDI). Here, transcription factor ESR1 and SP1 (the green bin in Fig.3 (a)) are added in to make the network more complete. Although they are not covered in Table 3, their important neighbors are ranked quite high through our analysis. Also, some existing biology literatures provide strong support to our computational results. Among all the selected TFs, E2F1, a key transcription factor involved in proliferation, differentiation and apoptosis has been observed to associate with breast cancer favorable outcome [14]. CREB1 is a co-activator of ESR1. ELK1 is a subclass of ETS family and a breast cancer study has shown over expression of ETS1 is indicative of poor prognostics [15]. JUN is associated with drug resistance [16] and CEBPB is already well known to be associated with breast cancer [17]. WT1 also has relationship with breast carcinoma [18]. Although two important TFs PPARA and PBX1 are not connected directly to the other components in our network, these two TFs have intense relationship with AP1. AP1 is a well known TF that has been identified in breast cancer research [19]. POU2F1 is a member of the POU family of TFs and is involved in hormonal signals [20].

Moreover, based on Eq. (11), the candidate target genes are sorted by the RSDG value. In the Fig. 3 (b), we show the expression pattern of the significant target genes identified by sNCA for several TFs appearing in the network. These genes are regulated by TFs with significant different activities between the early and late groups. So they also show significant and consistent pattern associating with their regulators. To certain extent, these genes have some relationship with the patient survival time. To test the prediction performance of these genes, support vector machine (SVM) is adopted here. 5-fold cross validation is applied to the 100 breast cancer samples in the early and late groups. The receiver operator characteristics (ROC) curve and the Kaplan–Meier (KM) survival time plot are shown in Fig. 4 (a) and (b), respectively.

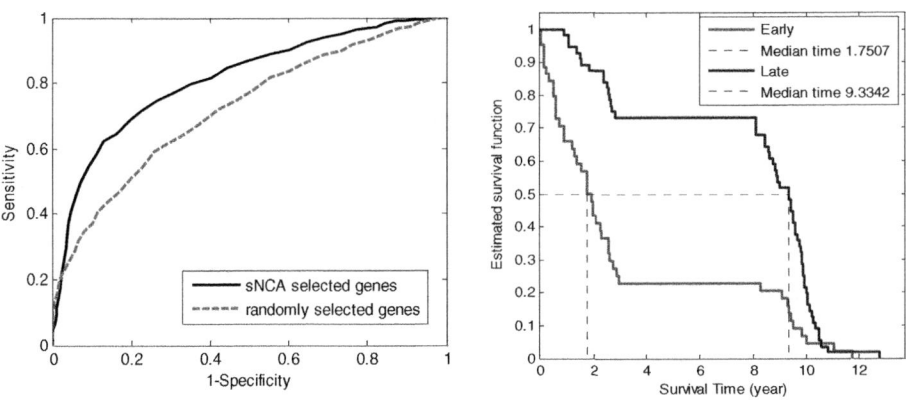

Fig. 4. Prediction performances of 200 downstream target genes. (a) Comparison of ROC performance, (b) KM plot of the genes selected by SA.

From the cross validation ROC curve in Fig. 4(a), the AUC value of the prediction performance of genes identified by sNCA is 0.8106, compared to 0.72 of the genes randomly selected from the ~4,000 genes (from t-test). Also, when the sensitivity (true positive rate) is fixed at 0.70, the specificity is 0.79 for sNCA identified genes and 0.66 for randomly selected genes. Considering Fig. 4(b), the KM plot, when the survival time is at 3 years, the late group survival portion is 0.72, compared to 0.22 for the early group. The division of the two groups is quite well. So from computational point of view, the genes that the sNCA scheme identifies are highly associated with the survival time of breast cancer patients.

Further, we selected the top 5% genes (~ 200) ranked by their RSDG value to do functional analysis and annotation. Based on Ingenuity Pathway Analysis (IPA) system [21], the functional annotation of these genes can be found in Table 4.

Table 4. Summary of functional groups of sNCA identified target genes

Functional annotation	Molecules
Cancer	81
DNA replication, repair	31
Cell death	46
Cellular growth and proliferation	39
Cell cycle	33

Table 4 shows that a high portion of the target gene can be functionally grouped together. Through the annotation, we know that these processes have intense relationship with cancer progression. Also, considering that our data set is estrogen-dependent breast cancer, many transcription factors included in estrogen-receptor signaling pathways (e.g., CREB1, STAT5, ATF2, JUN, ELK1) appear in our identified TFs and target genes. Also, from bottom to up, CREB1, CEBPA, CEBPB, JUN, ETS1 and E2F1 are highlighted by our identified target genes in IPS system. This further proves that our identified TFs and target genes support each other well.

4 Conclusions

In this paper, we have proposed a novel approach, stability-based network component analysis (sNCA), to prioritize TFs and their target genes simultaneously. The sNCA approach overcomes a major limitation in many existing methods as caused by the inconsistency between gene expression data and TF-gene binding knowledge. With a proper stability analysis procedure, the sNCA approach also overcomes the limitation of NCA due to the small sample size of gene expression data. Through simulation studies, the proposed sNCA approach outperforms both NCA and LS regression, especially when the network topology information is incomplete or with a high level of noise. The experimental results from an analysis of real breast cancer data have further demonstrated that sNCA can help identify biologically meaningful regulatory networks associated with the development and progression of breast cancer.

Acknowledgement. This work was supported by the National Institutes of Health (CA139246, CA149653 CA149147 and NS29525-18A).

References

1. Liao, J.C., Boscolo, R., et al.: Network component analysis: reconstruction of regulatory signals in biological systems. Proc. Natl. Acad. Sci. USA 100(26), 15522–15527 (2003)
2. Luscombe, N.M., Babu, M.M., et al.: Genomic analysis of regulatory network dynamics reveals large topological changes. Nature 431, 308–312 (2004)
3. Chen, L., Xuan, J., et al.: Multilevel support vector regression analysis to identify condition specific regulatory networks. Bioinformatics 26(11), 1416–1422 (2010)
4. Rverter, A., Hudson, N.J., et al.: Regulatory impact factors: unraveling the transcriptional regulation of complex traits from expression data 26(7), 896–904 (2010)
5. Nguyen, D.H., Haeseleer, P.D.: Deciphering principles of transcription regulation in eukaryotic genomes. Mol. Syst. Biol. 2 (2006)
6. Boscolo, R., Sabatti, C., et al.: A generalized framework for network component analysis. IEEE/ACM Trans. Comput. Biol. Bioinformatics 2(4), 289–301 (2005)
7. Brynildsen, M.P., Liao, J.C.: An integrated network approach identifies the isobutanol response network of Escherichia coli. Mol. Syst. Biol. 5, 277 (2009)
8. Ye, C., Galbraith, S.J., et al.: Using network component analysis to dissect regulatory networks mediated by transcription factors in yeast. PLoS Comput. Biol. 5(3), e1000311 (2009)
9. Loi, S., Haibe-Kains, B., et al.: Definition of clinically distinct molecular subtypes in estrogen receptor-positive breast carcinomas through genomic grade. J. Clin. Oncol. 25(10), 39–46 (2007)
10. Wang, C., Xuan, J., et al.: Motif-directed network component analysis for regulatory network inference. BMC Bioinformatics 9, S.1, S21 (2008)
11. Kel, A.E., Gossling, E., et al.: MATCH: A tool for searching transcription factor binding sites in DNA sequences. Nucleic Acids Res. 31(13), 3576–3579 (2003)
12. Matys, V., Kel-Margoulis, O.V., et al.: Transfac and its module TransCompel: transcriptional gene regulation in eukaryotes. Nucleic Acids Res., D108–D110 (2006)
13. Jolliffe, I.T., et al.: Principle component analysis, 2nd edn., pp. 167–198. Springer, NY (2002)
14. Vuaroqueaux, V., et al.: Low E2F1 transcript levels are a strong determinant of favorable breast cancer outcome. Breast Cancer Research 9, R33 (2007)
15. Buggy, Y., Maguire, T.M., et al.: Over expression of the Ets-1 transcription factor in human breast cancer. Br. J. Cancer 91(7), 1308–1315 (2004)
16. Daschner, P.J., Ciolino, H.P., et al.: Increased AP-1 activity in drug resistant human breast cancer MCF-7 cells. Breast Cancer Res. Treat. 53(3), 229–340 (1999)
17. Shackleford, T.J., et al.: Stat3 and CCAAT/enhancer binding protein beta (C/EBP-beta) regulate Jab1/CSN5 expression in mammary carcinoma cells. Breast Cancer Research 13, R65 (2011)
18. Domfeh, A.B., et al.: WT1 immunoreactivity in breast carcinoma: selective expression in pure and mixed mucinous subtypes. Modern Pathology 21, 1217–1223 (2008)
19. Shen, Q., et al.: The AP-1 transcription factor regulates breast cancer cell growth via cyclins and E2F factors. Oncogene 27, 366–377 (2008)
20. Kakizawa, T., et al.: Silencing mediator for retinoid and thyroid hormone receptors interact with octamer transcription factor-1 and acts as a transcriptional repressor. J. Biol. Chem. 276, 9720–9725 (2001)
21. Ficenec, D., et al.: Computational knowledge integration in biopharmaceutical research. Brief. Bioinformatics 4, 260–278 (2003)

A Transcript Perspective on Evolution

Yann Christinat and Bernard M.E. Moret

Laboratory of Computational Biology and Bioinformatics
EPFL, 1015 Lausanne, Switzerland

Abstract. Alternative splicing is now recognized as a major mechanism for transcriptome and proteome diversity in higher eukaryotes. Yet, its evolution is poorly understood. Most studies focus on the evolution of exons and introns at the gene level, while only few consider the evolution of transcripts.

In this paper, we present a framework for transcript phylogenies where ancestral transcripts evolve along the gene tree by gains, losses, and mutation. We demonstrate the usefulness of our method on a set of 805 genes and two different topics. First, we improve a method for transcriptome reconstruction from ESTs (ASPic), then we study the evolution of function in transcripts. The use of transcript phylogenies allows us to double the specificity of ASPic, whereas results on the functional study reveal that conserved transcripts are more likely to share protein domains than functional sites. These studies validate our framework for the study of evolution in large collections of organisms from the perspective of transcripts; we developed and provide a new tool, TrEvoR, for this purpose.

Keywords: alternative splicing, transcript, evolution, phylogeny, protein domain, transcriptome reconstruction.

1 Introduction

Gene duplication and loss are the main driving forces for transcriptome and proteome diversity. However, alternative splicing—a greatly underestimated mechanism twenty years ago—has now been shown to play a major role for diversity in higher eukaryotes [1,2]. In many genomes, most genes are thus split into introns and exons. The standard splicing scheme keeps all exons and removes all introns, but alternative splicing permits removal of alternative exons. Some mRNAs are further translated into proteins—named isoforms—and alternative proteins can therefore vary in large regions or may even not overlap.

Alternative splicing is limited in plants and fungi but quite common in vertebrates. Some researchers conjecture that 90% of human multi-exon genes are alternatively spliced [3]. Yet the study of evolution from a transcript perspective has not seen much work, while the evolution of the mechanism itself is poorly understood. Several articles focus on the evolution of the gene structure—exons and introns—but none address the problem of transcript evolution [3]. The few studies on this matter are limited to mouse and human and agree on the fact that alternative splicing is a fast evolving mechanism [4,5]. For instance, Nurtdinov *et al.* showed that only three quarters of the human isoforms have an ortholog in mouse [6].

The number of alternative isoforms is specific to species but also unevenly documented. For instance, the Ensembl database [7] reports numerous transcripts for human, mouse, rat, several apes, and a few fishes but almost no alternative splicing for dogs or cats. Some gene families also display a very different rates of alternative splicing across their species; a member can be extensively spliced in a species and have only one transcript in another species. The available data comes mainly from experiments and it is expected to be incomplete. Current automated pipelines for transcriptome reconstruction are not trusted, so that large-scale multi-species analysis is not doable at present. Confronted with the incompleteness of the data and the presence of extensively spliced genes, some researchers conjecture that all alternative transcripts are possible and that the observed set reflects a regulated distribution of all transcripts. Nonetheless, some transcripts are conserved among homologous genes. The question remains to be answered whether the function—be it gene ontologies, tissue or sub-cellular localization, or developmental stages—is also conserved or if they just represent noise in the splicing machinery.

In this paper, we extend our previous work [8] and present a transcript evolution framework where ancestral transcripts evolve along the gene tree through transcript gains or losses, and exon gains or losses. This entire process is represented by a forest of transcript phylogenies that links the observed transcripts of a gene family. This framework has applications in many transcript-related fields. Among these last, we selected two to demonstrate the benefits gained through our method. In Section 3.1, we address the problem of transcriptome reconstruction, as it represents a core issue for transcript analysis. We refined the output of ASPic (the Alternative Splicing Prediction DataBase [9]): as measured using the RefSeq database [10], our method doubles the specificity of ASPic. In Section 3.2, we study the distribution of protein domains in transcript phylogenies, using many Ensembl tracks. Our results indicate that transcripts are more likely to conserve their domains than their functional sites through evolution.

These two studies establish the usefulness of our transcript phylogeny framework for large-scale biological analyses. The tool we developed for these analyses TrEvoR (Transcript Evolution Reconstruction software), is publicly available and can be downloaded at http://lcbb.epfl.ch/trevor.

2 A Model of Transcript Evolution

Our model of transcript evolution is a refinement of the extended model we presented in [8]. Given the set of transcripts of the most ancient gene, a transcript evolves along the gene tree through three simple events: mutation (gain or loss of exons), fork (creation of new transcripts), and death (loss of transcripts). This process is represented in a set of transcript trees, one for each ancestral

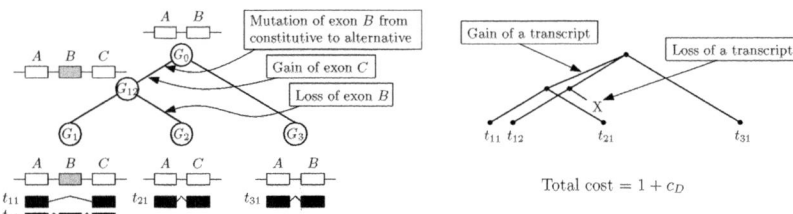

Fig. 1. Illustration of the two-level model. The first level is represented in A where the gene evolution happens. In B, one can see the transcript phylogeny. Transcripts t_{12} and t_{31} differ by exon C which was gained during the evolution from G_0 to G_{12}. This event belong to the first level and thus has zero cost in the transcript phylogeny. There one transcript loss (cost $= c_D$) and there is a new splicing pattern in G_{12} which cannot be explained by the evolution of the gene structure. The total edge costs is thus $1 + c_D$.

transcript. Mutations happen along the branches and affect the content of transcripts whereas forks and deaths affect the structure of the transcript trees.

In the absence of prior work, we opted for a maximum parsimony framework. Every event is assigned a unit cost, except for transcript death, which is the sole event of the model to be parameterized (c_D). Other frameworks such as maximum likelihood or Bayesian networks can be used, but they tend to yield higher computational costs.

Our model aims at reflecting the cost of transcript evolution alone; thus the cost of gene evolution is discarded. This implies a two-level model, where the evolution of the gene structure serves as a basis for the evolution of the transcriptome. For instance the loss of an exon at the gene level implies that all transcripts lose this exon. In a classical maximum parsimony framework, these events would add their cost to the score. In our model, however, they do not since they are the unavoidable consequence of a gene event. This concept is illustrated through an example in Figure 1.

3 Results

3.1 Transcriptome Reconstruction

Next generation sequencing methods yield an increasing amount of data and reconstructing transcripts from short reads is a complex problem [11]. Once ESTs are mapped on the genome and splice junctions are identified, a splice graph can be constructed—nodes represent exons and edges splice junctions. Any path on this graph is thus a potential transcript. The remaining problem is to identify the "true" transcripts within this graph.

Several methods exist to predict transcripts from ESTs. ESTGene, which is part of the Ensembl pipeline, reconstructs the minimal set of transcripts that cover the splice graph [12,7]. ECGene, another method, parses the splice graph and clusters transcripts based on the nature of the splice sites [1]. ATP, the algorithm behind the ASPic database, is similar to ESTGene but includes additional rules to predict transcripts [11,9]. Other methods such as Scrip- ture [13], Cufflinks [14], or the EM algorithm by Xing et al. [15] also aim at transcriptome reconstruction but have no associated database. However, none of these methods make use of phylogenies.

Methods. We selected from the ASPic database 805 human genes that have an ortholog in rat, mouse, opossum, chimpanzee, marmoset, and macaque. These six species had the highest rate of alternative splicing in the Ensembl database while being the closest to human. Exons and transcripts were collected from the ASPic database for human and from Ensembl for the remaining species. Genes were aligned using MAFFT and orthologous exons were established when reciprocally overlapping at 70%. Alternative 3'- or 5'-end exons were assigned when the overlap was not reciprocal.

We refine the prediction of ASPic through a simple algorithm. For each human transcript present in the ASPic database, we collected the Ensembl transcripts of the homologous genes and reconstructed a transcript phylogeny on this set plus the ASPic transcript. The total cost of this transcript phylogeny—as defined in our model—is assigned as the score of the ASPic transcript. Once every ASPic transcript is assigned a score, the algorithm discards all transcripts that have an unreasonable evolutionary score within the transcript set of a given gene. Since the evolutionary score is dependent of the number of exons and may vary greatly between genes, we considered the score ratio. Consequently the algorithm searches for the two groups of transcripts that maximize the ratio of their respective mean scores—a 2-mean clustering algorithm. If that ratio is larger than t, then the high-cost group is discarded. Therefore, if $t = \infty$ the performance of the refinement algorithm will be equal to the original algorithm since no transcript is removed.

Results. We tested the performance of ECGene and ASPic by matching their predicted transcripts (from their online database) to the RefSeq database [10]— a gold standard for RNA sequences. A true positive was defined as an exact sequence match between the query and a sequence in RefSeq. A false positive is thus a transcript predicted by ASPic that could not be matched in RefSeq.

As shown in Table 1, both methods performed quite poorly. The best method, ASPic, can only recover 7% of all refseq transcripts. The use of transcript phylogenies yielded a significant increase in specificity while withstanding a minor decrease in sensitivity. Note that since we filter the ASPic's transcripts, sensitivity cannot increase. The Ensembl database had a specificity of 97% on the same set of genes. It is however impossible to assess the part played by ESTGene as the Ensembl database includes all sequences from RefSeq. The ECGene results could not be refined as exon information was not available; only transcript RNA sequences are available for download.

Table 1. Results of the different algorithms using exact matches on the RefSeq database. Our refinement on the ASPic prediction yielded a slight decrease in sensitivity and a 2-fold increase in specificity. Both refined results were achieved with the same threshold t but different transcript loss cost.

	Sensitivity	Specificity
ECGene	0.9%	1.2%
ASPic	7.23%	5.2%
Ref. ASPic ($c_D = 1$)	6.4%	11.5%
Ref. ASPic ($c_D = 10$)	7.0%	9.7%

Following Bonizzoni et al. [11], we opted for a gentler setup where exact matches are not required at the end of the sequences. That is, a predicted transcript is a true positive if it is a substring of a sequence in RefSeq and if no additional exons could have been added by the algorithm. That is, there exists no predicted transcript that contains the missing ends. Under this setup, ASPic performed better—13.7% specificity and 22.8% sensitivity—and our refinement method pushed the specificity to 29.8%, again a two-fold increase, while lowering the sensitivity to 19.7%.

We then varied the threshold t in our refinement algorithm and investigated the influence of different models and phylogenies on the performance of the refinement step. We tested two evolutionary models,

- two-level model: Our standard model of transcript evolution where gene events have zero cost.
- simple model: Our standard model of transcript evolution but with gene cost. For instance, the cost of losing an exon at the gene level is passed onto each transcript.

and three different phylogenies.

- Best tree: The reconstructed phylogeny output by our algorithm.
- Random tree: Random transcript trees that still agree with the gene tree. (Average on 100 runs)
- Star tree: All transcripts are directly linked to the root.

The ROC curves in Figure 2 show that the two-level model performs better than the simple model. The star phylogeny outperforms random trees. In random trees, the signal is completely lost as orthologous transcripts may not be on the same tree, thus conveying wrong information, whereas, on a star phylogeny, no information is provided beyond the contents of the transcripts. It performs then a simple comparison of the human transcript to all non-human transcripts. This is enough to gain some specificity but it cannot reach the performance of the best tree setup—be it on the two-level or the simple model. If a random subset of ASPic transcripts was selected, the sensitivity would simply diminish without any gain in specificity.

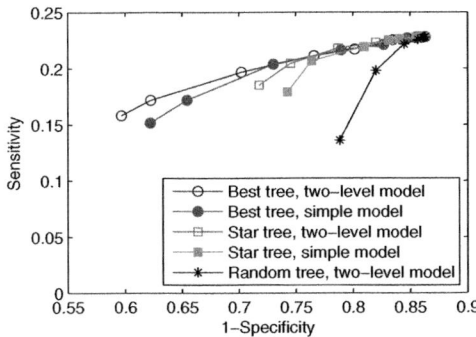

Fig. 2. Loose matches on RefSeq under the different setups and thresholds. All curves converge towards the ASPic performance at the top right corner. An optimal curve would go horizontally from the ASPic performance towards the left.

The previous experiment tested how a single transcript fits within the phylogeny. However, transcriptome reconstruction is about sets of transcripts. Therefore, we computed the cost of evolution for the Ensembl transcripts in the homologous genes plus different sets of human transcripts: all ASPic transcripts (ALL), no human transcripts (NO), and ASPic transcripts that were matched in RefSeq under the exact setup ($EXCT$) and loose setup (LOO). In this experiment, TrEvoR is only used to score different sets of human transcripts; our refinement algorithm is not applied. Note that human is the only species to be affected by these changes since we selected only human genes from the ASPic predictions. The NO setup corresponds to an evolutionary scenario where all human transcripts were lost in this gene family. The expected result is that the set of exactly matched transcript should have the lowest cost. We expect the loose match setup to sometimes have the lowest cost, as the RefSeq may not be exhaustive. We ran the algorithm on the 213 genes that had at least one exact match and observed that with a transcript loss cost of 1, removing all transcripts yielded the minimum cost in 76% of the cases. Nonetheless with a transcript loss cost of 10, 94% of the 213 genes had a exact match score lower than the loose match score and lower than the empty set. Table 2 summarizes these results and shows that the choice of the transcript loss cost (c_D) is an important matter.

To conclude, we demonstrate that transcript phylogenies can enhance transcriptome reconstruction from ESTs. Focusing on transcript evolution and discarding the cost of gene evolution yielded better results. A direction for future work is to integrate the transcript phylogenies directly into the reconstruction method. That way, both sensitivity and specificity could be increased.

3.2 Functional Study on Transcripts

To demonstrate the broad scope of our framework, we applied transcript phylogenies to the study of function in transcripts. We inquired whether transcript

Table 2. Statistics on the evolutionary cost of different subsets of the predicted transcripts. The first three conditions look at the expected behavior. (The *LOO* and *EXCT* setups should have the lowest score.) The last condition tests whether it is always profitable to remove transcripts. Values indicate the percentage of genes that fit the condition.

Condition	$c_D = 1$	$c_D = 10$
$s_{ALL} > s_{LOO} < s_{NO}$	21%	95%
$s_{ALL} > s_{EXCT} < s_{NO}$	23%	98%
$s_{ALL} > s_{EXCT} < s_{LOO}$	23%	94%
s_{NO} is min	76%	2%

phylogenies carry any functional information and, in the positive case, if two different transcript trees vary in functions. We thus studied the correlation between protein domains—structurally stable regions that often correspond to specific functions—and transcript phylogenies. Interestingly, conserved exons also displayed high correlation with protein domain boundaries [4].

Methods. We used the same dataset, 7 species and 805 genes, as for the transcriptome reconstruction problem. Transcripts and domain annotations were retrieved from the Ensembl database and transcript phylogenies were reconstructed with our algorithm. The sole parameter, the cost of transcript loss, was set as a proportion of the average number of exons in a set of homologous genes, denoted as $c_D = \alpha E[exs]$, and three values of α were tested: 0.1, 1, and 10. Different domain annotation databases, all available as tracks in Ensembl, were selected:

- *InterPro*: An integrated resource for protein families, domains, regions, and functional sites. It combines data from several databases such as *PROSITE, PRINTS, SMART, SUPERFAMILY, Pfam* and many others.
- *Pfam*: A database of protein families identified by sequence alignments and hidden Markov models.
- *PROSITE*: A collection of protein domains, families, and functional sites identified through patterns and profiles.
- *SMART*: A database of well annotated protein domains.
- *SUPERFAMILY*: A set of hidden Markov models that represent domains at the superfamily level in SCOP (structure-based classification of proteins).
- *SEG*: A software that divides the sequence into low- and high- complexity regions.
- *Transmembrane*: Identification of transmembrane helices with TMHMM.

We tested the transcript phylogenies for robustness and found 135 gene families where transcripts were grouped under the same parents across the three different transcript loss costs (c_D). A higher c_D may force two trees to be reunited under the same tree which is the reason why we only look at the last two levels of the transcript phylogenies. These genes are good candidates for having more than one well-conserved ancestral transcript.

We want to test whether the ancestral transcripts had the same functional content or not. If equal then the distribution of the domain content in each tree should be roughly equal to the distribution of the domains across all transcripts. Therefore, for each gene family F, we computed the probability of a domain d to appear in a transcript. This is our background probability,

$$P_F[d] = \frac{\text{Nb of transcripts in } F \text{ that contain } d}{\text{Nb of transcripts in } F} \qquad (1)$$

A similar value was computed for each tree, $P_t[d]$. Note that we only account for the presence of the domain. The number of occurrences per transcript does not matter.

We selected phylogenies with multiple trees and computed, for all domains, the deviation of each tree from the background probability. For a given domain, we have thus

$$Dev[d] = \frac{1}{|T|} \sum_{t \in T} (P_t[d] - P_F[d])^2 \qquad (2)$$

where T represent the set of transcript trees. The mean value over all domains, $E[Dev]$, gives us an indication of the global deviation of the domain content in the trees from the domain content in all transcripts.

In order to test if the 135 "stable" genes had a different deviation from the rest, we performed a random sampling among the 805 genes. The sampling was repeated a hundred times then averaged. We refer to this sampling as the "random set".

Results. As can be seen in Figure 3, the deviation between the stable and the random sets differ mainly for low values of c_D. The stable set, except for *PROSITE*, has always a lower deviation than the average on the random set. As the cost increases, the stable set converges towards the random set. Remarkably, *Transmembrane* and *SEG* did not display any significant differences between random and stable sets for any values of c_D. A study by Cline et al. showed that transmembrane regions are not likely correlated with alternative splicing [16]—a finding that endorses our results. Interestingly, the *PROSITE* database is the only one to exhibit a higher deviation for the stable set (at $c_D = E[exs]$). It is also the only database to include functional sites. The *InterPro* database does include the *PROSITE* database but its behavior resembles the other databases. *InterPro* collects data from 11 databases. The *PROSITE* functional sites annotations may thus be a minority and have little influence on the global result.

To test if the difference in *PROSITE* for $c_D = E[exs]$ was indeed significant and not a visual artifact, we performed a one-way ANOVA on *PROSITE*, *Pfam*, *SMART*, and *SUPERFAMILY* and another on the same databases but without *PROSITE*. The first analysis returned a p-value of $1.5036 \cdot 10^{-28}$ while the second returned a value of 0.1632. Consequently we can confidently reject the null hypothesis—all means are equal—in the first case but not in the second. This indicates that the *PROSITE* database has a significantly larger deviation than the other databases between the two sets. Note that these databases may

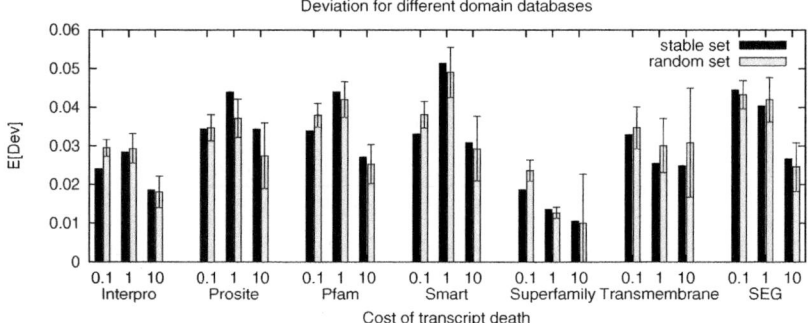

Fig. 3. Deviation from the expected domain presence for different domain annotation databases. For each database, three setups for the transcript loss cost were tested for the stable and the random sets: $c_D = 0.1E[exs]$, $c_D = 1E[exs]$, and $c_D = 10E[exs]$. The *PROSITE* is the only database to display a significantly higher deviation for the stable set.

cover similar domains and consequently may not be independent, which poses a problem for statistical analyses.

All databases, except *PROSITE*, have a lesser deviation than the random set for low c_D value. The *PROSITE* exception can be explained by an averaging effect. Proteins domains tend to yield a smaller deviation while functional sites create a larger deviation. The combination of both averages the score and results in a deviation similar to the random set. The *SUPERFAMILY* has, globally, a lower deviation than any other database. That is to be expected as the *SUPERFAMILY* database clusters domains with a higher abstraction level than families. That is, two different domain families may be reunited under a single superfamily and thus lower the deviation.

We also tested the stable set against the 805 genes for GO enrichment with GOrilla [17] but could not find any over- or under-represented ontologies.

Based on these results, one could conjecture that well conserved transcripts contain similar sets of domains but different functional sites. A deeper study could focus on the functional sites and potentially identify unknown functions in some transcripts.

4 Reconstruction of Transcript Phylogenies

Reconstructing transcript phylogenies is nontrivial: the problem is at least as hard as the standard phylogeny reconstruction problem, which is NP-hard. In a standard tree reconstruction, the tree structure is unknown but we know that only one tree exists. In a transcript phylogeny reconstruction, the tree structure is partially known, as it has to be a subtree of the gene tree, but the number of trees is unknown. Our previous algorithm did not scale well, hence we designed a heuristic based on neighbor-joining and packaged it into a convenient tool:

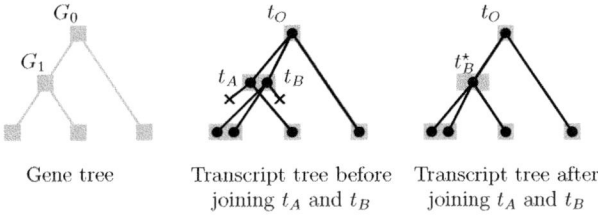

Fig. 4. A join operation on a simple transcript tree. JOIN(t_A, t_B) is valid because they share a common ancestor and belong to the same gene, G_1. Note that the two transcript deaths are lost after the operation.

TrEvoR (Transcript Evolution Reconstruction). The latter is available online and can be downloaded at http://lcbb.epfl.ch/trevor. A manual and some toy examples are also present.

4.1 TrEvoR Algorithm

Our two-level model of transcript evolution depends on the gene structure and, similar to our previous algorithm, the first step is thus to reconstruct the exons of the ancestral genes. Sankoff's algorithm for the small parsimony problem is applied and backtracking yields the ancestral states [18]. Our algorithm searches then for the most parsimonious forest of transcript trees.

Parsimony methods for the "standard" phylogeny reconstruction problem use a specific operation to search the tree space—nearest neighbor interchange, subtree pruning and regrafting, or tree bisection and reconnection. In our case, we define the "join" operation, which given an ancestral gene, merges two of its transcripts. The join operation simply assigns all children of transcript t_A to transcript t_B, deletes transcript t_A, and updates transcript deaths in t_B. Note that a join operation is only possible if the two transcripts share a common ancestor. Figure 4 illustrates the join operation on a simple example.

In the initialization step of the algorithm, every current transcript has its own ancestor (trivial trees). A neighbor-joining algorithm with the "join" operation is then applied at the root of the gene tree. At each iteration all possible join operations on two trees are tested, the best candidates are retained, their root are joined, and the algorithm moves to the next iteration. Similar to the leaf assignment procedure of the first algorithm, the score of a tree is tested by propagating the join operation from the root to the leaves. For each possible join operation on two roots, we apply a recursive neighbor-joining algorithm to find the best transcript tree (Algorithm 1). Note that any join operation that was done when computing the score of the transcript tree is undone before passing to the next iteration.

Algorithm 1. A recursive Neighbor-Joining algorithm using the JOIN operation.

```
 1: procedure RECNJ(Transcript t)
 2:     if no children of t can be joined then
 3:         return COMPUTESCORE(t)            ▷ Compute the score of the tree containing t.
 4:     s_best = ∞
 5:     while some children of t can be joined do
 6:         s* = ∞
 7:         for (a, b) s.t. a, b ∈ CHILDRENOF(t) do
 8:             JOIN(a,b)                      ▷ Assume that joining a and b is feasible.
 9:             s = RECNJ(b)
10:             UNJOIN(a,b)                    ▷ Revert to the situation before joining a and b.
11:             if s < s* then
12:                 s* = s
13:                 a* = a and b* = b          ▷ Save the best join.
14:         JOIN(a*,b*)                        ▷ Apply the best join.
15:         s = RECNJ(b*)                      ▷ Iterate on the "new" node.
16:         if s < s_best then
17:             s_best = s
18:     return s_best
```

5 Conclusion

We presented a model of transcript evolution and an associated tool, TrEvoR, to reconstruct transcript phylogenies. The model represents the evolution of transcripts as a second layer above the exon evolution.

On 805 genes from the ASPic database, we demonstrated that transcript phylogenies can enhance transcriptome reconstruction from ESTs. The use of transcript phylogenies doubled the specificity while retaining a similar sensitivity. Results also showed that our two-level model performed better than a gene-centric model. This implies that a transcript-focused approach is more powerful for this particular task.

Additionally, we broadened the scope of transcript phylogenies by correlating them with the protein domains of their isoforms. It turned out that transcript trees indeed contain useful functional information and may be used in studies on function evolution. Domain information was gathered from different tracks in Ensembl and results revealed that conserved transcripts show a greater variability in functional sites than in protein domains.

Future work can be directed in several directions. Different models—for instance, a model based on splice sites and not exons—and different hypotheses can be tested through TrEvoR. The accuracy of automated pipelines for transcriptome reconstruction could be improved by developing a method that includes transcript phylogenies of model organisms. Deeper studies on functional sites within a transcript phylogeny framework could shed some light on the evolution of functions.

In previous work, we proposed the concept of transcript phylogenies and demonstrated its feasibility. Here, we applied this concept to two large-scale analyses, demonstrated good improvements on transcriptome reconstruction, new findings on the evolution of function in transcripts, and consequently validated the usefulness of our method for transcriptome studies.

References

1. Kim, N., Shin, S., Lee, S.: ECgene: genome-based EST clustering and gene modeling for alternative splicing. Genome Research 15(4), 566–576 (2005)
2. Modrek, B., Lee, C.: A genomic view of alternative splicing. Nature Genetics 30(1), 13–19 (2002)
3. Keren, H., Lev-Maor, G., Ast, G.: Alternative splicing and evolution: diversification, exon definition and function. Nature Reviews Genetics 11(5), 345–355 (2010)
4. Artamonova, I.I., Gelfand, M.S.: Comparative genomics and evolution of alternative splicing: the pessimists' science. Chemical Reviews 107(8), 3407–3430 (2007)
5. Harr, B., Turner, L.M.: Genome-wide analysis of alternative splicing evolution among Mus subspecies. Molecular Ecology 19(suppl.1), 228–239 (2010)
6. Nurtdinov, R.N.: Low conservation of alternative splicing patterns in the human and mouse genomes. Human Molecular Genetics 12(11), 1313–1320 (2003)
7. Flicek, P., et al.: Ensembl 10th year. Nucleic Acids Research 38(suppl.1), D557–D562 (2010)
8. Christinat, Y., Moret, B.: Inferring transcript phylogenies. In: Proc. of IEEE International Conference on Bioinformatics and Biomedecine, pp. 208–215 (2011)
9. Martelli, P., et al.: ASPicDB: a database of annotated transcript and protein variants generated by alternative splicing. Nucleic Acids Research 39(suppl.1), D80 (2011)
10. Pruitt, K., et al.: NCBI Reference Sequences: current status, policy and new initiatives. Nucleic Acids Research 37(suppl.1), D32–D36 (2009)
11. Bonizzoni, P., et al.: Detecting alternative gene structures from spliced ESTs: a computational approach. Journal of Computational Biology 16(1), 43–66 (2009)
12. Eyras, E., et al.: ESTGenes: alternative splicing from ESTs in Ensembl. Genome Research 14(5), 976–987 (2004)
13. Guttman, M., et al.: Ab initio reconstruction of cell type-specific transcriptomes in mouse reveals the conserved multi-exonic structure of lincrnas. Nature Biotechnology 28(5), 503–510 (2010)
14. Trapnell, C., et al.: Transcript assembly and quantification by rna-seq reveals unannotated transcripts and isoform switching during cell differentiation. Nature Biotechnology 28(5), 511–515 (2010)
15. Xing, Y., et al.: An expectation-maximization algorithm for probabilistic reconstructions of full-length isoforms from splice graphs. Nucleic Acids Research 34(10), 3150 (2006)
16. Cline, M., et al.: The effects of alternative splicing on transmembrane proteins in the mouse genome. In: Pac. Symp. Biocomput. 2004, pp. 17–28 (2004)
17. Eden, E., et al.: GOrilla: a tool for discovery and visualization of enriched GO terms in ranked gene lists. BMC Bioinformatics 10(1), 48 (2009)
18. Sankoff, D.: Minimal Mutation Trees of Sequences. SIAM Journal on Applied Mathematics 28(1), 35–42 (1975)

A Fast Algorithm for Computing the Quartet Distance for Large Sets of Evolutionary Trees

Ralph W. Crosby and Tiffani L. Williams

Department of Computer Science and Engineering
Texas A&M University
{rwc,tlw}@cse.tamu.edu

Abstract. We present the QuickQuartet algorithm for computing the all-to-all quartet distance for large evolutionary tree collections. By leveraging the relationship between bipartitions and quartets, our approach significantly improves upon the performance of existing quartet distance algorithms. To explore QuickQuartet's performance, sets of biological data containing 20,000 and 33,306 trees over 150 taxa and 567 taxa, respectively are analyzed. Experimental results show that QuickQuartet is up to 100 times faster than existing methods. With the availability of QuickQuartet, the use of quartet distance as a tool for analysis of evolutionary relationships becomes a practical tool for biologists to use in order to gain new insights regarding their large tree collections.

1 Introduction

Phylogenetic analysis consists of reconstructing the best estimate of the evolutionary history for a set of organisms. Typically, the result is represented as an unrooted binary tree, with living organisms (taxa) as leaves on the tree, extinct ancestors as interior nodes, and the evolutionary relationships between them as edges. Figure 1 shows two unrooted evolutionary trees for the big cats: Jaguar, Leopard, Tiger, Snow Leopard, and Lion [2]. Additionally, a single phylogenetic study can also produce multiple trees. For example, Bayesian analysis techniques such as MrBayes [3] can produce tens of thousands of trees. Given t trees, how are they related to each other? Ultimately, relatedness is dependent on our measure of comparison. The Robinson-Foulds (RF) distance [6] is the most common approach used by biologists for comparing trees and its popularity continues as a result of the availability of software (such as PAUP* [12] and HashRF [11])for computing it. While other tree distances exist (such as the quartet distance [1,10]), their use in practice is non-existent. The lack of practical implementations are barriers to the use of these distances in the phylogenetics community. This paper presents the QuickQuartet algorithm for the practical computation of the quartet distance between large sets of trees.

The QuickQuartet algorithm rapidly computes the matrix of quartet distances between sets of phylogenetic trees. A quartet is a partitioning of the tree involving four of the n taxa. That is, if we remove an edge e from the tree, we have two sets X and Y of taxa. A quartet is formed by selecting two taxa from each

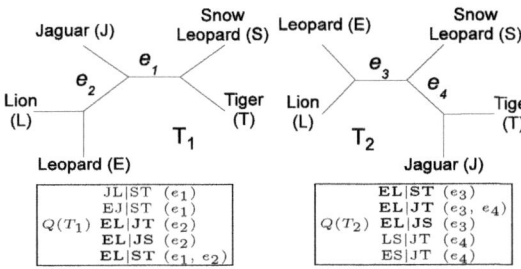

Fig. 1. Two evolutionary trees and their set of quartets. Quartets in bold are shared by both trees. Tree T_1 is the current best estimate of the big cats.

of the sets. A binary tree is uniquely defined by it's set of $\binom{n}{4}$ quartets. The quartet distance is the number of quartets that differ between two trees. Consider Figure 1. Tree T_1 contains two internal edges e_1 and e_2. Removing edge e_1 from the tree generates the quartets $JL|ST$ and $EJ|ST$. Edge e_2 generates additional quartets $EL|JT$, $EL|JS$ and $EL|ST$. Let $Q(T_1)$ and $Q(T_2)$ represent the set of quartets in trees T_1 and T_2, respectively. Then, the quartet distance between the two trees is $\frac{|Q(T_1)-Q(T_2)|+|Q(T_2)-Q(T_1)|}{2}$, where $Q(T_1) - Q(T_2)$ is the set of quartets in T_1 that are not in T_2. In Figure 1, the quartet distance between T_1 and T_2 is $\frac{2+2}{2} = 2$. QuickQuartet computes the all-to-all distance between t trees, where each cell (i, j) in the $t \times t$ matrix is the quartet distance between trees T_i and T_j.

While scientists have studied the quartet distance theoretically [9], its use is invisible in practice as biologists prefer to use the Robinson-Foulds (RF) distance [6] when they compare trees. One of the arguments for quartet distance is that it is more robust to minor edge rotations between the two trees—especially when such rotations lead to the maximum RF distance between the trees. While few papers use quartets as a measure of distance, quartets have been used in phylogenetic methods to reconstruct evolutionary trees [7]. To the best of our knowledge, QDist [5] is the only software that biologists can use for computing the quartet distance between a collection of trees. QDist is based on Stissing et al.'s [10] all-to-all quartet distance algorithm. However, its CPU and memory requirements preclude its use on the large collections of trees ($> 10,000$ trees over hundreds of taxa) that are returned from current large-scale phylogenetic analyses. Furthermore, given that the grand challenge of phylogenetic is to reconstruct the Tree of Life, which is estimated to contain between 10 million and 100 million taxa, tree collections will continue to grow in size. Thus, new algorithms such as QuickQuartet are needed that can scale to the current and future needs of biologists.

Our contributions. QuickQuartet was developed using algorithm engineering techniques to improve upon Stissing et al.'s algorithm (as implemented in QDist) for computing the all-to-all quartet distance. More specifically, Stissing et al. construct a directed acyclic graph (DAG) based on the shared bipartitions (or internal edges) in the set of t phylogenetic trees. QuickQuartet improves upon the work of Stissing et al. in several fundamental ways. First, for faster algorithmic performance, QuickQuartet strengthens the relationship between quartets and bipartitions with the use of color strings. The difference between the two is that the quartet is a partitioning of four taxa and a bipartition of n taxa. Second, QuickQuartet reduces the memory requirements substantially by calculating the lifespan of a node in the DAG. Third, while Stissing et al. analyzed the performance of their algorithm using 100 artificial trees over 1,000 taxa, QuickQuartet was used to perform experimental studies on two large collections of biological trees (20,000 trees over 150 taxa and 33,306 trees over 567 taxa).

Overall, experimental results show that QuickQuartet is over 100 times faster (with a 10 fold reduction in memory requirements) than QDist on the biological datasets studied in this paper. By providing a practical quartet distance algorithm, biologists can potentially discover new relationships and gain novel insights regarding their evolutionary trees.

2 Previous Work: QDist

Our QuickQuartet algorithm builds upon Stissing et al.'s all-pairs quartet distance algorithm [10]. Since QDist is an implementation of Stissing et al. approach, we use the terms interchangeably.

Ordered quartets. The computation of the quartet distance between two trees is based on an insight related to ordered quartets, which allows for faster computation time than $O(n^4)$. Brodal et al. [1] observed that the number of quartets for a tree can be viewed as twice the number of ordered quartets (where the direction of the center, connecting, edge is significant). Each inner node in a binary tree has three edges: one pointing upward in the tree and two pointing downward. The direction of the quartets is defined as pointed downward toward the inner node. Let A represent the set of taxa (leaves) based on the edge pointing upward. B and C represent the sets of taxa from the other two downward pointing edges. Equation 1 defines the total number of ordered quartets can be evaluated as the sum of the ordered quartets present at each of the inner nodes of a tree.

$$\binom{|A|}{2}|B||C| + \binom{|B|}{2}|A||B| + \binom{|C|}{2}|A||B| \tag{1}$$

When comparing two trees, the set of shared ordered quartets represented by a node in the first tree and a node in the second tree can be determined for a particular pair of edges [10]. By summing the values for all pairs of edges ($3 \times 3 = 9$) incident to the two nodes, the count of directed quartets at the pair of nodes can be determined. This value, when divided by 2, gives the number of shared, unordered quartets exposed by the pair of nodes.

An overview of QDist. To compute the $t \times t$ distance matrix, QDist computes a single row of the matrix at a time. For row i of the matrix, source tree T_i

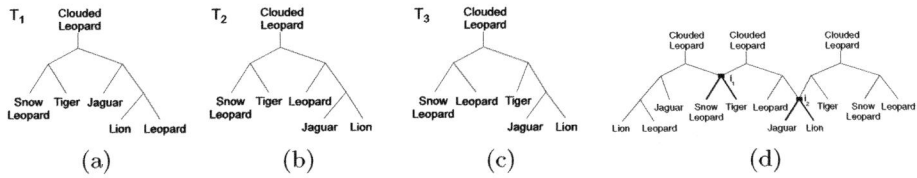

Fig. 2. Generation of the DAG structure. The first taxa encountered in the input is used as the common root (e.g. Clouded Leopard). Each of the three trees shown in (a), (b) and (c) have their own copies of the root node. Common subtrees (i_1 and i_2) are detected and stored only once in the DAG as shown in (d).

is compared to all target trees T_j, where $1 \leq j \leq t$ and $i \neq j$. QDist uses a DAG data structure as shown in Figure 2 to facilitate faster computation of quartets. Sets of trees are input in newick format as unrooted binary trees and the first taxa encountered is selected as the common root for the DAG. The trees are then organized in the DAG based on the location of the common root. If multiple trees contain common subtrees, the inner nodes are not replicated resulting in a substantial improvement in running time and memory.

Lastly, QDist uses the concept of coloring nodes wherein taxa connected to each of the three edges of an inner node are set to one of three colors: Red, Blue, and Green denoted by R, B, and G, respectively. The taxa are colored with respect to a source tree. Then, for each target tree, the set of ordered quartets in common with it and an inner node of the source tree can be determined by counting colored taxa with respect to the target tree. A complete traversal of a target tree is required for each inner node in the source tree. Once all of a source tree's inner nodes have been traversed, the quartet distance between the source and target trees can be computed.

An illustration of the algorithm. Stissing et al. discussed two different algorithms for computing the all-to-all quartet distance: (i) a DAG/DAG algorithm, where both the source and target trees are processed from the DAG data structure and (ii) a DAG/Tree algorithm, where only the target trees are processed from the DAG. In practice, Stissing et al. found that the DAG/DAG performed the best so we focus on that algorithm here.

Figure 3 shows the steps of the DAG/DAG algorithm in QDist. The number of ordered quartets shared between the source and target trees can be determined by first setting all leaf nodes to the color R as shown in Figure 3a. The root node is then colored B and a depth-first traversal of the source tree (tree T_1) is performed where the larger subtree is processed first. At each inner node u of a source tree, the leaves reachable from the smaller child subtree are colored G as shown in Figure 3b. After coloring the leaves, a full depth-first traversal of each target tree T is performed. Let V_T be the set of inner nodes from target tree T. For each inner node $v \in V_T$, let X_e be the number of taxa with color $X \in \{R, B, G\}$ attached to edge $e \in \{e_1, e_2, e_3\}$ of the target node v. In order to

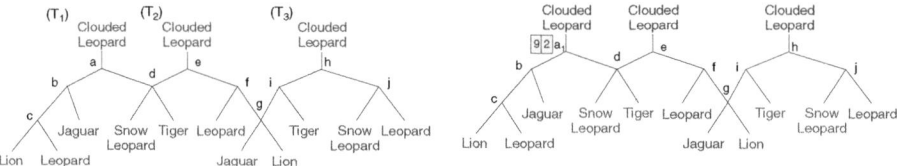

(a) Initial coloring of the DAG. (b) Step 1, compute down vector for node a.

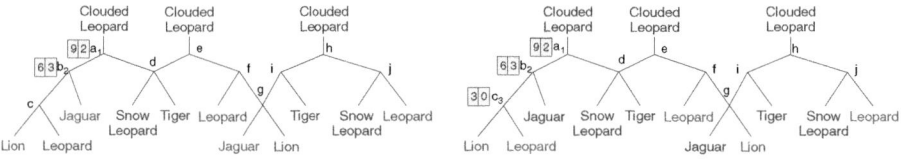

(c) Step 2, compute down vector for node b. (d) Step 3, compute down vector for node c.

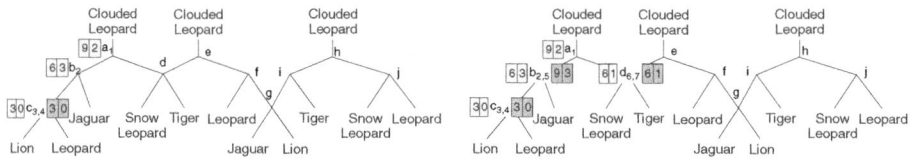

(e) Step 4, compute up vector for node c. (f) Step 7, compute up vector for node d.

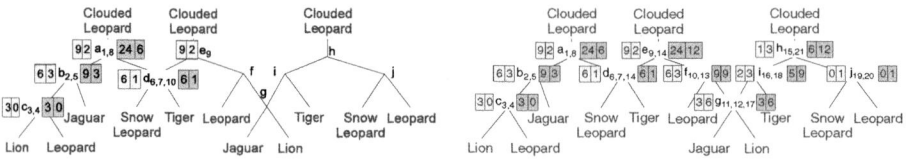

(g) Step 10, reuse down vector for node d. (h) QDist algorithm finished.

Fig. 3. The DAG/DAG QDist algorithm. The yellow and gray boxes at a source tree's inner node represent the down and up vectors computed during the downward upward tree traversals, respectively. The up vector at node k in the source tree is the sum of the its down vector and its children's down vectors. Each box contains an entry for each target tree.

compute the number of shared directed quartets between u and v, Equation 1 is modified to accommodate the coloring of leaf nodes. For example, the shared quartet count for color R associated with an edge (e.g., e_1) leading down into node v is shown in Equation 2.

$$\text{count}(R, e_1) = \binom{R_{e_1}}{2} \times (G_{e_2} \times B_{e_3} + G_{e_3} \times B_{e_2}) \tag{2}$$

To compute the total number of shared directed quartets between inner nodes u of the source tree and v of the target tree, a similar calculation is performed for all combinations of colors and edges as shown by Equation 3.

$$\text{shared}(u, v) = \sum_{X \in \{R, G, B\}} \sum_{Y \in \{e_1, e_2, e_3\}} \text{count}(X, Y) \tag{3}$$

During the downward traversal of the source tree S, we store the total number of shared quartets between node u in the source tree S and the target tree T. Thus, $\text{down}(u, T) = \sum_{v \in V_T} \text{shared}(u, v)$.

As a result, in Figure 3b, the number of shared quartets between the inner node a in source tree T_1 and the target trees T_2 and T_3 as shown by the down vector are 9 and 2, respectively. Continuing to node b in step 2 shown in Figure 3c, the taxa reachable from the smaller subtree are colored G and the taxa reachable through the upward edge are colored B. The larger subtree is then traversed as illustrated in Figure 3c and 3d. Once a larger or smaller subtree contains a single taxa, the taxa is colored B and a up vector initialized to all zeroes is returned.

After traversing the larger and smaller subtrees for an inner node of the source tree, the contents of the down vector are summed with the up vectors for the larger and smaller subtrees to produce the up vector for this inner node. This process is illustrated in Figures 3e and 3f. The up vector at the topmost inner node is divided by 2 (to convert from ordered quartets to unordered quartets) giving the number of quartets in common between the source and each of the targets. Subtracting these values from the maximum possible quartets $\binom{n}{4}$ gives the quartet distance for each source/target pair. Thus, at step 8, the quartet distance between source tree T_1 and trees T_2 and T_3 are 3 ($15 - \frac{24}{2}$) and 12 ($15 - \frac{6}{2}$), respectively.

In Figure 3g at step 10, tree T_2 has become the source tree. For each new source tree, the down vectors are cleared and the color of all taxa is initialized to R. The down vector for node d was rebuilt during the first traversal of T_1 as the target tree and the algorithm wants to now traverse down to node d. Since the down vector for node d already exists, and it can be reused to save computational time. The complete results of the QDist algorithm are shown in Figure 3h, where for example, node e shows that the quartet distance between source tree T_2 and target trees T_1 and T_3 is 3 and 9, respectively.

Since each vector is of length $t-1$ and generated for every unique inner node, the memory requirement for vector storage is significant. In the worst case, there are no common inner nodes, giving a space complexity of $O(t^2 n)$.

Analysis of the Stissing et al. algorithm. The $O(n^2)$ (DAG/DAG) and $O(n \log^2 n)$ (DAG/Tree) quartet distance algorithms for two trees were implemented by Mailund and Pedersen in the QDist application [5]. Building on the single pair algorithms, Stissing et al. produced several variations of the DAG/DAG and

DAG/Tree algorithms for the computation of the quartet distance between all pairs of trees [10]. The worst case time complexity of these algorithms for all-to-all computation is time $O(t^2n^2)$ and $O(t^2n\log^2 n)$ respectively where n and t are the number of taxa and trees, respectively. By loading all of the trees to be compared into a common data structure and retaining the vectors produced during traversal of the trees Stissing et al. was able to improve the best case complexity from time $O(t^2n^2)$ to $O(t^2 + n^2)$ and from $O(t^2n\log^2 n)$ to $O(t^2n + tn\log^2 n)$ for the two algorithms. The worst case complexity was unaffected in either case.

While the computational complexity of the DAG/Tree algorithm is better than the complexity of the DAG/DAG algorithm, the DAG/DAG algorithm performs significantly better in practice due to the presence of complex polynomials in the DAG/Tree algorithm.

3 The QuickQuartet Algorithm

QuickQuartet is based on QDist's implementation of the Stissing et al. DAG/DAG algorithm for single tree pair comparison but with a new all-to-all pairs algorithm based on algorithm engineering techniques. When computing the quartet distance between two trees, QuickQuartet identifies additional common bipartitions in the DAG with the use of color strings. These shared bipartitions are not recognized by the Stissing et al. approach and results in QuickQuartet's performance improvement by up to two orders of magnitude over QDist.

Relationship between bipartitions and quartets. Similarly to a quartet, a bipartition is based on an internal edge that when removed splits the taxa into two sets X and Y. All of the taxa in set X and in set Y compose the bipartition and we represent it as $X|Y$. For example, in Figure 1 the bipartition $EJL|ST$ is created from edge e_1. For each internal edge, there is a single bipartition and many quartets. That is, a bipartition embeds a set of quartets. Given a bipartition B, the number of quartets contained in it is $\binom{p}{2} \times \binom{q}{2}$, where p and q are the number of taxa in the sets X and Y, respectively. For the bipartition $EJL|ST$, the number of embedded quartets is $\binom{3}{2} \times \binom{2}{2}$ or 3. In this example, if another tree had bipartition $EJL|ST$, then we could automatically say that they had 3 quartets in common based on exploring a single bipartition. This is a significant source of cost savings leveraged by QuickQuartet with the use of color strings.

Creating color strings. Let A be the set of unique taxa names in any arbitrary order. $|A| = n$, which is the number of taxa. More specifically, $A = \{a_1, a_2, \ldots, a_n\}$, where a_1 is the first taxa name, a_2 is the second taxa name, etc. We can represent a coloring of a node i compactly by using a color string S, where $|S| = |A|$. That is $S = \{s_1, s_2, s_3, .., s_n\}$, where each s_i is the color assigned to taxa a_i. As discussed in Section 2, the possible colors of a node are Blue, Green, and Red represented by B, G, and R, respectively. Consider the trees in Figure 3b. Let $A=$ {Lion, Leopard, Clouded Leopard, Snow Leopard, Tiger, Jaguar}. The coloring for node a in Figure 3b is $S = \{RRBGGR\}$ for tree T_1. Node e for tree

T_2 has the same color scheme even though the trees T_1 and T_2 have different topologies and are represented as distinct nodes in the DAG. Finding a color string at a previously computed source tree inner node eliminates traversing the remaining target trees resulting in significant running time improvements.

The QDist algorithm creates cost savings by identifying nodes in the trees that have the exact same subtrees (e.g., node d in Figure 3b). However, in Figure 3b, QDist would redo the calculation for nodes a and e even though both nodes represent the same bipartition and induce the exact same coloring. Quick-Quartet, on the other hand, identifies those nodes as being identical bipartitions and retrieves a previously calculated and saved down vector. This eliminates the need to perform the traversal of any of the target trees at that specific source tree inner node. Thus, by saving the down vectors associated with bipartitions (color strings) a significant number of tree traversals can be eliminated.

Color strings and hash tables. A map (implemented as a hash table) keyed by the color string is used to store the down vector associated with the coloring. The hash key is generated by a standard universal hashing algorithm operating on the color string. Prior to performing the traversal of a target tree, the hash table is queried using the current color string. If the down vector has already been computed for the color string, it is retrieved from the hash table and the set of target tree traversals is not performed. If the color string is not in the hash table, the down vector is computed and inserted in the hash table.

Additional improvements. While the identification of shared bipartitions using color strings is responsible for a majority of the performance optimization implemented in QuickQuartet, there were additional improvements that we incorporated into the algorithm. First, we only compute the upper triangle of the symmetric quartet distance matrix. Second, the fundamental equations used for quartet counting were refactored, which produced a 6% overall improvement in CPU time (not shown). To reduce the consumption of memory, down vectors are released when they are no longer required. The life of a down vector is defined between the first and last trees in which the vector's associated inner node appears. Since inner nodes are not duplicated in the DAG, by keeping track of the last tree that uses an inner node, we can free down vectors after the last tree containing the node is processed as the source tree. As a result, the vector storage increases to an upper limit during the middle of the quartet distance computation but then decreases as vectors were released. Thus, reducing the algorithm's, overall memory requirements.

Analysis. Although QuickQuartet outperforms QDist in practice, the overall best and worst case time complexities of the algorithms are the same. In the worst case, there would be no bipartitions in common and there would be no common DAG nodes. A common DAG node is a sufficient but not a necessary condition for the existence of a common bipartition. Without any common DAG nodes, each source tree inner node will force the traversal of each other target tree with the resulting worse case time complexity of $O(t^2 n^2)$ for t trees over

n taxa. In the best case, all bipartitions would be in common (i.e., all t trees are identical) and discovered during the first and only traversal of a source tree. During this one source tree traversal, only the first target tree traversal set would produce unique results. No subsequent target traversals would be required as the counts for all DAG nodes would have already been computed. This yields the same best case time complexity as QDist of $O(t^2 + n^2)$.

The advantages of the QuickQuartet algorithm occur in biologically interesting cases where there is medium to high similarity between the trees. In a typical phylogenetic inference the trees are generated though perturbations of the tree topology using operations such as nearest neighbor interchange (NNI) or tree bisection and reconnection (TBR). Even when these transformation operators are applied, there remains a significant amount of similarity between the trees. Finding a shared color string, at a source tree inner node, eliminates traversals of the target trees. For each color string found, there is a total savings of n^2 target tree traversals. Our experimental results show that this substantial reduction of tree traversals leads to significantly faster performance.

4 Experimental Methodology

Biological datasets. We constructed our benchmarks based on two tree collections from published phylogenetic analyses.

1. Dataset #1: 20,000 trees obtained from the Bayesian analysis of an alignment of 150 taxa consisting of 23 desert taxa and 127 from other habitats [4].
2. Dataset #2: 33,306 trees obtained from an analysis of a three-gene, 567 taxa dataset consisting of 560 angiosperms and seven outgroups [8].

For each dataset of t trees, different subsets were used to study the performance profile of the application. Our experiments used 10%, 20%, ..., 100% of the t trees available. Due to the memory requirements of QDist, a maximum of 16,000 trees from Dataset #1 and 10,000 from Dataset #2 were used. For QuickQuartet, the complete datasets were used.

Experimental platform. Our benchmarks were performed on the Texas A&M Brazos Cluster (http://brazos.tamu.edu), where each node consisted of dual 2.5GHz Intel quad core processors and 32GB of memory. Although QDist and QuickQuartet are sequential algorithms, we used a high-performance computing cluster to run our experiments in parallel across the nodes. On a computing node, there is a single experiment which executes sequentially on a single core. Finally, both QuickQuartet and QDist were written in C++ and compiled with gcc 5.3 with the -O3 compiler option.

Validation and reporting of results. Experimental results were compared across the competing algorithms for experimental validation. Plots show the average performance over five runs.

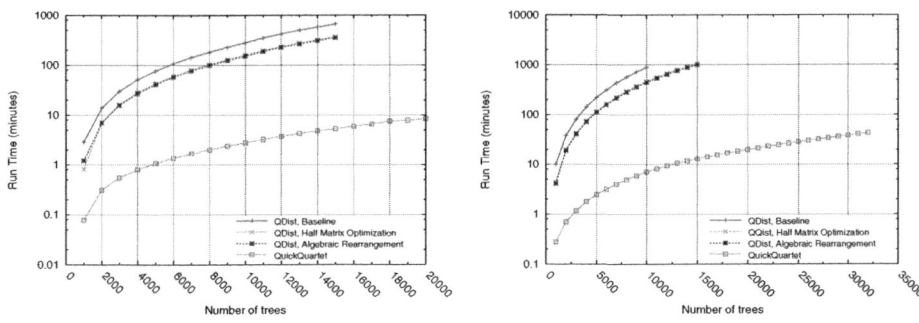

(a) Dataset #1: 20,000 trees of 150 taxa. (b) Dataset #2: 33,306 trees of 567 taxa.

Fig. 4. Running time of the algorithms. The scales differ across the plots. The y-axis is on a logarithmic scale.

5 Experimental Results

Figures 4 and 5 compare the QDist and QuickQuartet algorithms in terms of running time and memory usage. The CPU performance of the algorithms is shown in Figure 4. The top three curves show the baseline QDist application followed by our optimizations to QDist, which includes computing the lower triangle of the distance matrix and algebraic rearrangement of equations. The lowest curve in the plots shows the running time of our QuickQuartet algorithm, which includes the use of color strings to leverage the relationship between bipartitions and quartets and reduce redundant computations. In Figure 4b, at 10,000 trees, the running time for QuickQuartet and QDist (with algebraic rearrangements) is 7 minutes and 524 minutes. This leads to QuickQuartet having a 98.7% performance improvement over QDist. We note that QDist ran out of resources to complete the computation beyond 15,000 trees.

In QDist, repeated traversals of the target trees during computation of directed quartet counts consumes the majority of the running time. By saving the down vectors associated with unique color strings (bipartitions), the number of traversals of the target trees required is reduced significantly as illustrated in Figure 5a. In QuickQuartet, each time a duplicate color string is found, $O(n^2)$ target traversals are eliminated. For the 150 taxa dataset, at 15,000 trees, 654,566 unique inner nodes existed in the DAG. QDist performed a set of target tree traversals for each of these inner nodes giving a total of 4,840,798,085 target traversals. QuickQuartet requires 55,465,669 traversals eliminating the need for 638,401 redundant target traversals. By reducing tree traversals by nearly two orders of magnitude, running time improves from 362.2 minutes (QDist) to 5.2 minutes (QuickQuartet). Dataset #2 shows similar trends (not shown).

Figure 5b shows running time and memory performance for QuickQuartet. Running time and memory usage are better on our smaller dataset (Dataset #1). However, as the dataset size increases, memory requirements grow quite large

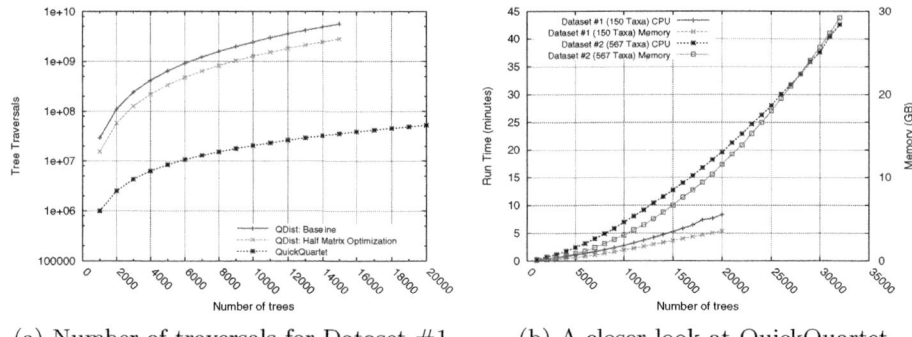

(a) Number of traversals for Dataset #1. (b) A closer look at QuickQuartet.

Fig. 5. The impact of traversals and memory. The scales differ across the plots.

up to 28GB. While QuickQuartet requires substantially less memory than QDist (not shown), its memory requirements can be further improved. For the 567 taxa dataset, the curves for memory and running time converge. However, for the 150 taxa dataset, the memory and CPU performance are slightly divergent. This behavior may be a result of the number of taxa (and consequently inner nodes) on the memory efficiency of the algorithm. However, additional experimentation is needed to study this relationship in more detail.

6 Conclusions

We have presented the QuickQuartet algorithm, a new technique for the computation of the all-to-all pairs quartet distance between phylogenetic trees. QuickQuartet strengthens the relationship between the bipartitions of a tree and its quartets in order to obtain improved performance on large-scale biological tree collections in comparison to QDist, which is currently the only available software for computing the all-to-all quartet distance. Our experiments show that QuickQuartet is two orders of magnitude better than QDist in terms of running time. When applied to 10,000 tree subset of the 33,306 trees over 567 taxa dataset (the maximum number of trees that could be processed with QDist on our platform), the running time improved by a factor of 126 from 14.5 hours to 7 minutes. QuickQuartet required 43 minutes to compute the complete dataset (a $33,306 \times 33,306$ matrix). Furthermore, the memory usage of QuickQuartet was reduced by an order of magnitude in comparison to QDist.

Our QuickQuartet algorithm allows biologists to add the quartet distance to the their workbench of software tools. But, more importantly, by having a practical implementation of the quartet distance, it is now possible to study what such a distance tells us about how our large tree collections are related to each other. Future work will focus on improving the QuickQuartet algorithm and comparing the quartet distance to other tree measures such as the RF distance.

Acknowledgment. We would like to thank M. Gitzendanner, P. Lewis, D. Soltis for providing us with the tree collections used in this paper. This work was supported by the National Science Foundation under grants DEB-0629849, IIS-0713618, and IIS-1018785. Finally, the authors acknowledge the Texas A&M University Brazos HPC cluster that contributed to the research reported here.

References

1. Brodal, G.S., Fagerberg, R., Pedersen, C.N.S.: Computing the Quartet Distance between Evolutionary Trees in Time $O(n \log^2 n)$. In: Eades, P., Takaoka, T. (eds.) ISAAC 2001. LNCS, vol. 2223, pp. 731–742. Springer, Heidelberg (2001)
2. Davis, B.W., Li, G., Murphy, W.J.: Supermatrix and species tree methods resolve phylogenetic relationships within the big cats, panthera (carnivora: Felidae). Molecular Phylogenetics and Evolution 56(1), 64–76 (2010)
3. Huelsenbeck, J.P., Ronquist, F.: MRBAYES: Bayesian inference of phylogenetic trees. Bioinformatics 17(8), 754–755 (2001)
4. Lewis, L.A., Lewis, P.O.: Unearthing the molecular phylodiversity of desert soil green algae (chlorophyta). Syst. Bio. 54(6), 936–947 (2005)
5. Mailund, T., Pedersen, C.N.S.: QDist–quartet distance between evolutionary trees. Bioinformatics 20(10), 1636–1637 (2004)
6. Robinson, D.F., Foulds, L.R.: Comparison of phylogenetic trees. Mathematical Biosciences 53, 131–147 (1981)
7. Schmidt, H.A., Strimmer, K., Vingron, M., von Haeseler, A.: Tree-puzzle: maximum likelihood phylogenetic analysis using quartets and parallel computing. Bioinformatics 18(3), 502–504 (2002)
8. Soltis, D.E., Gitzendanner, M.A., Soltis, P.S.: A 567-taxon data set for angiosperms: The challenges posed by bayesian analyses of large data sets. Int. J. Plant Sci. 168(2), 137–157 (2007)
9. Steel, M.A., Penny, D.: Distributions of tree comparision metrics—some new results. Systematic Biology 42(2), 126–141 (1993)
10. Stissing, M., Mailund, T., Pedersen, C., Brodal, G., Fagerberg, R.: Computing the all-pairs quartet distance on a set of evolutionary trees. Journal of Bioinformatics & Computational Biology 6(1), 37–50 (2008)
11. Sul, S.-J., Williams, T.L.: An Experimental Analysis of Robinson-Foulds Distance Matrix Algorithms. In: Halperin, D., Mehlhorn, K. (eds.) ESA 2008. LNCS, vol. 5193, pp. 793–804. Springer, Heidelberg (2008)
12. Swofford, D.L.: PAUP*: Phylogenetic analysis using parsimony (and other methods), Sinauer Associates, Underland, Massachusetts, Version 4.0 (2002)

Error Propagation in Sparse Linear Systems with Peptide-Protein Incidence Matrices

Peter Damaschke and Leonid Molokov

Department of Computer Science and Engineering
Chalmers University, 41296 Göteborg, Sweden
{ptr,molokov}@chalmers.se

Abstract. We study the additive errors in solutions to systems $Ax = b$ of linear equations where vector b is corrupted, with a focus on systems where A is a 0,1-matrix with very sparse rows. We give a worst-case error bound in terms of an auxiliary LP, as well as graph-theoretic characterizations of the optimum of this error bound in the case of two variables per row. The LP solution indicates which measurements should be combined to minimize the additive error of any chosen variable. The results are applied to the problem of inferring the amounts of proteins in a mixture, given inaccurate measurements of the amounts of peptides after enzymatic digestion. Results on simulated data (but from real proteins split by trypsin) suggest that the errors of most variables blow up by very small factors only.

Keywords: protein mixture inference, linear system, error propagation, bipartite graph, shortest path.

1 Introduction

Suppose that we are given an analytical chemistry problem that obeys the following informal "axioms".

- An unknown mixture of *compounds* is given.
- The compounds can be split into *constituents* by a certain chemical reaction triggered by an external compound E.
- The compounds react only with E and decay independently of each other. No other chemical reactions take place.
- We have a database telling which constituents result from the splitting of each compound.
- The compounds are hard to measure directly.
- The amount of each constituent can be measured directly.

Now we would like to infer which compounds are in the mixture, along with the amount of each one. Obviously this problem can be formulated as a system $Ax = b$ of linear equations, where A is the incidence matrix of constituents (one per row) and compounds (one per column), vector b is the measured mixture of constituents, and x is the unknown vector of compounds.

Only nonnegative vectors make sense here: $b \geq 0$ and $x \geq 0$. Therefore we can immediately set to zero all variables that appear in rows j where $b_j = 0$, thus leaving a system $Ax = b$ with a (perhaps) smaller matrix A where all entries of b are strictly positive.

However, reality dictates that we have to take noise in account. We cannot measure b exactly. In the following we adopt the assumption that all entries of b have an additive error of at most some parameter value ϵ. Instead of the true b_j we therefore measure some amount between $b_j - \epsilon$ and $b_j + \epsilon$. This involves a certain simplification because one may be able to measure some constituents more accurately than others, however the assumption is not too artificial and allows for nice mathematical analysis, as we will see. Note also that one may disregard entries known to be unreliable right from the beginning and restrict the linear system to selected rows.

We may observe $b_j = 0$ instead of some true $b_j \leq \epsilon$, thus discarding some variables that should actually be positive. But since we have $x \geq 0$, affected variables are in this case bounded by ϵ, too, hence it is not a bad mistake to ignore them. In the following we are aim for upper bounds on the additive errors of the variables in general. Note that only the structure of matrix A matters.

Our aim is to classify variables by error bounds in a specific application where A has only entries 0 and 1, and A is sparse, i.e., has very few 1 entries: We are particularly interested in the question of absolute quantification of proteins in a protein mixture by means of shotgun proteomics. The general idea of shotgun proteomics is to split large protein molecules into smaller peptides, which later can be observed by mass spectrometry. That is, the compounds are proteins, and constituents are peptides after enzymatic digestion. The problem of just inferring which proteins are present leads to the combinatorial problem of Set Cover or, equivalently, Hitting Set. But we may also want to quantify proteins, e.g., in order to detect changes of protein expression. See, e.g., [11,2,13,9] for more background.

Evidence of protein presence is established through analysis of peptide observations, as each protein is believed to be generating consistent peptide observations. Many proteins contain unique peptides, what means that after splitting they will produce some peptide that other proteins would not. Current quantification procedures use this fact and attempt to quantify proteins expression by measuring unique peptides' quantities with one or another lab technique, e.g., iTRAQ. In the system described in [3], relative or absolute quantification is done by mass spectrometry. Also error propagation is addressed. However it is said that for "protein quantification, only unique peptides are taken into consideration, whereas peptides belonging to more than one protein sequence are only used for proving the identification of the corresponding proteins" although they are ubiquitous.

A natural question is whether the linear system of equations could be used to include also ambiguous peptides directly in the inference process (assuming bounded errors), and how measurement errors would then propagate. Intuitively, unique peptides' quantities will not affect noise level, while the more shared a peptide is, the more noisy are the inferred protein quantities. We stress that error propagation solely depends on the incidence matrices, and the present paper focuses on this aspect. In particular, this work is not concerned with measurement technologies but only with the structure of the protein-peptide incidence matrices. We observe that many peptides appear in only two candidate proteins, which also allows a graph-theoretic view of the problem.

Apparently the work most related to ours is "the first attempt to use shared peptides in protein quantification" made in [6]. Non-unique peptides can also contribute to quantification, specifically to quantify those proteins which either lack unique peptides themselves or unique peptide observations. The suggested LP model incorporates shared peptides' relative quantities and attempts to explain them by minimizing some error measure. Different detectability of peptides can be modelled by individual scaling factors applied to the entries of vector b before giving it to the linear system, whereas matrix A is not affected. While error minimization is quite natural, it might still not yield the proper prediction of actual quantitites. Despite similarities to [6], the present paper explicitly focuses on the aspect of controlled errors and their propagation and also makes some graph-theoretic contributions.

Not surprisingly, error bounds in linear systems is an extensively studied subject, see e.g. [1,4,12]. However, here we want elementary, easily applicable bounds for the case of $0, 1$-matrices A with sparse rows and perturbations limited to b, and we aim at conclusions for the protein inference problem (using simulated digestion and mixture data, however from a real protein database.) A related problem is to explain the measurements by minimal mixtures (following the parsimony principle), or algebraically, finding and enumerating the solutions with minimal sets of nonzero variables in linear systems. In [5] we proved fixed-parameter tractability of that problem, parameterized by the solution size and the number of variables per equation. Another, less related problem with data mining applications is the reconstruction of a low-rank factorization of a given matrix under extra assumptions, as in [8] where an MILP formulation is used to attack the problem.

The paper is organized as follows. Section 2 gives the necessary definitions and basic facts. The central definition is a measure of error propagation as the solution to some linear program (LP). A relevant special case is linear systems where A is the incidence matrix of a graph. In Section 3 we obtain a complete characterization of the optimal error measure in this case. Section 4 briefly discusses the sparsity of optimal combinations of measures, and a method to obtain optimality certificates. After perturbation of vector b, a system $Ax = b$ may have no solution at all, and then we would like to satisfy all equations as good as possible. Section 5 gives a combinatorial characterization of the optimal solution in the case of bipartite graphs. (Due to space limitations we could not insert figures to illustrate the graph-theoretic results and proofs.) In Section 6 we report some simulation results and draw conclusions.

2 Error Bounds through Linear Programming

Vectors x are always understood to be column vectors, therefore the transposed x^T is always a row vector. For two vectors x and y of equal lengths we mean by $x \leq y$ that $x_i \leq y_i$ for all i, similarly $x < y$ means that $x_i < y_i$ for all i. Symbol $|x|$ denotes a vector that consists of the entries $|x_i|$ for all i. The notation should not be confused with any norm of x. Instead, by $||.||_1$ we mean the ℓ_1-norm, i.e., the sum of absolute values of all entries of a vector. Symbol $\bar{1}$ denotes a vector where all entries are 1, with a length that will always be clear from context. Observe that the ℓ_1-norm can be written as the inner product $||x^T||_1 = |x|^T \bar{1}$. The incidence vector of a subset V of variables is the vector where all variables in V are set to 1, and all others are 0. A unit vector is an incidence vector of a single variable: $(0, \ldots, 0, 1, 0, \ldots, 0)$.

The following is the central concept of the paper. We first state it formally and discuss it later after having proved some properties.

Definition 1. *Let A be a fixed $m \times n$-matrix and z^T an n-vector. With respect to A we define $e(z^T) := \min\{||y^T||_1 : y^T A = z^T\}$. If no such y^T exists, $e(z^T)$ remains undefined.*

Theorem 1. *Consider an admissible linear system $Ax = b$ with $m \times n$-matrix A. We also denote by x a specific solution vector. Let z^T be any n-vector for which $e(z^T)$ is defined, and $\epsilon > 0$ some scalar. Furthermore let b' be an m-vector with $|b - b'| \leq \epsilon \bar{1}$. Then there exists an n-vector x' with $|Ax' - b'| \leq \epsilon \bar{1}$. Moreover, any such x' satisfies $|z^T x' - z^T x| \leq 2e(z^T)\epsilon$. If $Ax' = b'$ is admissible, too, i.e., has an exact solution x' then $|z^T x' - z^T x| \leq e(z^T)\epsilon$.*

Proof. The first assertion is trivial: In particular, $x' := x$ yields

$$|Ax' - b'| = |Ax - b'| \leq |b - b'| \leq \epsilon \bar{1}.$$

To show the second assertion, let y^T be a vector as in the definition of $e(z^T)$, that is, $y^T A = z^T$ and $e(z^T) = ||y^T||_1$. Then we get:

$$|z^T x' - z^T x| = |y^T A x' - y^T b| = |y^T(Ax' - b)| = |y^T(Ax' - b' + b' - b)|$$

$$\leq |y^T(Ax' - b')| + |y^T(b' - b)| \leq |y|^T |Ax' - b'| + |y|^T |b' - b|$$

$$\leq 2|y|^T \epsilon \bar{1} = 2\epsilon |y|^T \bar{1} = 2\epsilon ||y^T||_1 = 2e(z^T)\epsilon.$$

The third assertion follows from the same analysis, since one term disappears. □

In particular, if z^T is the incidence vector of a set V, Theorem 1 says that the uncertainty in the inferred total amount of compounds in V is, at most, proportional to the uncertainty of measured values and proportional to $e(z^T)$. Hence $e(z^T)$, which depends only on A, serves as an important measure of accuracy of the inferred results. If z^T ist the unit vector with the ith entry being 1, then $e(z^T)\epsilon$ bounds the deviation of the value of the ith variable, since $z^T x = x_i$.

Here some discussion is in order. In the mentioned application the uncorrputed system $Ax = b$ always has a solution, for trivial reasons. System $Ax = b$ may lack a unique solution, but in our examples this is often the case only due to undistinguishable variables that always appear together. We can merge such variables (declare their sum a new variable). That is, subsets of variables may still have uniquely determined sums. Moreover, Definition 1 and Theorem 1 obviously do not become more complicated for general vectors z^T, therefore we consider also subsets of variables, even though we primarily want the values of single variables. The disturbed system $Ax' = b'$ usually becomes inconsistent, because some rows of A are linearly dependent, but the errors in b' are "uncorrelated".

We refer to the problem in Definition 1 as BOUND MINIMIZATION. By a well-known standard trick, minimizing an ℓ_1-norm $\sum_i |y_i|$ under linear constraints can be written as an LP, hence we can compute $e(z^T)$ for any given z^T by an LP solver. (Introduce a new variable s_i for each $|y_i|$ and new constraints $s_i \geq y_i$ and $s_i \geq -y_i$, then minimize $\sum_i s_i$. Then any optimal solution satisfies $s_i = |y_i|$ for all i.) However in relevant special cases we can apply simpler combinatorial methods to BOUND MINIMIZATION, as we will see.

For non-zero vectors z^T we always have $e(z^T) \geq 1$, and $e(z^T) = 1$ holds if and only if z^T is a row of A. We omit the proof of this simple observation.

3 Unit Vectors and Rows with Two Variables

In the following, all matrices A have only entries 0 and 1. In this section we consider the speical case of BOUND MINIMIZATION when all rows of A have at most two entries 1. (Experimental results suggest that this case is of particular interest in our application, see the example later on.) We define a graph G such that A is the incidence matrix of G, that is, columns and rows of A correspond to vertices and edges, respectively, of G, and an entry 1 means that the vertex belongs to the edge. In particular, a row with exactly one entry 1 corresponds to a loop. Therefore, when z^T is a unit vector, we can obviosuly reformulate BOUND MINIMIZATION as an edge labeling problem in graphs.

EDGE LABELING: We are given a graph G, possibly with loops, and a distinguished vertex v. The problem is to label the edges (including the loops) of G with real numbers, thereby minimizing the sum of *absolute values* of all edge labels, under the following constraint: Let $S(u)$ denote the sum of labels of all edges incident to vertex u. Then $S(v) = 1$, and $S(u) = 0$ for all $u \neq v$.

If several edge labelings satisfy the above constraints, then any convex linear combination of them, i.e., linear combination with positive coefficients that sum up to 1, is also a feasible solution to EDGE LABELING.

Assume that $Ax = b$ has a solution x. Also assume that G is connected, otherwise the following reasoning holds in every connected component. It is folklore in linear algebra that x is uniquely determined if G has a loop or an odd cycle. Otherwise G is bipartite, and $Ax = b$ has a 1-dimensional solution space. In the following we extend this result to error propagation, in the sense that we characterize the optimal solutions to BOUND MINIMIZATION. Here, a *tour* is a path in a graph, where edges may be traversed several times in any direction.

Theorem 2. *If graph G has neither loops nor odd cycles (bipartite graph), then* EDGE LABELING *has no solution. Otherwise there exists a labeling. Furthermore, there exists an optimal labeling of one of the following two types, where edges not mentioned get label 0:*

(i) Take a shortest path from v to some loop, and assign labels $+1$ and -1 alternatingly to its edges, including this loop.
(ii) Take a shortest tour of odd length, and without loops, from v to itself, thereby adding alternatingly $+1/2$ and $-1/2$ to the labels of traversed edges.
Moreover, a tour of type (ii) consists of a path P (maybe empty), followed by an odd cycle C and the reverse of P.

Proof. For the moment we relax the constraint in EDGE LABELING and call a labeling *valid* if $S(v) > 0$, and $S(u) = 0$ for all $u \neq v$. Denote by $s > 0$ the smallest absolute value in a given labeling. Any valid labeling can then be changed into a correct labeling by dividing all labels by $S(v)$. We shall see that successive subtraction of type (i) and (ii) labelings from a given labeling eventually produces a zero labeling, i.e., where all labels are zero.

So, consider a given optimal labeling. We start at v a tour T through, alternatingly, positively and negatively labeled edges. At the same time we subtract s from the absolute value of the label of every traversed edge. Observe that there cannot exist an even cycle where positive and negative labels alternate, since then subtraction would not alter any $S(u)$, contradicting optimality of the initial labeling. Hence a tour T with the following properties exists:

– T terminates at a loop or at v.
– In the former case, T has no other loops.
– in the latter case, T is free of loops and has odd length.
– Subtraction of labels (as described) leaves a new valid labeling.

Hence we find another tour in the graph equipped with a new valid labeling, and so on, until we get the zero labeling. In the event that v gets isolated, that is, all edges incident to v have labels reduced to 0, property $S(u) = 0$ for all $u \neq v$ implies that all labels in the graph are already 0 as well, since otherwise the initial labeling was not optimal. It follows that any given optimal labeling is a convex linear combination of labelings of type (i) or (ii). It also follows that loop-free bipartite graphs have no labeling at all.

In the set of valid labelings obtained in the above construction, every edge has a fixed sign: either positive or negative or zero. Moreover the sum of absolute values of labels is a linear function on the restricted set of labelings with those fixed signs. Since a linear function on a convex polytope achieves its optimum at some extremal point, it follows that some optimal labeling is of type (i) or (ii). Now, $s = 1$ in case (i) and $s = 1/2$ in case (ii) is enforced by the constraint in the problem statement. Finally, optimality yields that T is a shortest tour of the respective type.

Consider type (ii). If a vertex w appears at least twice in T and the subtour between two copies of w has even length, we can delete this subtour from T, which contradicts optimality. If w appears at least three times in T, then some subtour between two of the copies has even length, which was already excluded. Thus w appears at most twice in T. Now let w be any vertex that actually appears twice. Such w exists, since at least v appears twice. Now we choose w such that all vertices on the subtour between the two copies of w are distinct; note that such w exists. Let P, C, Q denote the subtours from v to w, w to w, and w to v. Then C is an odd cycle. Since T has odd length, P and Q are both even or both odd. If P is no longer than Q, we can replace Q with the reverse of P without increasing the length of T, while keeping the total length odd. The case where P is longer than Q is symmetric. □

In the case of bipartite graphs, $e(z^T)$ for a unit vector z^T is undefined, however: Consider any two vertices u and v from different partite sets of G, and let z^T be the incidence vector of $\{u, v\}$. Then $e(z^T)$ exists and is the distance of u and v. More generally we have in any graph G, not necessarily bipartite:

Theorem 3. *Let u and v be any two vertices of G, and let z^T be the incidence vector of $\{u, v\}$. Then $e(z^T)$ is the minimum length of an odd path between u and v, and $e(z^T)$ is undefined if no such odd path exists.*

We omit the proof which is very similar to Theorem 2. Optimal tours in Theorems 2 and 3 can be found by adaptations of standard algorithms for shortest paths.

We conclude the section with a few remarks.

We can get upper bounds on errors also by elimination: In the following we use $e(x_i)$ as a shorthand for $e(z^T)$ where z^T is the indicator vector of x_i. As observed earlier, $e(x_i) = 1$ if x_i appears as the only variable in some row. By induction and the triangle inequality we obtain: If some row has the form $x_i + \sum_{k \in K} x_k$, where $\sum_{k \in K} e(x_k) \leq t$, then $e(x_i) \leq t + 1$. This can be extended straightforwardly to subsets of variables.

We emphasize that our results give worst-case upper bounds, while the actual error propagation should be much better. We illustrate this for case (i) of Theorem 2 (which is responsible for the error bounds of most variables in our simulation examples). The worst case appears only if the edges on the path are alternatingly labeled $+1$ and -1. To be specific, consider the following sequence of edge labels (beginning with the loop): $+1, -1, +1, -1, +1, -1 \ldots$ Then Gaussian elimination along this path yields the following error factors of variables: $+1, -2, +3, -4, +5, -6, \ldots$ (i.e., these are the factors in front of ϵ). Now the point is that, in real data, the errors in vector b which yield the edge labels will barely form such alternating paths. Assuming, for instance, that measurement errors are random and independent, the sequence of error factors will be a random walk that is unlikely to move very far away from 0.

4   Properties of Optimal Combinations of Rows

We come back to BOUND MINIMIZATION in general, that is, the computation of $e(z^T) = \min\{||y^T||_1 : y^T A = z^T\}$ for a given matrix A and vector z^T. In our sparse matrices we observe that optimal linear combinations are almost always built from a few rows, and often by just applying Theorem 2 (i). Hence optimal y are easier to obtain than by running an LP solver on relatively large matrices. The sparsity of optimal vectors y is partly explained by the fact that rows of A are sparse, and there always exists an optimal y where the set of rows of A with nonzero coefficients is linearly independent. (The latter claim follows by standard arguments from LP theory that we omit.)

On the other hand, the possible numerical values of $e(z^T)$ are dense, e.g., they can be arbitrarily close to 1. As an example that gives the idea, consider an $n \times n$ matrix with one row of only 1s, and $n-1$ other rows where x_1 appears together with any one of the variables $x_2, \ldots, x_n$. Then $e(1, 0, \ldots, 0) = 1 + \frac{2}{n-2}$. In general matrices A this means that linear combinations of rows that yield $e(z^T)$ (as in Definition 1) may involve rows with arbitrarily many 1 entries.

Therefore it would be nice to confirm optimality of a vector y obtained by a combinatorial method, without actually solving an LP. The natural idea to get an optimality certificate is to *dualize* the LP for BOUND MINIMIZATION. Straightforward calculation and algebraic manipulation yields the following dual problem: $\max z^T r$, such that $-\bar{1} \leq Ar \leq \bar{1}$ (where entries of r may be negative). To confirm a value $e(z^T)$ in the case of unit vectors z^T, we only have to set the corresponding entry of r to the alleged value and choose other entries so as to keep all entries of Ar in the range from -1 to $+1$. Since y usually involves only a few rows, some sparse r is often quickly found in an ad-hoc way, without running an LP solver on the entire dual problem.

5 Approximate Solutions with Two Variables Per Row

Theorem 1 and the subsequent discussion suggests a problem of independent interest. Given a 0, 1-matrix A and a vector b, we actually want to compute x such that ϵ is minimized in $|Ax - b| \leq \epsilon \bar{1}$.

If A has at most two entries 1 per row, this yields another graph labeling problem, on the incidence graph G of A:

VERTEX LABELING: Given a graph with edges labeled by real numbers, label the vertices by real numbers such that the label of every edge differs from the sum of labels of its vertices by at most ϵ.

The problem can be transformed into an LP, but again we would like to have simpler combinatorial algorithms and structural insights.

The case of bipartite incidence graphs G is of special interest, since already the exact problem has no unique solution. In the following we give a combinatorial characterization of an optimal vertex labeling. Given a solution, i.e., some vertex labeling, where the maximum deviation is ϵ, we call an edge high (low) if the sum of its vertex labels equals the edge label plus (minus) ϵ.

Theorem 4. *For* VERTEX LABELING *in a bipartite graph we have: If there exists an alternating cycle of high and low edges, then the vertex labeling is optimal.*

Otherwise there exists a vertex incident with some high edge but not incident with any low edge, or vice versa.

Proof. Let C be an alternating cycle. In order to improve the labeling we must get rid of all high and low edges, in particular those in C. Let e be any high edge in C. We must decrease the label of some vertex v of e by some $\delta > 0$. But since v is also incident with a low edge, we must increase the label of its other vertex by more than δ, and so on. In this way we go round the cycle and eventually arrive at v whose label must be decreased, but now by more than δ, a contradiction.

By definition, there always exists some high or low edge. If every vertex incident with some high (low) edge is also incident with some low (high) edge, we can obviously form an alternating cycle. Now both assertions are proved. □

Theorem 4 suggests a simple algorithm for VERTEX LABELING in bipartite graphs: Take a vertex incident with some high (low) edge but no low (high) edge and decrease (increase) its label. This removes some high or low edge, or improves ϵ. This is repeated until an alternating cycle appears. Of course, worst-case time analysis would require some more work, along the lines of similar algorithms for flow problems.

6 Some Experimental Results and Conclusions

Remember from Theorem 1 that $e(z^T)$ of a unit vector z gives an error bound for the corresponding variable. To get an idea of the actual distribution of these errors we studied 100 matrices obtained from mixtures of 20 proteins randomly picked from Swissprot database, with trypsin digestion simulated in silico. Input matrices A are then constructed as explained in the Introduction. Instead of peptide identities (sequences) we distinghuished them only by their masses. Clearly, peptide identity information would have further improved the results.

We evaluated the $e(z^T)$ values for each variable by using both EDGE LABELING (Theorem 2) and the LP considering all rows of the matrix, and it seems that in almost all cases where EDGE LABELING manages to find a bound, these results strongly agree. In spite that for few variables $e(z^T)$ are undefined or bounds are scaringly high, the average values are quite reassuring. For the majority of variables $e(z^T)$ is still small, taking values at most 4, with an average below 3. The low $e(z^T)$ values persist also for larger mixtures; we tried some with 50 proteins in the mixture, and with further candidate proteins inserted to account for undetected peptide masses.

As a more detailed illustration we report one typical example. To simplify notation we write i for x_i, every row is written as the set of variables appearing there, without parantheses, and separated by commas. We use symbol e to denote the error bound. Sets of variables with $e = 1$ are the given rows sorted by the number of variables. In particular, single variables with $e = 1$ are the proteins identified from unique peptides. Then we list the variables for every bound $e > 1$ that occured. The example contains 42 variables (candidate proteins).

After merging of variables that appear only together, and therefore cannot have uniquely determined individual values, there remain 40 distinct variables.

e=1: 1, 2, 10, 14, 15, 16, 21, 22, 27, 28, 29, 32, 33, 34, 35, 38, 40, 41, 42,
1 21, 1 28, 2 15, 2 22, 2 24, 2 29, 3 21, 4 15, 5 28, 7 29, 8 33, 9 21, 9 27, 12 27,
12 28, 12 42, 13 17, 14 17, 14 22, 14 27, 15 17, 15 35, 16 30, 16 31, 20 28, 22 36,
23 29, 26 29, 27 29, 29 34, 29 40, 33 35, 37 41,
1 6 23, 1 11 37, 2 18 19, 2 28 41, 9 29 41, 10 29 41, 18 19 41, 23 24 41, 25 28 31,
27 33 40, 1 23 24 42, 23 24 35 41,
14 27 28 35 36, 15 23 25 26 42, 23 25 27 41 42,
20 21 24 25 27 37, 11 14 24 25 27 35 41,
1 2 10 12 14 21 23 24 27 29 33 34 35 41 42,
1 2 14 15 16 22 23 24 27 28 29 31 35 37 38 39 40 41 42

e=2: 3, 4, 5, 7, 8, 9, 12, 17, 18, 20, 23, 24, 26, 30, 31, 36, 37

e=3: 13; **e=3.67:** 11; **e=3.83:** 25; **e=4:** 6

The solution to this system turns out to be uniquely determined, subject to the few merged variables. (By elementary linear algebra, this fact depends only on A but not on b.) Remarkably, almost half of them have $e = 1$ and $e = 2$, respectively, and only a small rest have $e \leq 4$. We emphasize that many rows consist of one or two variables, and the optimal linear combinations almost exclusively use these graph edges and loops. The graph also exhibits some cycles, but only case (i) from Theorem 2 takes effect in this example, because loops are abound. Only a few optimal error factors are "hidden more deeply". For instance, $e \leq 4$ for variables 11 and 25 is easy to find without computer help, but the optimal result detected by GLPK is slightly better: $e = 3.67$ for variable 11 is a result of the following rows:
1, 21, 28, 35, 20 28, 37 41, 1 11 37, 20 21 24 25 27 37, 11 14 24 25 27 35 41,
with coefficients -2,+1,-1,-1,+1,-1,+2,-1,+1, all divided by 3.

We observe that the matrix structure is rich enough to give almost all variables uniquely determined values (where the only exceptions we found were sets of proteins with equal peptide mass spectrum), and at the same time their accumulated errors are small. Moreover, optimal combinations of measurements are mostly built from very sparse equations using a bit of elementary graph theory. (Thus the restriction to rows with at most two 1s is not as narrow as it may seem.) In some cases the LP gave somewhat better combinations, but this degree of accuracy does not seem to be very important, as the $e(z^T)$ merely have the role of pessimistic worst-case bounds; see the earlier remark.

We conclude that quantitative mixture reconstruction is possible when sufficiently accurate measures are available. At least this does not fail already for intrinsic mathematical reasons, i.e., by the structure of the incidence matrices. Although we could test, at this stage, only simulated (i.e., random) mixtures, we expect similar matrix structures also for real mixtures. It must be admitted that our simulations did not take undetectable peptides into account, i.e., the set of available rows in the matrices may be more limited. However we also notice that most unit vectors have several alternative representations as "small" linear combinations of rows, and it suffices that some of these row sets is present. It was also pointed out in [6] that improved accuracy of the measurement technology will increase the power of mathematical inference methods. Another type of errors not considered here is that some proteins could very well be in the mixture but are ignored because their masses are not listed among the observed ones, although some of the masses they share with other proteins are detected. We are planning to address this issue in the future by extending the selection of columns included in the input matrix A. However, current quantification procedures consider only unique peptides, and thus the first results for shared peptides are a step forward.

Other directions of further work include: to examine larger simulated mixtures and real mixtures, to derive both simple upper bounds and combinatorial optimality criteria for BOUND MINIMIZATION also when rows have some more 1s entries (however the resulting hypergraph problems could be intrinsically more complicated), to give a theoretical explanation of the observed limited error propagation by "helpful" properties of the matrices, and to study probabilistic error models. In our study we adopted a simple error model to start with, assuming a uniform error bound ϵ in b. Due to practical considerations we may want to allow for different errors of the b_i, a "vector-ϵ" so to speak, which may be known or unknown in advance. On the other hand, our exact systems $Ax = b$ typically have uniquely solutions (after removal of duplicate columns) and are even largely overdetermined. These two points give rise to the following type of problem:

Given a matrix A with full column rank, and a corrupted vector b' coming from an unknown $b = Ax$, can we efficiently compute x' such that in Ax' most entries are close to the given ones in b', with only few outliers? In particular, can we recognize which b_i had small and large errors in a particular instance? What global error assumptions on b are sufficient to enable such an approximate reconstruction of x? It is clear that some error assumptions must be made. Also, this is not a single clear-cut problem but rather a research direction. The work in [6] belongs to this direction and proposes some error norm, but can also think of alternative ones. For an $m \times n$ matrix A this problem amounts to a geometric one: reporting and clustering the intersection points of subsets of n linearly independent hyperplanes in an arrangement of $m > n$ hyperplanes in n dimensions, perhaps under additional assumptions how the arrangement was obtained. The problem can also be formulated as finding maximal feasible subsystems of linear equations. Negative complexity results as in [7,10] do not rule out the possibility of efficient parameterizations.

Acknowledgments. This work has been supported by the Swedish Research Council (Vetenskapsrådet), grant no. 2010-4661, "Generalized and fast search strategies for parameterized problems". Early stages of the second author's work have also been supported by Devdatt Dubhashi through a Chalmers Bioscience Initiative grant. The authors thank Azam Sheikh Muhammad for some discussions and for valuable technical help in preparing data and scripts for GLPK.

References

1. Arioli, M., Demmel, J.W., Duff, I.S.: Solving Sparse Linear Systems with Sparse Backward Error. SIAM J. Matrix Analysis and Appl. 10, 165–190 (1989)
2. Bantscheff, M., Schirle, M., Sweetman, G., Rick, J., Kuster, B.: Quantitative Mass Spectrometry in Proteomics: A Critical Review. Anal. Bioanal. Chem. 389, 1017–1031 (2007)
3. Boehm, A.M., Pütz, S., Altenhöfer, D., Sickmann, A., Falk, M.: Precise Protein Quantification Based on Peptide Quantification Using iTRAQ. BMC Bioinformatics 8, 214 (2007)
4. Chandrasekaran, S., Ipsen, I.C.F.: On the Sensitivity of Solution Components in Linear Systems of Equations. SIAM J. Matrix Analysis and Appl. 16, 93–112 (1995)
5. Damaschke, P.: Sparse Solutions of Sparse Linear Systems: Fixed-Parameter Tractability and an Application of Complex Group Testing. In: Marx, D., Rossmanith, P. (eds.) IPEC 2011. LNCS, vol. 7112, pp. 94–105. Springer, Heidelberg (2012)
6. Dost, B., Bafna, V., Bandeira, N., Li, X., Shen, Z., Briggs, S.: Shared Peptides in Mass Spectrometry Based Protein Quantification. In: Batzoglou, S. (ed.) RECOMB 2009. LNCS, vol. 5541, pp. 356–371. Springer, Heidelberg (2009)
7. Feige, U., Reichman, D.: On the Hardness of Approximating Max-Satisfy. Inf. Proc. Lett. 97, 31–35 (2006)
8. Fritzilas, E., Milanic, M., Rahmann, S., Rios-Solis, Y.A.: Structural Identifiability in Low-Rank Matrix Factorization. Algorithmica 53, 313–332 (2010)
9. Gerber, S.A., Rush, J., Stemman, O., Kirschner, M.W., Gygi, S.P.: Absolute Quantification of Proteins and Phosphoproteins from Cell Lysates by Tandem MS. Proc. Nat. Academy of Sc. USA 100, 6940–6945 (2003)
10. Giannopoulos, P., Knauer, C., Rote, G.: The Parameterized Complexity of Some Geometric Problems in Unbounded Dimension. In: Chen, J., Fomin, F.V. (eds.) IWPEC 2009. LNCS, vol. 5917, pp. 198–209. Springer, Heidelberg (2009)
11. Nesvizhskii, A.I., Aebersold, R.: Interpretation of Shotgun Proteomic Data: The Protein Inference Problem. Mol. Cellular Proteomics 4, 1419–1440 (2005)
12. Yang, X., Dai, H., He, Q.: Condition Numbers and Backward Perturbation Bound for Linear Matrix Equations. Num. Lin. Algebra with Appl. 18, 155–165 (2011)
13. Zhang, B., Chambers, M.C., Tabb, D.L.: Proteomic Parsimony through Bipartite Graph Analysis Improves Accuracy and Transparency. J. Proteome Res. 6, 3549–3557 (2007)

Models and Algorithmic Tools for Computational Processes in Cellular Biology: Recent Developments and Future Directions

(Invited Keynote Talk)

Bhaskar DasGupta

Department of Computer Science, University of Illinois at Chicago, IL 60607
bdasgup@uic.edu

Over the last few decades, researchers in various fields have witnessed applications of novel computing models and algorithmic paradigms in many application areas involving biological processes, quantum computing, nanotechnology, social networks and many other such disciplines. Typical characteristics of these application areas include their interdisciplinary nature going beyond previous traditional approaches that were used, and often high-risk high-gain nature of resulting collaborations. Major research challenges on a macroscopic level for such collaborations include forming appropriate interdisciplinary teams, effective communication between researchers in diverse areas, using individual expertise for overall goal of the project, and collaboration with industry if necessary. In addition, one also faces the usual challenge that lies in collaboration between theory and practice, namely sometimes theory follows application, sometimes theory precedes application, and sometimes they walk hand-in-hand. Recent and not so recent developments on analysis of models of computational processes in biology, in the context of gene and protein networks that arise in organism development and cell signalling, have given rise to many types of discrete, continuous and hybrid models, and researchers have studied the inter-relationships, powers and limitations, computational complexity and algorithmic issues as well as biological implications and validations of these models. Such investigations have given rise to fascinating interplay between many diverse research areas such as biology, control theory, discrete mathematics and computer science.

In this talk, I will discuss some successful interdisciplinary collaborative projects of mine on cell signalling processes with other researchers from control theory, cell biology and computational complexity theory [1–11]. These projects involve interesting integration of concepts from control theory (dynamical systems) and computational complexity theory (approximation algorithms and inapproximability results) with the signalling mechanisms (signal transduction networks and other models) in cellular processes in biology. I will also discuss future research problems in these topics that may be of importance to new or current researchers. No prior knowledge of dynamical systems or other models for cell signalling mechanisms will be assumed.

Acknowledgements. These research projects in my talk could not have been completed without collaboration with a large number of collaborators from different research areas. I would like to thank all of them for their involvements. Special thanks go to my colleagues Réka Albert, Piotr Berman and Eduardo Sontag for their enormous patience and contribution during the collaboration phases. Last, but not the least, I would like to thank individually all the students and post-doctoral fellows involved in these projects (Anthony Gitter, Gamze Gürsoy, Rashmi Hegde, Gowri Sangeetha Sivanathan, Pradyut Pal, Paola Vera-Licona, Riccardo Dondi, Sema Kachalo, Ranran Zhang, Yi Zhang, Kelly Westbrooks and German Enciso); these collaborations would not have been so successful without their involvements. We also thankfully acknowledge generous financial support from the National Science Foundation through grants DBI-1062328, IIS-1064681, IIS-0346973, DBI-0543365, IIS-0610244, CCR-9800086, CNS-0206795, and CCF-0208749, and generous support from the DIMACS Center of Rutgers University during my Sabbatical leave through their special focus on computational and mathematical epidemiology.

References

1. Albert, R., DasGupta, B., Gitter, A., Gürsoy, G., Hegde, R., Pal, P., Sivanathan, G.S., Sontag, E.: A New Computationally Efficient Measure of Topological Redundancy of Biological and Social Networks. Physical Review E 84(3), 036117 (2011)
2. DasGupta, B., Vera-Licona, P., Sontag, E.: Reverse Engineering of Molecular Networks from a Common Combinatorial Approach. In: Elloumi, M., Zomaya, A. (eds.) Algorithms in Computational Molecular Biology: Techniques, Approaches and Applications, ch. 40. John Wiley & Sons, Inc. (2011)
3. Albert, R., DasGupta, B., Sontag, E.: Inference of signal transduction networks from double causal evidence. In: Fenyo, D. (ed.) Methods in Molecular Biology: Topics in Computational Biology, ch. 16, p. 673. Springer Science+Business Media, LLC (2010)
4. Berman, P., DasGupta, B., Karpinski, M.: Approximating Transitive Reductions for Directed Networks. In: Dehne, F., Gavrilova, M., Sack, J.-R., Tóth, C.D. (eds.) WADS 2009. LNCS, vol. 5664, pp. 74–85. Springer, Heidelberg (2009)
5. Albert, R., DasGupta, B., Dondi, R., Sontag, E.: Inferring (Biological) Signal Transduction Networks via Transitive Reductions of Directed Graphs. Algorithmica 51(2), 129–159 (2008)
6. Kachalo, S., Zhang, R., Sontag, E., Albert, R., DasGupta, B.: NET-SYNTHESIS: A software for synthesis, inference and simplification of signal transduction networks. Bioinformatics 24(2), 293–295 (2008)
7. Berman, P., DasGupta, B., Sontag, E.: Algorithmic Issues in Reverse Engineering of Protein and Gene Networks via the Modular Response Analysis Method. Annals of the New York Academy of Sciences 1115, 132–141 (2007)
8. Albert, R., DasGupta, B., Dondi, R., Kachalo, S., Sontag, E., Zelikovsky, A., Westbrooks, K.: A Novel Method for Signal Transduction Network Inference from Indirect Experimental Evidence. Journal of Computational Biology 14(7), 927–949 (2007)

9. DasGupta, B., Enciso, G.A., Sontag, E., Zhang, Y.: Algorithmic and Complexity Results for Decompositions of Biological Networks into Monotone Subsystems. Biosystems 90(1), 161–178 (2007)
10. Berman, P., DasGupta, B., Sontag, E.: Computational Complexities of Combinatorial Problems With Applications to Reverse Engineering of Biological Networks. In: Wang, F.-Y., Liu, D. (eds.) Advances in Computational Intelligence: Theory and Applications. Series in Intelligent Control and Intelligent Automation, vol. 5, pp. 303–316. World Scientific Publishers (2007)
11. Berman, P., DasGupta, B., Sontag, E.: Randomized Approximation Algorithms for Set Multicover Problems with Applications to Reverse Engineering of Protein and Gene Networks. Discrete Applied Mathematics 155(6-7), 733–749 (2007)

Identifying Rogue Taxa through Reduced Consensus: NP-Hardness and Exact Algorithms*

Akshay Deepak, Jianrong Dong, and David Fernández-Baca

Department of Computer Science, Iowa State University, Ames, Iowa, USA

Abstract. A rogue taxon in a collection of phylogenetic trees is one whose position varies drastically from tree to tree. The presence of such taxa can greatly reduce the resolution of the consensus tree (e.g., the majority-rule or strict consensus) for a collection. The reduced consensus approach aims to identify and eliminate rogue taxa to produce more informative consensus trees. Given a collection of phylogenetic trees over the same leaf set, the goal is to find a set of taxa whose removal maximizes the number of internal edges in the consensus tree of the collection. We show that this problem is NP-hard for strict and majority-rule consensus. We give a polynomial-time algorithm for reduced strict consensus when the maximum degree of the strict consensus of the original trees is bounded. We describe exact integer linear programming formulations for computing reduced strict, majority and loose consensus trees. In experimental tests, our exact solutions improved over heuristic methods on several problem instances.

1 Introduction

Consensus methods such as majority rule and strict consensus trees [13] are often used to summarize in a single tree, a *consensus tree*, the common information in a collection of trees over a common leaf set. Such collections are routinely produced by phylogenetic analyses by parsimony, maximum likelihood or Bayesian methods. Consensus trees can be greatly affected by *rogue taxa* (sometimes called *wandering taxa*); that is taxa whose positions can vary dramatically without having a strong effect on a tree's overall score. The presence of just a few such taxa can lead to poorly-resolved consensus trees, even when there is substantial agreement relative to the remaining taxa [16,22,21,19,14]. Figure 1 illustrates this for the case of strict consensus.

Here we consider the problem of finding a set of leaves (i.e., possible rogue taxa) to be removed so as to maximize the number of internal edges in the consensus tree on the reduced leaf set. Our main results are the following.

- Proofs that the underlying decision problem is NP-complete for three trees in the strict consensus case and for four trees in the case of majority rule trees.
- A polynomial-time algorithm for reduced strict consensus when the maximum degree of the strict consensus tree of the original collection of trees is fixed.
- An integer linear programming (ILP) formulation for obtaining exact solutions for reduced strict, majority and loose consensus.

---

* Supported in part by National Science Foundation grants DEB-0830012 and CCF-106029.

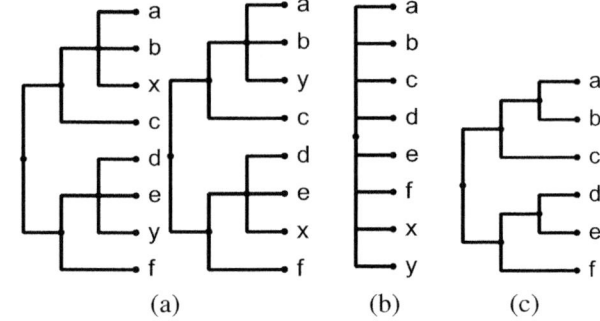

Fig. 1. The effect of rogue taxon on consensus trees. (a) Two input trees and (b) their strict consensus – a star-like tree. (c) The strict consensus of the trees, after removing the rogue taxon x and y from the input trees.

- An experimental comparison with the heuristic proposed in [15], showing that in several cases our reduced consensus formulations allow us to uncover increased common phylogenetic information without discarding many more leaves.

Relationship to Previous Work. There is a sizable literature on identifying rogue taxa (see, e.g., [22,21,15,5,24,23]). Agreement and frequent subtree approaches — where one seeks induced subtree common to all or some fraction of the input trees — have been suggested by some. For example, Cranston and Rannala use it to summarize the posterior distribution of a collection of trees resulting from Bayesian analysis [5]. Intuitively, since the placement of rogue taxa varies widely from tree to tree, these taxa will be excluded from the agreement subtrees. There are several papers on computing agreement subtrees (e.g., [9,2,8,12,20]) and frequent subtrees (e.g., [4,25,26,27]); however, agreement-based approaches can tend to be more conservative than consensus trees. Further, the underlying optimization problems are NP-hard [2,15].

Wilkinson [22,24,23] appears to have been the first to propose the use of reduced majority rule and strict consensus as a means to identify rogue taxa. While the underlying principle behind all proposed reduced consensus methods is to gain internal edges at the expense of dropping leaves [16], the precise cost measure can vary. The criterion we use in this paper is to maximize the resolution of the consensus tree, without taking into account how many leaves are eliminated. In contrast, Pattengale et al.'s objective function is a weighted sum of the total number of leaves retained and the number of clusters (actually, bipartitions) obtained [15]. Our NP-completeness proof shows that this problem is hard for strict and majority-rule consensus, when the weight assigned to the leaves is zero. To our knowledge, the complexity of the problem in the general case has not been determined. We conjecture that this case is also hard. Indeed, Pattengale et al. only offer a heuristic for it.

The use of integer linear programming in phylogenetics appears to be relatively new [10,18]. The formulations presented here are related to the authors' previous work on constructing majority-rule supertrees and conservative supertrees [7,6]. Nevertheless, there are some important differences from that work. While our ILP formulations can only be used for relatively small data sets, they offer a valuable benchmark against which to compare heuristics. They have also allowed us to identify certain interesting characteristics and limitations of the various consensus methods. In particular, our experiments suggest that assigning equal weight to the number of clusters and the number of taxa in the consensus tree might be too conservative an approach.

2 Preliminaries

A *phylogenetic tree* is an unordered rooted or unrooted tree whose leaves are in a bijection with a set of *labels* or *taxa*. We sometimes refer to a phylogenetic tree simply as a *phylogeny* or as a *tree*. A node is *internal* if it is not a leaf. An edge is *internal* if its two endpoints are internal. If a phylogeny T is rooted, we require that each internal node have at least two children; if T is unrooted, every internal node is required to have degree at least three. Here, we limit our discussion to collections of trees over the same set of taxa. Without loss of generality, we can view such trees as rooted, since, when the trees are unrooted, we can arbitrarily pick any taxon as an out-group, which is equivalent to fixing a common root [17,1].

Let T be a phylogenetic tree. We write V_T, E_T, and $\mathcal{L}_T$ to denote the set of vertices, the set of edges, and the set of leaf-labels of T respectively. For convenience, we refer to the set of leaf nodes by their labels in $\mathcal{L}_T$. For an internal node u, let $\text{Ch}(u)$ denote its set of children and let $d(u) = |\text{Ch}(u)|$. Two leaves form a *cherry* if they have a common parent. Let $\varepsilon(T)$ denote the number of internal edges in T.

To *contract* an internal edge is to delete the edge and identify its endpoints. To *suppress* a degree-two node is to delete the node and identify its neighbors. The *restriction* of T to set $X \subseteq \mathcal{L}_T$, denoted by $T|_X$, is the tree on leaf set X obtained from the minimal subtree of T spanning X by suppressing all degree-two nodes except the root. The root of $T|_X$ is the node that was originally closest to the root of T.

The *cluster* induced by a node u in tree T, denoted $\Gamma(u)$, is the set of all the leaf descendants of u. A cluster is *trivial* if it is induced by a leaf or the root and is *nontrivial* otherwise. Let A be a subset of $\mathcal{L}_T$. Tree T *contains* cluster A if there is a node u of T such that $A = \Gamma(u)$. We say that $A \subseteq \mathcal{L}_T$ is *compatible with* T if there exists a tree T' such that T' contains all the clusters of T, as well as cluster A.

Throughout the rest of the paper, $P = (T_1, \ldots, T_m)$ is a tuple of trees[1] over a common leaf set, which is denoted by $\mathcal{L}_P$. Let X be a subset of $\mathcal{L}_P$. Then, the *restriction* of P to X, denoted $P|_X$, is the tuple of trees $(T_1|_X, \ldots, T_m|_X)$.

Consensus Methods. A *phylogenetic consensus method* (or simply a *consensus method*) is a function C that maps a tuple of trees P to a tree $C(P)$ with leaf set $\mathcal{L}_P$. A variety of consensus methods have been defined [3]. Here we focus on the following three.

- The *majority-rule tree* of P, denoted $\text{Maj}(P)$ is the tree containing precisely those clusters that occur in more than half of the trees in P.

---

[1] We refer to a *tuple* of trees, rather than a set or collection, because, for the case of majority-rule consensus, it is necessary to allow repetitions.

- The *strict consensus tree* of P, denoted $\text{Str}(P)$ is the tree containing precisely those clusters that occur in every tree in P.
- The *loose consensus tree* of P, denoted $\text{Loose}(P)$, contains exactly those clusters that appear in some tree in P and that are compatible with every tree in P.

Reduced Consensus. Let C denote some consensus method. Given a collection of trees P, the *reduced C consensus problem* is to find a set $X \subseteq \mathcal{L}_P$ that maximizes $\varepsilon(C(P|_X))$. For example, the reduced majority-rule consensus problem aims at restricting the input trees to a set X such that $\text{Maj}(P|_X)$ has the maximum number of internal edges among all such possible Xs. Observe that $\varepsilon(C(P|_X)) \leq |\mathcal{L}_P| - 2$, regardless of the consensus method C.

We also consider a *decision version* of the reduced consensus problem. Given a tuple of trees P, a consensus method C, and an integer k between 1 and $|\mathcal{L}_P| - 2$, the question is whether there exists a set $X \subseteq \mathcal{L}$ such that $\varepsilon(C(P|_X)) = k$.

3 The Reduced Consensus Problem Is NP-Complete

Theorem 1. *The reduced strict consensus problem is NP-complete even for three trees.*

Proof. Clearly the problem is in NP. We use reduction from the *3-dimensional matching problem* (3DM), which is known to be NP-complete [11]. The input to 3DM consists of finite disjoint sets A, B and C, where $|A| = |B| = |C| = k$, and a set $M \subseteq A \times B \times C$. The question is whether there exists a set $M' \subseteq M$ of size k such that for any two distinct triplets (x_i, y_i, z_i) and (x_j, y_j, z_j) in M', $x_i \neq x_j$, $y_i \neq y_j$ and $z_i \neq z_j$. Such an M', if it exists, is called a *witness* for the 3DM instance.

Given an instance (A, B, C, M) of 3DM, construct a tuple of trees (T_A, T_B, T_C) as follows. Assume without loss of generality that every element of A, B, C is present in at least one triplet in M else a witness is not possible. For each $m \in M$, create two taxa $m^{(1)}$ and $m^{(2)}$, and let $\mathcal{L}_P = \bigcup_{m \in M} \{m^{(1)}, m^{(2)}\}$. Let T_A initially be an empty tree with only r_A as its root node. For each $x_i \in A$, add a child x_i to r_A. Now T_A is a star-like tree with k leaves. For each triplet $m = (x_i, y_i, z_i)$ in M, attach leaves $m^{(1)}, m^{(2)}$ at node x_i in T_A. At this point if any x_i in T_A does not have any children, remove it. Similarly, construct trees T_B and T_C corresponding to the input sets Y and Z. Clearly, $\varepsilon(T_A) = \varepsilon(T_B) = \varepsilon(T_C) = k$. This reduction is clearly polynomial.

We claim that instance (A, B, C, M) has a witness if and only if there exists an $X \subseteq \mathcal{L}_P$ for $P = (T_A, T_B, T_C)$ such that $\varepsilon(\text{Str}(P|_X)) = k$.

For the forward direction, let $M' \subseteq M$ be a witness for (A, B, C, M). Let $X = \bigcup_{m \in M'} \{m^1, m^{(2)}\}$ be the reduced leaf set. Let $S = \text{Str}(P|_X)$. Since for every two distinct triplets $m_i = (x_i, y_i, z_i)$ and $m_j = (x_j, y_j, z_j)$ in M', $x_i \neq x_j$, $y_i \neq y_j$ and $z_i \neq z_j$, the corresponding cherries $(m_i^{(1)}, m_i^{(2)})$ and $(m_j^{(1)}, m_j^{(2)})$ are not attached at the same node in any of the trees in (T_A, T_B, T_C). Thus, each such cherry corresponding to an $m \in M'$ is a distinct two-element cluster in $\text{Str}(P|_X)$. Therefore, there is an internal edge corresponding to each such size-two cluster in $\text{Str}(P|_X)$. Further $|M'| = k$ implies $\varepsilon(\text{Str}(P|_X)) = k$.

Now we prove the reverse direction. Suppose there exists a subset $X \subseteq \mathcal{L}_P$ such that $\varepsilon(\text{Str}(P|_X)) = k$. Note that $\text{Str}(P|_X)$ cannot have a cherry with leaves corresponding to two distinct triplets m_i and m_j in M as this will lead to $m_i = m_j$, a contradiction. Thus, the endpoints of each internal edge are the root and the parent of a cherry, and each such cherry corresponds to a unique element in M. Let $M' \subseteq M$ be the set of triplets such that for each $m_i = (x_i, y_i, z_i)$ in M', the corresponding cherry with leaves $m_i^{(1)}, m_i^{(2)}$ is a cluster in $\mathrm{Str}(P|_X)$. Clearly $|M'| = k$. Since $\varepsilon(\mathrm{Str}(P|_X)) = \varepsilon(T_A) = k$, each internal edge — whose endpoints are the root and the parent of the cherry consisting of $m_i^{(1)}, m_i^{(2)}$ — in $\mathrm{Str}(P|_X)$ corresponds to a unique internal edge attached to x_i in T_A. Thus, triplet m_i covers a unique x_i in A. The same applies with respect to B and C. Thus, for every two distinct triplets $m_i = (x_i, y_i, z_i)$ and $m_j = (x_j, y_j, z_j)$ in M', $x_i \neq x_j$, $y_i \neq y_j$ and $z_i \neq z_j$ holds. Thus M' is a witness for the given 3DM instance. □

Theorem 2. *The reduced majority rule consensus problem is NP-complete, even for four trees.*

Proof. Modify the construction used in the proof of Theorem 1 by adding a dummy star-like tree T_D with $\mathcal{L}_{T_D} = \mathcal{L}_{T_A} = \mathcal{L}_{T_B} = \mathcal{L}_{T_C}$ to the tuple (T_A, T_B, T_C). The reduced majority rule tree of this new collection of trees is the strict consensus of the original tuple, because the dummy tree contains no non-trivial clusters, so the clusters from the first three trees constitute a majority. □

4 Fixed Parameter Tractability

We now prove that the reduced consensus problem relative to strict consensus is fixed parameter tractable in the maximum out-degree of $\mathrm{Str}(P)$. Formally, we claim the following.

Theorem 3. *Let $P = (T_1, \ldots, T_m)$ be a tuple of phylogenetic trees over the same leaf set and let d denote the maximum out-degree of a node in $\mathrm{Str}(P)$. Then, the reduced strict consensus problem for P can be solved in $O(m \cdot |\mathcal{L}_P| \cdot 2^d)$ time.*

Note that the above result is practical only when the strict consensus of P does not contain high-degree nodes. Nevertheless, even if this is not the case, the algorithm can still be used in combination with heuristic approaches such as [15,5]. Through these methods, one can bring the maximum out-degree to within reasonable bounds, after which one can invoke Theorem 3.

Suppose X is a subset of $\mathcal{L}_P$. Let $S = \mathrm{Str}(P)|_X$ and $R = \mathrm{Str}(P|_X)$. Observe that R is a *refinement* of S. That is, S can be obtained from R by contracting zero or more internal edges of R (see Figure 2). Thus, for every $u \in V_S$, there exists a mapping $\vartheta_u : 2^{\mathcal{L}_S} \to 2^{E_R}$ such that $\vartheta_u(X)$ is the set of all edges in R that are contracted into u. If $\varepsilon(R) > \varepsilon(S)$, there must exist at least one internal node u in S such that $|\vartheta_u(X)| > 0$. Node u is said to be *refined* in R. For example, in Figure 2, the root of S is refined in R. Since R is a refinement of S, $V_S \subseteq V_R$. Let $\psi : V_S \to V_R$ denote the injective mapping from the nodes of tree S to the corresponding nodes in R.

The proof of Theorem 3 relies on the following result, which is proved at the end of this section.

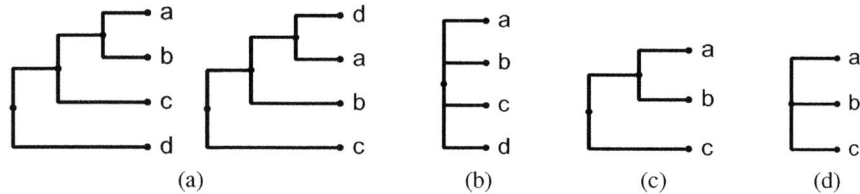

Fig. 2. (a) A pair P of trees on leaf set $\mathcal{L}_P = \{a, b, c, d\}$. (b) $\mathrm{Str}(P)$. (c) $R = \mathrm{Str}(P|_X)$ for $X = \{a, b, c\}$. (d) $S = \mathrm{Str}(P)|_X$.

Lemma 1. *Let u be a node in $\mathrm{Str}(P)$. Let X be a maximal set of taxa such that u is refined in $\mathrm{Str}(P|_X)$ and let $Y = \mathcal{L}_P - X$. Then the following hold.*

(i) For every $a \in \mathrm{Ch}(u)$ if $\Gamma(a) \cap Y \neq \emptyset$ then $\Gamma(a) \subseteq Y$
(ii) $Y \subseteq \bigcup_{a \in \mathrm{Ch}(u)} \Gamma(a)$

Lemma 1 shows that the refinement of node u must occur at the cost of removing one or more clusters corresponding to $\mathrm{Ch}(u)$ and that only clusters corresponding to $\mathrm{Ch}(u)$ need to be removed. Therefore, to refine any node u in $\mathrm{Str}(P)$ we only need to consider the clusters corresponding to $\mathrm{Ch}(u)$ in their entirety; i.e., either a cluster corresponding to $\mathrm{Ch}(u)$ is completely retained in $\mathrm{Str}(P|_X)$ or is completely absent from $\mathrm{Str}(P|_X)$.

For a given node u in $\mathrm{Str}(P)$, define $\mathrm{opt}(u)$ as follows

$$\mathrm{opt}(u) = \max_{X \subseteq \Gamma(u)} \{\varepsilon\left(\mathrm{Str}(P|_X)\right)\}. \qquad (1)$$

When u is a leaf, $\mathrm{opt}(u) = 0$. When u is the root node, $\mathrm{opt}(u)$ gives the value of the optimum solution to the reduced strict consensus problem. Note that when considering the optimum solution in (1), an $X \subset \Gamma(u)$ needs to be considered only if it refines u. By Lemma 1, such an X must contain one or more clusters corresponding to $\mathrm{Ch}(u)$ in entirety. Thus, $\mathrm{opt}(u)$ can be expressed as:

$$\mathrm{opt}(u) = \max_{U \subseteq \mathrm{Ch}(u)} \left\{ \sum_{v \in U} \mathrm{opt}(v) + \iota(U) + |\vartheta_u\left(\Gamma(U)\right)| \right\}, \qquad (2)$$

where $\iota(U)$ denotes the number of nodes in U that are internal and $\Gamma(U) = \bigcup_{v \in U} \Gamma(v)$. Here, $\sum_{v \in U} \mathrm{opt}(v)$ corresponds to the contribution of internal edges by the subtrees rooted at each of the children of u in U, $\iota(U)$ is the number of internal edges contributed by the internal nodes in U and $|\vartheta_u\left(\Gamma(U)\right)|$ refers to the internal edges gained due to the refinement of u.

Note that while (1) requires iteration over all subsets of $\Gamma(u)$, (2) only requires iteration over subsets of $\mathrm{Ch}(u)$. For a given $U \subseteq \mathrm{Ch}(u)$, $\vartheta_u\left(\Gamma(U)\right)$ can be computed in $O(m|U|)$ time. This can be done in the following way. For every node u in $\mathrm{Str}(P)$, we maintain an array of size m whose i-th entry array points to the node v in the i-th tree in P such that $\Gamma(u) = \Gamma(v)$. Thus, for a given $U \subseteq \mathrm{Ch}(u)$ and a tree T in P, we can find the subtree induced by $U + u$ in T in $O(|U|)$ time[2]. Observe that here U

---

[2] Given a set A and an element x, we write $A + x$ to denote $A \cup \{x\}$.

corresponds to the leaves of this subtree. For all the trees in P, this takes $O(m|U|)$ time. Computing the strict consensus of these subtrees takes $O(m|U|)$ time [1]. The number of internal edges in this strict consensus tree corresponds to $|\vartheta_u(\Gamma(U))|$.

Assuming $opt(v)$ is known for all $v \in \text{Ch}(u)$, expression (2) can be evaluated in $O(m|U|)$ time. Thus, $opt(u)$ can be computed in $O(m \cdot d(u) \cdot 2^{d(u)})$ time. Thus, in an inorder depth first traversal of $\text{Str}(P)$, opt values for all nodes can be calculated in $O(m \cdot |\mathcal{L}_P| \cdot 2^d)$. Theorem 3 follows.

One appealing feature of the algorithm just described is that it can easily be implemented so that, among all subsets X that maximize $\varepsilon(\text{Str}(P|_X))$, it returns a maximal one. That is, no proper superset of X yields an optimal solution or, equivalently, the set of dropped taxa is minimal.

Proof of Lemma 1. Let ℓ be any leaf from $Y = \mathcal{L}_P - X$. Since X is maximal, u is not refined in $\text{Str}(P|_{X+\ell})$. Let (u', v') be an edge in $\vartheta_u(X)$, where $u' = \psi(u)$ and u' is the parent of v' i.e. this edge does not exist in $\text{Str}(P|_{X+\ell})$ but exists in the refined strict consensus tree $\text{Str}(P|_X)$.

1. We use proof by contradiction. Suppose there exists an $a \in \text{Ch}(u)$ such that $\Gamma(a) \cap Y \neq \emptyset$ but $\Gamma(a) \not\subseteq Y$. Thus, the corresponding node $a' = \psi(a)$ exists in $\text{Str}(P|_X)$. Without loss of generality, assume $\ell \in \Gamma(a) \cap Y$. In the refined strict consensus tree $\text{Str}(P|_X)$, both a' and v' are children of u'. Based on the relative position of a' with respect to v', there are two possibilities.
   (a) a' is a descendant of v'. Since the edge (u', v') does not exists in $\text{Str}(P|_{X+\ell})$ but exists in $\text{Str}(P|_X)$, cluster $\Gamma(v') + \ell$ does not exist in all the trees in $P|_{X+\ell}$ but exists (as $\Gamma(v')$) in all the trees in $P|_X$. Thus, at least one tree $T \in P|_{X+\ell}$, must contain cluster $A = \Gamma(v')$. However, T also contains clusters $B = \Gamma(a') + \ell$ and $C = \Gamma(u') + \ell$. Given that v' is a child of u' and a' is a descendant of v', the existence of clusters A, B, and C in a tree is not possible. Hence this case leads to a contradiction.
   (b) a' is not a descendant of v'. Again, since the edge (u', v') does not exist in $\text{Str}(P|_{X+\ell})$ but exists in $\text{Str}(P|_X)$, cluster $\Gamma(v')$ does not exist in all the trees in $P|_{X+\ell}$ but exists in all the trees $P|_X$. Thus, at least one tree $T \in P|_{X+\ell}$, must contain cluster $A = \Gamma(v') + \ell$. T also contains cluster $B = \Gamma(a') + \ell$. Given that neither of v', a' is a descendant of the other, the existence of clusters A and B in a tree is not possible. Hence this case leads to a contradiction.
   Since each case leads to a contradiction, we have $\Gamma(a) \subseteq Y$.
2. Suppose $Y \not\subseteq \bigcup_{a \in \text{Ch}(u)} \Gamma(a)$. We can assume without loss of generality that $\ell \notin \bigcup_{a \in \text{Ch}(u)} \Gamma(a)$. Thus, $\ell \notin \Gamma(u')$. Again, at least one tree $T \in P|_{X+\ell}$, must contain cluster $A = \Gamma(v') + \ell$. This tree also contains cluster $B = \Gamma(u')$. Given that v' is the child of u' and $\ell \notin \Gamma(u')$, the existence of clusters A and B in a tree is not possible, so we have a contradiction. Thus, $Y \subseteq \bigcup_{a \in \text{Ch}(u)} \Gamma(a)$. □

5 ILP Formulations

We present ILP formulations for computing reduced strict, loose, and majority-rule consensus trees. The formulations share some characteristics with those developed elsewhere for supertrees [7,6]; however, they differ in important aspects. Unlike the supertree case, there is no need to model the fill-in process, where each input tree is augmented with the taxa that it is missing. Ensuring that this is done legally for supertree computation necessitates many constraints. In contrast, for reduced consensus, reduced trees are completely determined by the selection of rogue taxa. Moreover, clusters that are identical or compatible prior to taxon reduction remain identical or compatible.

We now describe the main constants, variables, and constraints of the ILP formulations. Unless mentioned otherwise, variables are binary. For brevity, we will simply express the constraints as logical expressions involving set operations. These can easily be transformed into linear inequalities — see, e.g., [7]. As before, let $P = (T_1, \ldots, T_m)$ be a tuple of rooted trees over the same leaf set; let $n = |\mathcal{L}_P|$.

For $j = 1, \ldots, m$, we represent T_j by a $n \times g_j$ matrix $M(T_j)$ whose columns correspond to the nontrivial clusters of T_j that do not appear in T_i, for any $i < j$. Let $g = \sum_{j=1}^{m} g_j$. Suppose column i of $M(T_j)$ corresponds to cluster A in T_j. Then, $M_{xi}(T_j) = 0$ if taxon x is in A, and $M_{xi}(T_j) = 1$ otherwise. Tuple P is represented by the matrix $M(P)$ obtained by concatenating matrices $M(T_1), \ldots, M(T_m)$.

To represent the subset $X \subseteq \mathcal{L}(P)$ from which we obtain the reduced collection of trees $P|_X$, define variables $S_1, \ldots, S_n$, where $S_r = 1$ precisely if the r-th taxon is included in X. For every pair i, j of columns of $M(P)$, and each taxon r such that $S_r = 1$, we can encounter one of four patterns: 00, 01, 10, or 11. The presence or absence of these patterns is indicated by the settings of variables $B_{ij}^{(ab)}$, where $a, b \in \{0, 1\}$, $i = 1, \ldots, g_1$ (or g), $j = 1, \ldots, g$, and $i < j$. $B_{ij}^{(ab)} = 1$ precisely when there is a taxon r such that $S_r = 1$, $M_{ri}(P) = a$ and $M_{rj}(P) = b$. We have that $B_{ij}^{(ab)} \Leftrightarrow \bigcup_{q=1}^{p} S_{r_q}$ when ab are 01 and 10 for strict and majority-rule reduced consensus trees, ab are 00, 01 and 10 for loose reduced consensus trees.

Variables $\delta_1^0, \ldots, \delta_g^0$ and $\delta_1^1, \ldots, \delta_g^1$ represent the nontrivial clusters in the reduced consensus tree. Nontrivial clusters must have at least two 0s and two 1s. Hence, for cluster j, the sum of the S_r's corresponding to $M_{r,j}(P) = 0$ must be at least 2, and the sum of S_r corresponding to $M_{r,j}(P) = 1$ must be at least 1. For $1 \leq j \leq g_1$ (or g), we have $\delta_j^0 \Leftrightarrow \sum_{q=1}^{p} S_{r_q} \geq 2$. Similarly, we have $\delta_j^1 \Leftrightarrow \sum_{q=1}^{p} S_{r_q} \geq 1$ where $1 \leq r_q \leq n$. Note that for strict consensus we can use fewer δ^0 and δ^1 variables. This is because tree T_1 (in fact, any input tree) must contain every cluster of the reduced strict consensus tree. Thus, we only need g_1 cluster variables, and correspondingly fewer constraints. As seen below, similar savings can be achieved for other variables and constraints.

Variables E_{ij}, where $i = 1, \ldots, g_1$ (or g), $j = 1, \ldots, g$ and $i < j$, indicate identity between distinct reduced clusters. Observe that $E_{ij} \Leftrightarrow \neg B_{ij}^{(01)} \wedge \neg B_{ij}^{(10)}$. We also define g_1 (or g) D variables to indicate duplicate clusters. For $2 \leq p \leq g_1$ (or g), $D_p \Leftrightarrow \bigcup_{i=1}^{p-1} E_{ip}$ and $D_1 \equiv 1$.

For strict and majority reduced consensus trees, define variables U_{ij}, where $U_{ij} = 1$ precisely if the i-th reduced cluster is in the j-th reduced tree. $U_{ij} \Leftrightarrow \bigcup_{\ell \in M(T_j)} E_{i\ell} \cup \neg \delta_i^0 \cup \neg \delta_i^1$ (if $i > \ell$, $E_{\ell i}$ is replaced by $E_{i\ell}$).

Next, we define W-variables, that indicate which reduced input tree clusters appear in the reduced consensus tree. The number of these variables and the constraints that define them depend on the particular consensus method used.

For reduced strict consensus trees, define variables $W_1, \ldots, W_{g_1}$ where $W_i = 1$ precisely if the ith cluster is in every other tree. That is, $W_i \Leftrightarrow \sum_{j=2}^{m} U_{ij} = m - 1$.

For reduced majority-rule consensus trees, define variables $W_1, \ldots, W_g$ where $W_i = 1$ precisely if the i-th cluster is in more than half of the trees. That is, $W_i \Leftrightarrow \sum_{j=1}^{m} U_{ij} \geq \lceil \frac{m+1}{2} \rceil$.

The clusters of the reduced loose tree are in at least one tree and are compatible with every other cluster. Note that, even if input tree clusters i and j are incompatible, they could be compatible when restricted to the selected taxa. Thus, for reduced loose consensus trees, we define variables C_{ij} where $i = 1, \ldots, g$, $j = g_1 + 1, \ldots, g$, and $i < j$ to represent pairwise compatibility among reduced clusters. Observe that $\neg C_{ij} \Leftrightarrow B_{ij}^{(00)} \wedge B_{ij}^{(01)} \wedge B_{ij}^{(10)} \wedge B_{ij}^{(11)}$. When $1 \leq i, j \leq g_1$, $C_{ij} \equiv 1$. Note that $C_{ij} = C_{ji}$. Define variables $W_1, \ldots, W_g$ where $W_i = 1$ precisely if the ith cluster is compatible with every other cluster. That is, $W_i \Leftrightarrow \sum_{j=1, j \neq i}^{g} C_{ij} = g - 1$.

Define g_1 (or g) V variables to represent unique nontrivial clusters in the reduced consensus tree. For $1 \leq p \leq g_1$ (or g), we have $V_p \Leftrightarrow W_p \wedge \neg D_p \wedge \delta_p^0 \wedge \delta_p^1$. The objective function is the sum of the V_is.

It can be shown that the number of variables and constraints is $O(m^2 n^2)$ for the reduced majority-rule and loose consensus tree ILPs and $O(mn^2)$ for the reduced strict consensus tree ILP.

6 Experiments and Results

Our data sets were derived from those used in a previous study by Pattengale et al. [15]. There are 16 sets of bootstrapped trees constructed from single-gene and multi-gene DNA sequences, with 125-2,554 taxa. For each set of trees, we extracted a set of 50 trees on 20 leaves. For this, we randomly selected the required number of trees and restricted them on a random set of leaves of the required size.

We compared our ILP formulation with the rogue taxon solution proposed by Pattengale et al. [15]. The latter tries to uncover new internal edges by merging bipartitions that are sufficiently similar that the merged bipartition becomes frequent enough to be included in the consensus tree on the reduced leafset. Though quite fast, the heuristic cannot guarantee an optimal solution. The default solution in [15] tries to balance the gain in internal edges with the accompanying loss in the number of leaves. However, the authors also give a straightforward modification that attempts to maximize only resolution; i.e., the gain in internal edges. We use this modified approach to compare with our exact solutions. We use the original implementation of the authors — a Python script — for the purpose of comparison.

We computed the strict and majority reduced consensus trees with 16 20-taxon-50-tree input files. All experiments were run on an Intel Core 2 64 bit quad-core processor (2.83GHz) with 8GB RAM. The ILPs were solved using CPLEX[3]. We used MATLAB to generate the input files for CPLEX and to post-process the output files. Table 1 shows the nine cases where the ILP approach improves over the heuristic method. For other cases not shown in the table, ILP tied with the heuristic method in terms of the new

---

[3] CPLEX is a trademark of IBM.

Table 1. ILP and Heuristic Results for Strict and Majority-rule Reduced Consensus Trees. The columns represent data set number (Dataset), total running time in seconds (Time (sec)), original internal edges (OIE), new internal edges after rogue taxon removal (NIE), and total leaf loss (LL) for exact solution, followed by the heuristic NIE (HNIE) and LL (HLL) results.

	Strict						Majority-rule					
Dataset	Time(sec)	OIE	NIE	HNIE	LL	HLL	Time(sec)	OIE	NIE	HNIE	LL	HLL
1	334	4	**7**	4	10	0	1366	15	15	15	0	0
2	78	5	**8**	7	6	3	1762	15	**16**	15	1	0
3	243	5	**8**	6	8	3	62	18	18	18	0	0
4	79	7	**9**	8	7	1	6929	14	16	16	1	1
5	1897	6	**7**	6	3	0	7659	15	**16**	15	1	0
6	20	7	**10**	9	5	2	1417	15	**16**	15	2	0
7	22	8	**9**	8	2	0	28	18	18	18	0	0
8	9	9	9	9	0	0	46	15	**16**	15	1	0
9	163	8	8	8	1	0	30550	14	**15**	14	1	0

internal edges after rogue taxon removal. The columns represent the data set number (Dataset), total running time in seconds (Time (sec)), original internal edges (OIE), new internal edges after rogue taxon removal (NIE), and total leaf loss (LL) for the exact solution, followed by the heuristic NIE (HNIE) and LL (HLL) results.

Based on our experimental results, we can make some preliminary observations. First, the local search heuristics does not seem as effective when the removed rogue taxa account for a high percentage of the total number of taxa. On the other hand, in most cases, the improvement achieved by the exact algorithm over the heuristic method is at most one, showing that the heuristic method is quite accurate.

Second, since our ILP only maximizes NIE, it seems to discard more leaves than the heuristic method, which attempts to minimize the leaf loss as well as maximize NIE. It is straightforward to use an ILP solver to find all optimal solutions and choose the one with minimum leaf loss. Alternatively, it is easy to modify the objective function of our ILPs to make it a weighted sum of the number of leaves and the number of internal edges. No new variables need to be introduced, since the sum of the S_i's gives us the size of the selected set X. We did this an noticed an interesting phenomenon when the number of leaves and the number of internal edges have the same weight, which is effectively what is done in [15]. Here, we find that the ILP is reluctant to remove taxa to make gain on internal edges. Indeed, it would appear that the method is too conservative in this case.

Third, we found that reduced majority consensus trees tend to lose many fewer leaves than the reduced strict consensus trees; sometimes no leaves are lost at all. This might be explained as follows. For reduced majority-rule consensus, as long as a taxon is correctly placed in more than half of the input trees it will not be removed. In contrast, for strict reduced consensus, if a taxon is placed incorrectly in just one tree but correctly in others, it often has to be removed from all the input trees to achieve high resolution. Further, there are instances where two clusters that were non-majority originally, become identical after the removal of rogue taxa and their combined frequency exceeds the threshold value of 50%. However, the combined frequency is still not high enough (i.e., 100%) for the reduced cluster to appear in reduced strict consensus. In passing, we should note that in general the majority-rule reduced consensus ILP takes longer to solve than the strict reduced consensus ILP.

Finally, the ILP is much slower than the heuristic method. The latter typically ran within a fraction of a seconds in most of the cases we studied. In contrast, as the numbers of taxa and trees increase, the exact algorithm quickly becomes impractical. Despite their drawbacks, however, our exact solutions give a good benchmark against which to evaluate and explore the limitations of heuristic methods.

References

1. Amenta, N., Clarke, F., John, K.S.: A Linear-Time Majority Tree Algorithm. In: Benson, G., Page, R.D.M. (eds.) WABI 2003. LNCS (LNBI), vol. 2812, pp. 216–227. Springer, Heidelberg (2003)
2. Amir, A., Keselman, D.: Maximum agreement subtree in a set of evolutionary trees. SIAM Journal on Computing 26, 758–769 (1994)
3. Bryant, D.: A classification of consensus methods for phylogenetics. In: Janowitz, M., Lapointe, F.-J., McMorris, F., Mirkin, B.B., Roberts, F. (eds.) Bioconsensus. Discrete Mathematics and Theoretical Computer Science, vol. 61, pp. 163–185. American Mathematical Society, Providence (2003)
4. Chi, Y., Muntz, R.R., Nijssen, S., Kok, J.N.: Frequent subtree mining — an overview. Fundamenta Informaticae 66(1-2), 161–198 (2004)
5. Cranston, K.A., Rannala, B.: Summarizing a posterior distribution of trees using agreement subtrees. Systematic Biology 56(4), 578 (2007)
6. Dong, J., Fernández-Baca, D.: Constructing Large Conservative Supertrees. In: Przytycka, T.M., Sagot, M.-F. (eds.) WABI 2011. LNCS, vol. 6833, pp. 61–72. Springer, Heidelberg (2011)
7. Dong, J., Fernández-Baca, D., McMorris, F.R.: Constructing majority-rule supertrees. Algorithms in Molecular Biology 5(2) (2010)
8. Farach, M., Przytycka, T.M., Thorup, M.: On the agreement of many trees. Inf. Process. Lett. 55(6), 297–301 (1995)
9. Finden, C.R., Gordon, A.D.: Obtaining common pruned trees. Journal of Classification 2(1), 255–276 (1985)
10. Gusfield, D., Frid, Y., Brown, D.: Integer Programming Formulations and Computations Solving Phylogenetic and Population Genetic Problems with Missing or Genotypic Data. In: Lin, G. (ed.) COCOON 2007. LNCS, vol. 4598, pp. 51–64. Springer, Heidelberg (2007)
11. Karp, R.M.: Reducibility among combinatorial problems. In: Miller, R.E., Thatcher, J.W. (eds.) Complexity of Computer Computations. Plenum, New York (1972)
12. Lee, C.-M., Hung, L.-J., Chang, M.-S., Shen, C.-B., Tang, C.-Y.: An improved algorithm for the maximum agreement subtree problem. Information Processing Letters 94(5), 211–216 (2005)
13. Margush, T., McMorris, F.R.: Consensus n-trees. Bulletin of Mathematical Biology 43(2), 239–244 (1981)
14. Nadler, S.A., Carreno, R.A., Mejía-Madrid, H., Ullberg, J., Pagan, C., Houston, R., Hugot, J.P.: Molecular phylogeny of clade III nematodes reveals multiple origins of tissue parasitism. Parasitology 134(10), 1421–1442 (2007)

15. Pattengale, N., Aberer, A., Swenson, K., Stamatakis, A., Moret, B.: Uncovering hidden phylogenetic consensus in large datasets. IEEE/ACM Trans. Comput. Biol. Bioinformatics 8-4(99), 1 (2011)
16. Redelings, B.: Bayesian phylogenies unplugged: Majority consensus trees with wandering taxa (2009)
17. Semple, C., Steel, M.: Phylogenetics. Oxford Lecture Series in Mathematics. Oxford University Press, Oxford (2003)
18. Sridhar, S., Lam, F., Blelloch, G.E., Ravi, R., Schwartz, R.: Mixed integer linear programming for maximum-parsimony phylogeny inference. IEEE/ACM Trans. Comput. Biol. Bioinformatics 5(3), 323–331 (2008)
19. Sullivan, J., Swofford, D.L.: Are guinea pigs rodents? The importance of adequate models in molecular phylogenetics. Journal of Mammalian Evolution 4(2), 77–86 (1997)
20. Swenson, K.M., Chen, E., Pattengale, N.D., Sankoff, D.: The Kernel of Maximum Agreement Subtrees. In: Chen, J., Wang, J., Zelikovsky, A. (eds.) ISBRA 2011. LNCS, vol. 6674, pp. 123–135. Springer, Heidelberg (2011)
21. Thomson, R.C., Shaffer, H.B.: Sparse supermatrices for phylogenetic inference: taxonomy, alignment, rogue taxa, and the phylogeny of living turtles. Systematic Biology 59(1), 42 (2010)
22. Wilkinson, M.: Common cladistic information and its consensus representation: reduced Adams and reduced cladistic consensus trees and profiles. Systematic Biology 43(3), 343 (1994)
23. Wilkinson, M.: More on reduced consensus methods. Systematic Biology 44(3), 435 (1995)
24. Wilkinson, M.: Majority-rule reduced consensus trees and their use in bootstrapping. Molecular Biology and Evolution 13(3), 437 (1996)
25. Xiao, Y., Yao, J.F.: Efficient data mining for maximal frequent subtrees. In: Proc. IEEE International Conference on Data Mining, pp. 379–386. IEEE (2003)
26. Zaki, M.J.: Efficiently mining frequent trees in a forest: Algorithms and applications. IEEE Trans. on Knowl. and Data Eng. 17(8), 1021–1035 (2005)
27. Zhang, S., Wang, J.T.L.: Discovering frequent agreement subtrees from phylogenetic data. IEEE Trans. on Knowl. and Data Eng. 20, 68–82 (2008)

Analytics Approaches for the Era of 10,000* Genomes
*and Counting
(Invited Keynote Talk)

Cynthia J. Gibas

Department of Bioinformatics and Genomics, University of North Carolina at Charlotte,
Charlotte, North Carolina
cgibas@uncc.edu

Abstract. What does it mean to make data available in a world where not everyone is a bioinformatician, but anyone can sequence a genome? As low-cost sequencing democratizes genome sequencing, it is also democratizing data analysis, with predictably uneven results. Can analytics approaches save us from everything from poorly documented analysis protocols to increasingly compressed and confusing visualizations?

Keywords: Genomics, analytic provenance, visualization.

1 The Impact of High-Throughput Sequencing

In the small world of microbial genomics, an explosion in genome data has taken place since second-gen sequencing became widely available in the mid-00s. With nearly 2000 completed microbial genomes and over 7000 sequencing projects now appearing in NCBI, and the growing popularity of multi-strain and phylogenomic studies, this trend shows no sign of leveling off. As the collection of microbial genomes grows towards 10,000, it illustrates, in microcosm, the issues that will no doubt plague us as the collections of euykaryotes, of vertebrates, and of individual human genomes reach that point in the not too distant future. Genome data are available to the scientific community and to the public, but what does availability mean when data sets are this large? How do we meaningfully view and interrogate data sets on this scale?

The discipline of analytics lives at the intersection of computing, operations research, and statistics. It is still primarily focused around enterprise data with the purpose of informing decision making. Analytics concepts and tools, including provenance models, online analytical processing (OLAP) and visual analytics interfaces show promise for providing access to large genome collections, for supporting reasoning about genomic data sets, and for intuitive interaction with data via summary visualizations.

2 The Role of Analytic Provenance in Genomics

Computational analysis of even a single genome's sequence is a branching set of choices leading to a proliferation of possible outcomes. In microbial analysis alone, the bioinformatics community has generated several well-regarded assembly methods for second-gen sequencing data, each of which will produce a slightly different result when applied to the same data set. In the subsequent annotation phase of the analysis, the investigator can choose from several possible pipelines or construct a pipeline of their own based on a consensus of common analysis steps.

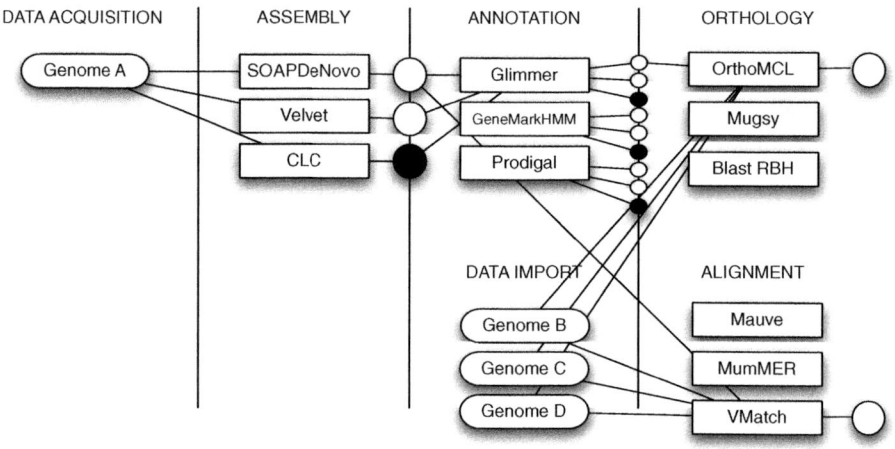

Fig. 1. The proliferation of analysis choices underlying a simple comparative genomic analysis

For each analysis step, each pipeline, the outcome of analysis is somewhat different. Not vastly different, for microbial genome analysis is reasonably well-served by automated analysis methods, but sufficiently different to, perhaps, change by a few genes the result of a query such as "which genes are present in genome A, but not in genome B". This is often the very first line of questioning with which an investigator approaches a pair or group of genome sequences, and the interest – the diagnostic sequence, the Discussion section of the paper -- lies in the content differences (Morrison et al., in review). These can often be quite small, and significantly impacted by the outcomes of choices earlier on in the bioinformatics analysis.

Therefore it is very important that we know the provenance of the data, and not just the information about strain and geographical origin and isolation that is covered by existing genomic data standards, but the provenance of the sequence presented to us complete or in contigs, the provenance of the annotations.

There is increasing pressure, though, from journal editors and reviewers, to keep publications describing microbial genome sequences extremely compact. Detailed reporting of bioinformatics analyses and parameters is often actively discouraged.

GenOLAP, an online analytical processing system for comparative genomic analysis, aims to provide methods for preserving analytic provenance – the investigator's process of analysis actions taken in the course of a comparative genome analysis.

3 Visual Analytic Approaches to Fast-Growing Data

A second emerging challenge for comparative genomic analysis is the linear genome alignment view. The dominant visualization concept for genomic data has been and continues to be a linear map of the genome, decorated with various information tracks. This visualization is useful when a single genome is being examined. Linear browser systems tend to assume that the user has *a priori* knowledge of genomic coordinates or gene names of interest, and wishes to see all information associated with a specific locus. But linear visualizations and the databases that support them generally do not provide adequate support for genomic data mining over multiple genome data sets. Especially as genomic data sets grow, literal linear visualizations of the genome and its dependent data become very compressed and difficult to interpret, even before comparative analysis is considered. Could other genomic data summarizations work better, and in fact better address the types of questions that biologists really intend to ask about genomic sequence data?

Visual analytic interfaces generally consist of multiple, coordinated views on the data. They allow the user to interact with and select data in real time based on direct interaction with the visualization. GenoSets[2] is an alternative genome browsing tool that offers a different visualization paradigm. Supported by the GenOLAP data warehouse, GenoSets summarizes descriptive data associated with a genome in visualizations that allow the data to be interactively subdivided and filtered. Within the system, questions about content and properties of sequences can be formulated on an abstract level, absent *a priori* knowledge of any specific gene locus or genomic coordinate. The visual analytic system facilitates searching based on combinations of properties that suggest a locus of interest, and then guides the user to the object of their search.

References

1. Morrison, S.S., Williams, T., Cain, A.A*, Froelich, B., Taylor, C., Verner-Jeffreys, D., Hartnell, R., Oliver, J.D., Baker-Austin, C., Gibas, C.J.: Pyrosequencing-based comparative genome analysis of Vibrio vulnificus environmental isolates. PLoS One (in review)
2. Cain, A.A., Kosara, R., Gibas, C.J.: GenoSets: Visual Analytic Methods for Comparative Genomics. PLoS One (in review)

GTP Supertrees from Unrooted Gene Trees: Linear Time Algorithms for NNI Based Local Searches

Paweł Górecki[1], J. Gordon Burleigh[2], and Oliver Eulenstein[3]

[1] Institute of Informatics, University of Warsaw, Poland
gorecki@mimuw.edu.pl
[2] Department of Biology, University of Florida, USA
gburleigh@ufl.edu
[3] Dept. of Computer Science, Iowa State University, USA
oeulenst@cs.iastate.edu

Abstract. Gene tree parsimony (GTP) problems infer species supertrees from a collection of rooted gene trees that are confounded by evolutionary events like gene duplication, gene duplication and loss, and deep coalescence. These problems are NP-complete, and consequently, they often are addressed by effective local search heuristics that perform a stepwise search of the tree space, where each step is guided by an exact solution to an instance of a local search problem. Still, GTP problems require rooted input gene trees; however, in practice, most phylogenetic methods infer unrooted gene trees and it may be difficult to root correctly. In this work, we (i) define the first local NNI search problems to solve heuristically the GTP equivalents for unrooted input gene trees, called unrooted GTP problems, and (ii) describe linear time algorithms for these local search problems. We implemented the first NNI based local search heuristics for unrooted GTP problems, which enable analyses for thousands of genes. Further, analysis of a large plant data set using the unrooted NNI search provides support for an intriguing new hypothesis regarding the evolutionary relationships among major groups of flowering plants.

1 Introduction

The deluge of genomic data from next-generation sequencing technologies has created much interest in phylogenetic methods that can exploit large-scale genomic data sets [9]. Such methods must be computationally tractable for data sets with thousands of genes and account for processes of genomic evolution that can confound species tree inference. For example, processes such as gene duplication, gene duplication and loss, incomplete lineage sorting, and lateral gene transfer can result in gene trees that differ from the true species phylogeny [20]. Simply concatenating the conflicting genes in a single matrix for maximum parsimony, or maximum likelihood phylogenetic analysis may produce erroneous species trees [22,18,4]. One alternate phylogenetic approach that can reconcile conflicting gene trees is gene tree parsimony (GTP). Given a collection of rooted gene trees, GTP (supertree) problems seek a species tree that implies the fewest evolutionary events (e.g., duplication, duplication and losses, and deep coalescences), or reconciliation costs [12,14,20,8,29]. In contrast to other supertree problems [6], GTP problems explicitly account for the evolutionary processes that cause incongruence among the given gene trees. GTP analyses have produced credible species trees using large-scale genomic data sets [26,8], However, while GTP problems require rooted gene trees, most standard phylogenetic inference methods produce unrooted gene trees, and correctly rooting the gene trees can be difficult if not impossible [12,14,20,8]. Therefore, the applicability of GTP problems is severely limited in practice.

All of the GTP problems are NP-hard [30,19], but they have been effectively addressed by search heuristics that employ highly efficient local search algorithms for rooted input gene trees [1,3]. In this work, we introduce the first local search problems to address versions of the GTP problems for a collection of unrooted gene trees under gene duplication cost and gene duplication and loss cost. Furthermore, we describe linear time algorithms for these local search problems. In absence of other local search algorithms for unrooted GTP problems we can not provide a comparative study, but we demonstrate the performance of our algorithms using a data set of more than 1000 unrooted gene trees for flowering plants. While our study confirms many of the existing clades, it provides support for an intriguing new evolutionary hypothesis regarding the evolution of major groups of flowering plants.

Previous Related Work. Numerous search heuristics based on efficient local search algorithms have been developed for GTP problems [1,2,3]. These algorithms iteratively search the space of all possible candidate species trees, guided at each step by solutions to a local search problem. Given a rooted gene tree and a rooted species tree, local search problems for GTP problems seek a species tree in the neighborhood of the given species tree that implies the minimum reconciliation cost. Two trees are neighbours when they can be transformed into each other by some tree edit operation. More formally, the search space can be described by a connected graph, where the vertices of the graph represent the candidate species trees, and an edge is drawn between two vertices if the corresponding trees are neighboring each other. Each vertex is assigned the overall reconciliation cost of the species tree represented by the vertex for the given input gene trees. Given an initial vertex, the heuristics' task is to find a path of steepest descent in the cost of its vertices and to return the last vertex (that is, the vertex with minimum cost) on this path. The local search problem is typically solved thousands of times during one heuristic run, and thus, the efficiency of solving this problem dictates the runtime of the heuristic.

Nearest Neighbor Interchange (NNI) local search neighborhoods are desirable, since they can be more efficiently searched than local search neighborhoods like SPR or TBR. For example, the NNI local search algorithm from Bansal et. al. [3] is orders of magnitude faster then existing naïve NNI searches as well as searches using other SPR or TBR local search neighborhoods [2]. However, in practice, the applicability of GTP problems, no matter what search problem, is severely limited by the strict requirement of rooted input gene trees.

The most commonly used methods to build gene trees, such as maximum parsimony and maximum likelihood, usually produce unrooted gene trees. If a gene tree has a history of duplication or loss, it often is not clear how to identify the root. For example, if there have been duplications and losses, rooting with an outgroup taxon may not be accurate because duplications may precede the common ancestor of all species in the gene tree. Rooting the tree by assuming a molecular clock, or similarly using mid-point rooting, also can result in error when there is molecular rate variation throughout the tree [16,15]. Thus, the requirement for rooted gene trees represents a major impediment for GTP problems in general.

Our Contribution. Here, for the first time, we (i) formulate the unrooted local NNI search problems called ULN-D and ULN-DL to address heuristically the unrooted versions of the GTP problems under the gene duplication cost and gene duplication and loss cost respectively, and (ii) describe linear time algorithms for these problems. The ULN-D and ULN-DL problems are defined as natural extensions of their rooted counterparts. Given an unrooted and full binary gene tree G and a rooted and full binary species tree S, the ULN-D problem seeks under all ordered pairs consisting of a rooted version of G and a species tree in the NNI neighborhood of S a pair where the rooted gene tree implies the minimum reconciliation cost for gene duplications. Analogously, the ULN-DL problem is defined. For a given gene tree G and species tree S each of our algorithms runs in $O(|S| + |G|)$ time, which improves on the naive solution by a factor of $|S|$. In fact, our algorithms are able to update a single unrooted reconciliation cost between G and S after an NNI operation on S in $O(|G|/|S|)$ time on average, or, in other words, in constant time on average if both trees have the same size. Our novel algorithms were implemented in the prototype software fasturec as a part of local search heuristics for the unrooted versions of the gene duplication and the gene duplication and loss supertree problems. We demonstrate that our software allows us to estimate species trees from data sets with over 1000 unrooted gene trees.

2 Definitions

Here, we introduce the fundamentals of the duplication-loss model. Our definitions are mostly adopted from [13]. A more detailed introduction can be found in [14,23,11].

Let $\mathcal{I}$ be the set of species. The *unrooted gene tree* is an undirected acyclic connected graph in which each node has degree 1 (leaves) or 3 (internal nodes), and the leaves are labeled by the elements from $\mathcal{I}$. A *species tree* S is a rooted binary tree with leaves uniquely labeled by the elements from $\mathcal{I}$. A *rooted gene tree* is a rooted binary tree with leaves labeled by the elements from $\mathcal{I}$. The internal nodes of a tree T we denote by $I(T)$. The set of leaf labels in a tree T will be denoted by $L(T)$. A subtree of a rooted tree T rooted at $t \in T$ will be denoted by $T(t)$.

Let $S = \langle V_S, E_S \rangle$ be a species tree. S can be viewed as an upper semilattice with $+$ a binary least upper bound operation and $\top$ the top element, that is, the root. In other words, $a + b$ is the least common ancestor of a and b in S. Additionally, for $a, b \in V_S$, $a < b$ means that a and b are on the same path from the root, with b being closer to the root than a. Let T be a rooted tree such that $L(T) \subseteq L(S)$. Let $M \colon V_T \to V_S$ be the *least common ancestor (lca) mapping*, that preserves the labeling of the leaves. Formally, if v is a leaf in T then $M(v)$ is the leaf in S labeled by the label of v. If v is internal node in T with two children a and b, then $M(v) = M(a) + M(b)$, that is, $M(v)$ is the least common ancestor of $M(a)$ and $M(b)$ in S.

Now, we provide the definition of costs based on duplication events [13]. Let T be a rooted gene tree and S a species tree such that $L(T) \subseteq L(S)$. A node $g \in I(T)$ is called a *duplication* if $M(g) = M(a)$ for a child a of g. Each non-duplication node of T will be called a *speciation node*. The total number of duplication nodes in T is called *the duplication cost* and is denoted by $D(T, S)$. The total number of gene losses required to reconcile T and S can be defined by the following formula: $L(T, S) = 2 * D(T, S) - \sum_{g \in I(T), a, b \text{ children of } g}(||M(a), M(b)|| - 2)$, where $||a, b||$ is the number of edges on the path connecting a and b in S. Finally, we can define the duplication-loss cost of reconciling a rooted gene tree T and a species tree S as follows: $DL(T, S) = D(T, S) + L(T, S)$.

2.1 Unrooted Reconciliation

Here we highlight some definitions and results from [13]. From now on, we assume that $G = \langle V_G, E_G \rangle$ is an unrooted gene tree. We define a rooting of G by selecting an edge $e \in E_G$ on which the root is to be placed. Such a rooted tree will be denoted by G_e. To distinguish between rootings of G, the symbols defined in previous section for rooted gene trees will be extended by inserting index e. Please observe, that the mapping of the root of G_e is independent of e.

Let G be an unrooted gene tree and S a species tree. The *unrooted duplication-loss cost* is defined as follows: $urDL(G, S) = \min\{DL(G_e, S) \mid e \text{ is an edge in } G\}$. An edge (or rooting) of G with minimal $DL(G_e, S)$ cost is called *optimal*. The set of all optimal edges in G is denoted by $\mathbf{Min}_G$. Similarly, we can define $urD(G, S)$ - the *unrooted duplication cost* (by replacing $DL(G, S)$ with $D(G, S)$).

Without loss of generality, the following is assumed: S and G have at least one internal node and the root of every rooting of G is mapped into the root of S. First, G is transformed into a directed graph $\widehat{G} = \langle V_G, \widehat{E_G} \rangle$, where $\widehat{E_G} = \{\langle v, w \rangle \mid \{v, w\} \in E_G\}$. In other words each edge $\{v, w\}$ in G is replaced in $\widehat{G}$ by a pair of directed edges $\langle v, w \rangle$ and $\langle w, v \rangle$. Edges in $\widehat{G}$ are labeled by nodes of S as follows. If $v \in V_G$ is a leaf labeled by a, then the edge $\langle v, w \rangle \in \widehat{E_G}$ is labeled by a. When v is an internal node in $\widehat{G}$ we assume that $\langle w_1, v \rangle$ and $\langle w_2, v \rangle$ are labeled by b_1 and b_2, respectively. Then the edge $\langle v, w_3 \rangle \in \widehat{E_G}$, such that $w_3 \neq w_1$ and $w_3 \neq w_2$ is labeled by $b_1 + b_2$. Such labeling will be used to explore mappings of rootings of G.

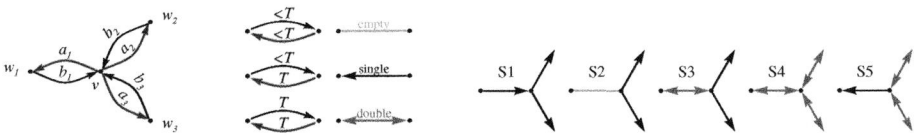

Fig. 1. Left: A star created after the transformation of gene tree edges incident to a node v. Middle: simplified notation of edges used in stars. Right: S1-S5 - all possible types of stars that can be present in the transformed gene trees.

Every internal node v and its neighbors in $\widehat{G}$ define a subtree of $\widehat{E_G}$, called a *star* with a center v, as depicted in Fig. 1. The edges $\langle v, w_i \rangle$ are called *outgoing*, while the edges $\langle w_i, v \rangle$ are called *incoming*. We will refer to the undirected edge $\{v, w_i\}$ as e_i, for $i = 1, 2, 3$.

The are several types of possible star topologies based on the labeling: **S1** a star has one incoming edge labeled by ⊤ and two outgoing edges labeled ⊤ and these edges are connected to the three siblings of the center, **S2** a star has exactly two outgoing edges labeled by ⊤, **S3** a star has all outgoing edges and exactly one incoming edge labeled by ⊤, **S4** a star has all edges labelled by ⊤, and **S5** a star has all outgoing edges and exactly two incoming edges labeled by ⊤. The star topologies are depicted in Fig. 1.

In summary, stars are atomic units that can be assembled into unrooted gene trees. Now, we briefly overview the main result of [13] (see Theorem 7). Let S be a species tree and G be unrooted gene tree. Then **(M1)** if $|\mathbf{Min}_G| > 1$, then $\mathbf{Min}_G$ consists of all edges present in all stars of type S4 or S5, **(M2)** if $|\mathbf{Min}_G| = 1$, then $\mathbf{Min}_G$ contains exactly one empty or double edge that is present in star of type S2 or S3. From the above statements and star topologies we can easily determine the set of optimal edges $\mathbf{Min}_G$, by the following observation: edges of stars outside $\mathbf{Min}_G$ are single (see Fig. 1) and share the same direction. Thus, to find an optimal edge it is sufficient to follow the reversed direction of single edges in $\widehat{G}$. An example is depicted in Fig. 2.

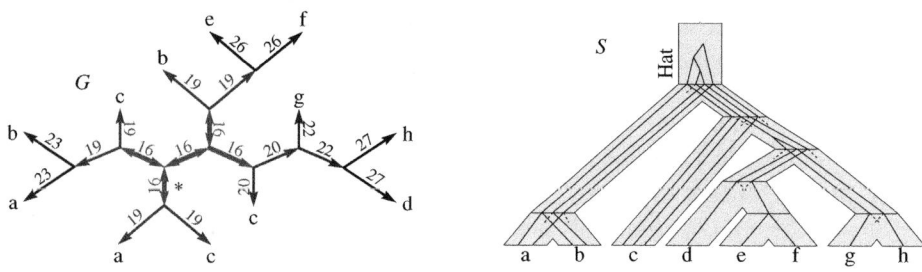

Fig. 2. Left: G - an unrooted gene tree shown with stars based on mappings from S. Each edge e of G is decorated with $DL(G_e, S)$. There are five optimal rootings with cost 16. Right: Embedding of the optimal rooting of G_e, where e is the lowest optimal edge in G (marked with $*$), into a rooted species tree S. It follows from [13] that the embeddings of optimal rootings differ only in the hat of S. Here, any optimal rooting has 4 duplications, 3 of them located in the hat, and 12 gene losses.

It follows from [5] that a single lca-query (that, is $a + b$ for nodes a and b in S) can be computed in constant time after an initial preprocessing step requiring $O(|S|)$ time. Other structures like $\widehat{G}$ with the labeling can be computed in $O(|G|)$ time. The same complexity has the procedure of finding an optimal edge in G. Thus, an optimal edge/rooting and the minimal cost can be computed in linear time.

3 Methods

First we describe the algorithm for computing $urDL$ cost and the set of optimal edges after one NNI operation performed on a rooted species tree (see Def. 1 and Fig. 3). Then we extended it to Alg. 1 to solve *the ULN-DL problem* defined as follows: for a given an unrooted gene tree G and a species tree S, compute $urDL(G, S')$ for each species tree S' that can be obtained from S by one NNI operation. Then, we briefly describe the solution for the ULN-D problem, that is, computing the unrooted duplication costs urD instead of the cost $urDL$.

Definition 1. *(NNI operation) Given a species tree S, an NNI operation transforms S into S' by replacing a subtree $((A, B), C)$ of S with $((C, A), B)$. We assume that the root of A is a, etc. as indicated in Fig. 3. Let x be the root of (A, B) and t be the root of $((A, B), C)$.*

The notation introduced in the above definition (see Fig. 3) will be used in the remaining part of this work. Note that there is an additional NNI operation, when $((A, B), C)$ is replaced with $((B, C), A)$. However, this operation can be easily defined; therefore, it is omitted for brevity. It should be clear, that the NNI operation can be performed in constant time. Note, that the number of elements in the 1-NNI neighborhood of S equals $2(|I(S)| - 1)$.

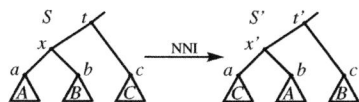

Fig. 3. An NNI operation that transforms a species tree S into S'

3.1 Algorithms

Preliminaries. First, let us assume that the following is given: (a) a species tree S with the root $\top$, (b) an unrooted gene tree G with the roots of rootings mapped into $\top$, (c) $\widehat{G}$ with the labelling of edges, that is, G with the lca-mappings of subtrees G into S, and (d) $urDL(G, S)$. Recall that (a)-(d) can be computed in $O(\max(|S|, |G|))$ time. First, we show how the mappings of $\widehat{G}$ can be updated. Then, we consider two cases of the NNI operations, depending on whether t equals $\top$.

Update of mappings. We start with the observation related to the labels of $\widehat{G}$. The proof follows easily from the definition of $\widehat{G}$:

Proposition 1 (Update of $\widehat{G}$). *If $e \in \widehat{G}$ is labeled x, then after NNI e is labeled t'. If $e \in \widehat{G}$ is labeled t, then after NNI e is labeled t' or x'. Other labels remain unchanged.*

NNIs below the root. Assume that $t \neq \top$. First, we show the following property:

Proposition 2. *The set of optimal edges is preserved by the NNI operation, when* $t \neq \top$.

The proof follows immediately from Proposition 1 and the fact that all stars are preserved by the NNI operation of this kind. Now, we can present the change of the cost, by referring to the result from [13]: "the embeddings of the optimal rootings differ only in the hat of the species tree". Moreover, by Prop. 2 the root of an optimal gene tree G_e, is preserved by the NNI operation. Thus, it is sufficient to show how the cost is changed for the parts of G_e, that are crossing the edge $\langle x, t \rangle$ in the embedding. In general, we have two cases: **(N1)** nodes of G_e that are mapped into x or t or **(N2)** edges $\langle p, q \rangle$ from G_e, such that $M_e(p) < x < t < M_e(q)$, that is, the edges that are partially intersecting $\langle x, t \rangle$ in the embedding. Such a case can be found in Fig. 2; for example, the right gene lineage located between nodes $defgh$ (t) and def (x) of S.

Let g be node from G_e, such that $s = M_e(g) \leq t$. Then $\mathcal{L}(S(s)) \subseteq \mathcal{L}(A) \cup \mathcal{L}(B) \cup \mathcal{L}(C)$, where $\mathcal{L}(T)$ is the set of all leaves from a rooted tree T. From now on, we use a simplified notation, called *a signature*, composed of at most three letters to indicate which sets among $\mathcal{L}(A)$, $\mathcal{L}(B)$ and $\mathcal{L}(C)$ have members in $\mathcal{L}(S(s))$. For instance, if the set of mappings of the leaves of g has nodes from A and B only, this is indicated by the signature AB.

Case (N1), $t \neq \top$. Next, let us assume that $M_e(g) \in \{t, x\}$. Then $g \in I(G_e)$ has two children g' and g''. The cost update is depicted in Fig. 4 with all variants of the mappings for the children of g expressed in the form (α, β), where α and β are the signatures of the children of g. For example, in $t4$, (A, AC) denotes $M_e(g') \in V_A$, $M_e(g'') \in V_A \cup V_C$ (here $M_e(g) = t$). Computing costs using this figure should be rather simple (we skip formal details for brevity). Note that one additional loss may occur, when the NNI operation changes the mapping of g from t into x' and the parent of g is mapped into an ancestor of t/t'. This condition has to be checked in $t1, t4, t8$ or $t11$. We refer to this additional case by **(N1a)**.

Case (N2), $t \neq \top$. We start with several definitions related to the embeddings and rooted gene trees. The number of duplication and speciation nodes mapped into s will be denoted by $\delta(s)$ and $\sigma(s)$, respectively. For a node s in S, let $\bar{s}$ denote the edge in S connecting s with its parent. This notion will be also used with $s = \top$, where the edge $\bar{s}$ is the rooting edge representing the hat of S. Let $\iota(s)$ be the total number of gene lineages that enter the edge $\bar{s}$ in the embedding and reach s without a loss. For instance, in Fig. 2, $\iota(b) = 2$, $\iota(ef) = 1$, $\iota(defgh) = 2$, $\delta(defgh) = 1$, $\sigma(\top) = 3$ and $\delta(\top) = 3$. Formally, $\iota(s)$ can be defined by the following formula:

$$\iota(s) = \begin{cases} \sigma(s) - \delta(s) & \text{if } s \text{ is a leaf,} \\ \iota(s') + \iota(s'') - \sigma(s) - \delta(s) & \text{if } s \in I(S) \text{ with two children } s' \text{ and } s''. \end{cases} \quad (1)$$

Note that $\iota(\bar{\top}) = 1$. The formula for the loss change is below (proof omitted):

$$\Delta L = \delta(x) - \rho(x) - \iota(b) - \rho(b) + \rho(c) + \iota(c), \quad (2)$$

where $\rho(s)$ denotes the total number of gene losses removed from $\bar{s}$ after the NNI operation in cases from **(N1)**. Observe, that the reversed case of **(N1a)**, when a potential gene loss from $\bar{c}$ is removed (cases $x1$-$x4$), is already incorporated in (2).

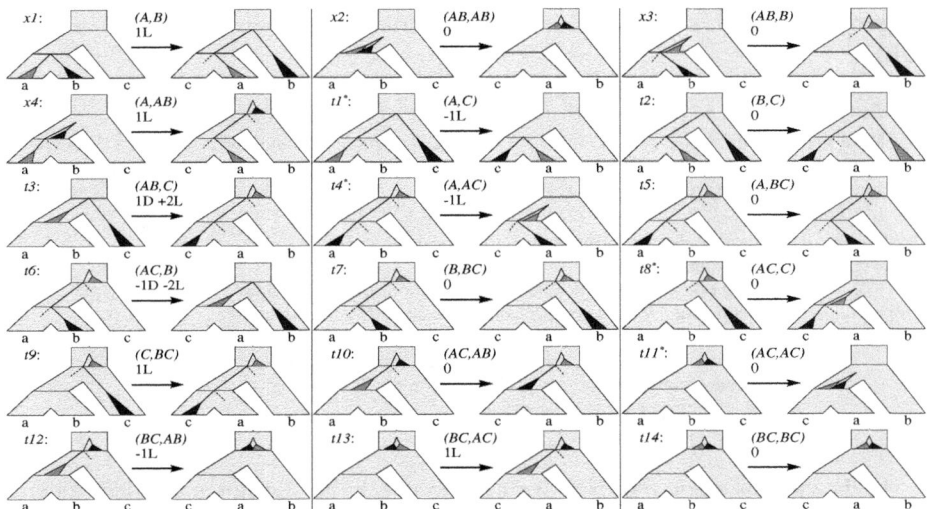

Fig. 4. Local transformations of the embedding for an optimal rooting G_e, when $g \in G_e$ is mapped into t or x. The cost update is expressed in the form $\alpha D + \beta L$. For instance, $1L$ denotes $+1$ gene loss, $-1D$ denotes -1 duplication, etc. In above variants, the signature BC can be replaced by ABC. (*) Additional losses can be introduced in $t1$, $t4$, $t8$ or $t11$.

NNI with the root. Now, we analyse the remaining case of NNI, when $t = \top$. In this case, we have at most 4 NNI neighbours of S. Moreover, it is the only case when optimal edges may be changed. In our algorithm we use the following formula for the cost computation:

$$c(s) = \begin{cases} \delta(s) & \text{if } s \text{ is a leaf,} \\ \delta(s) + \iota(s') + \iota(s'') - 2 \cdot \sigma(s) & \text{if } s \in I(S), s' \neq s'' \text{ the children } s, \end{cases} \quad (3)$$

It can be proved that $urDL(G_e, S') = c(r)$, where r is the root of S and $c(r)$ is computed for the tree S and $\widehat{G}$ after performing the NNI operation.

Data structures of the algorithm. Now, we can summarize required data structures. To compute efficiently the update of the cost after the NNI operation, we need two main data structures: the list of speciation nodes and the list of duplication nodes for some optimal rooting G_e. Internally the elements of these lists can be represented by the edges of $\widehat{G}$. In general, we have $2|V_S|$ lists, with the total size equal to $2(|V_G| - 1)$. The lengths of these lists are naturally used to access the values of δ and σ. For example, $\sigma(s)$ is the length of the speciation list for s, etc. Additionally, with each node s of S we store the value of $\iota(s)$. It should be clear that these data structures can be created in $O(|S| + |G|)$ time.

Each NNI operation requires traversal of the speciation and duplication lists for x and t, and moving some of the nodes between the lists according the potential change of the mapping and the type of a node. This can be done in the number of steps proportional to the length of the lists.

Algorithm 1. urDL costs for all species tress from 1-NNI neighborhood of S (ULN-DL)

1: **Input** A species tree S, an unrooted gene tree G, $L(G) \subseteq L(S)$.
2: **Output** $urDL(G, S')$ for all species trees S' from the 1-NNI neighborhood of S.
3: **Data preparation.** Let $\sigma = urDL(G, S)$. Compute lca query structure for S and $\widehat{G}$ by the algorithm from [13]. Create lists of duplication and speciation nodes. Compute ι and the DFS intervals in S.
4: **For** each edge $\langle x, t \rangle$ of S such that $x \in I(S)$
5:    **For** cycle$= 1, 2, 3$ *(comment: execute 3 times the NNI operation on $\langle x, t \rangle$)*
6:        Transform S and update mappings of $\widehat{G}$ (in situ).
7:        **If** t is the root of S
8:            Update the lists of duplication and speciation nodes for x and t.
9:            Recompute ι (for all nodes). Compute cost by (3) and assign to σ.
10:       **Else**
11:           (N1) and (N1a): update the cost σ and the lists for x and t (cases $x1$-$x4$, $t1$-14).
12:           Update losses by (N2). Let $\iota(x) := -\sigma(x) - \delta(x) + \iota(a) + \iota(c)$.
13:       **If** cycle < 3: print the current cost for S.

In the case of a non-root NNI we need to check the signatures of nodes required to determine variants from Fig. 4. This operation can be reduced to at most 6 single queries of the following form: "is a given node descendant of a (or b, c)?". Queries of this type, can be answered in constant time if each node of S has a DFS interval[1]. Such intervals can be created during the preprocessing phase in $O(|S|)$ number of steps. There is one technical problem with these intervals: NNI changes the interval structure used in the query. However, in our algorithm there is no need to update them. Observe, that, for each edge in S, we perform 3 NNI operations (see Alg. 1). Thus, after exploring two neighbors of S, the algorithm returns to the original species tree after the 3rd NNI operation. Then, instead of recomputing the intervals after each NNI operation (which could be costly), it is better to modify the query to use the intervals from the original tree. We omit easy details. After the non-root NNI, ι has to be updated. It should be clear that only $\iota(x)$ has to be recomputed by using the updated lengths of duplication and speciation lists (see the formula in Alg. 1 based on (1)).

Now, we can summarize the complexity of Alg. 1. A single node s of S is incident to at most 3 edges. It means that the gene nodes that are mapped into s will be processed at most $3 \cdot 3$ times (3 cycles of ≤ 3 NNIs). That gives the total complexity of the update proportional to $9 \cdot |G|$ for the whole 1-NNI neighborhood of S. Additionally, in the case of the root NNI, the algorithm traverses at most 4 times the species tree in order to compute ι. Clearly, the preprocessing is linear in time $O(|G| + |S|)$. In summary, the time complexity of Alg. 1 is $O(|G| + |S|)$, and the space complexity is the same.

ULN-D: local search with duplication cost. Computing the duplication cost can be achieved by reducing Alg. 1 to a form where gene losses are ignored. In other words, the solution to the ULN-D problem is Alg. 1, where the cases from Fig. 4 reduce to $t3$ and $t6$. However, the same data structures are required; therefore, the complexity remains the same. A more detailed description is postponed to a full version of this article.

Software. Alg. 1 is implemented in a computer written in C/C++ called *fasturec*. It determines an optimal species tree from collections of unrooted gene trees. In

---
[1] A DFS interval is a pair of discovery and finish times created during Depth First Search algorithm. Then $s \leq s'$ iff the interval for s is included in the interval for s'.

particular, *fasturec* has the following algorithms (not described here for brevity): local search based on Alg. 1 to compute scores from a set of unrooted gene trees, generator of initial species tree and hill climbing method for inferring locally optimal species tree. Computer program and exemplary datasets are freely available through the following website: http://bioputer.mimuw.edu.pl/gorecki/fasturec.

4 Experimental Results and Discussion

In order to evaluate the performance of our fast local search algorithm, we utilized the unrooted NNI-local search algorithm implemented in *fasturec* to infer the phylogeny from two large genomic data sets from plants. Plants represent an especially interesting case for the gene duplication problem, as the high frequency of gene, and even whole genome, duplications has limited the use of nuclear genes in phylogenetics (see [8]).

We first generated a new set of gene trees using sequences taken from the Green-Phyl comparative genomic database [25]. We downloaded sequences from all 1356 gene families in GreenPhyl that contained at least one sequence from all 13 land plant species represented in the database. Each gene family contained between 15 and 1713 sequences (mean = 96.4; median = 52). The amino acid sequences from each gene family were aligned using MUSCLE [10]. We performed maximum likelihood (ML) phylogenetic analysis on each gene alignment using RAxML-VI-HPC version 7.0.4 [28]. All ML analyses used the JTT+CAT amino acid substitution model [17]. We used the unrooted ML trees as input for *fasturec*. A single heuristic search with *fasturec* took approximately 2 seconds on a Macintosh Powerbook with a 2 GHz Intel Core 2 Duo processor and 2 GB memory. *Fasturec* identified a single optimal species tree that implies 88, 348 gene duplications from the gene trees (see Fig. 5). For the duplication-loss cost, *fasturec* found the same topology with score 208, 081. The species tree is rooted between the angiosperms (flowering plants) and the two non-seed plants (*Selaginella* and *Physcomitrella*). Within angiosperms, the monocots and eudicots each form clades, and within monocots, the well-established PACCAD and BEP clades are resolved [7]. Interestingly, within the eudicots, *Malpighiales* are sister to *Arabidopsis* and *Carica* rather than the *Fabales* (the *Fabales+Malpighiales* topology was ranked 4th on the best-tree list with 88, 644 duplications and was ranked 6th with the score of 211, 710 for the duplication-loss cost). While this placement of *Malpighiales* (see Fig. 5) conflicts with many previous molecular analyses, mostly based on chloroplast genes [21,27], it is found in recent large-scale analyses of mitochondrial and nuclear data sets [24,8]. This may suggest a different evolutionary history for the chloroplast genome than the mitochondrial and nuclear genomes. Furthermore, it emphasizes the existence of conflicting signals among genes and the importance of examining and reconciling the histories of multiple, independent genes in order to infer the true species phylogeny.

Additional tests with the GreenPhyl data set, proved that *Fasturec* was able to perform approximately 4000 NNIs per one minute. In other words, for a single gene tree, it gives $\sim 10^5$ executions of Alg.1 per second.

Providing credible comparative evaluations for our new algorithm is difficult because GTP based algorithms for unrooted gene trees under gene duplication based costs have not been developed. Possibly the closest program that uses NNI neighborhoods for the

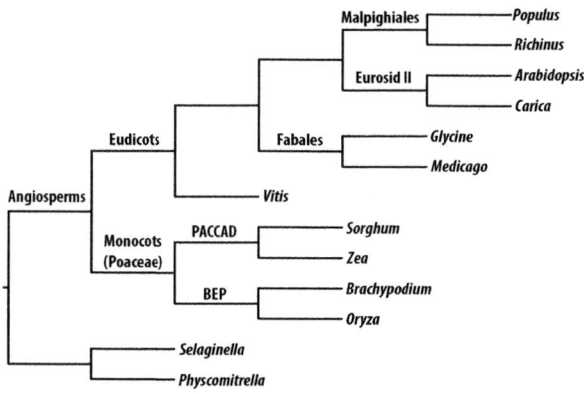

Fig. 5. Optimal (rooted) species tree with 88,348 duplications inferred from 1356 gene families of GreenPhyl database

local search is introduced in [3]. While there are similarities between our solution and [3] related to the general time complexity, this previous approach is defined for rooted gene trees only.

5 Outlook and Conclusion

There is an increased interest in inferring phylogenies from large genomic data sets. GTP problems are based on biologically motivated optimality criteria, and they allow scientists to incorporate complex processes of gene evolution into phylogenetic inference. Therefore, they may be more suitable for species tree construction than other problems that simply aim to maximize the phylogenetic information that is common to all of the input trees. However, most GTP approaches described in the literature are limited by their requirement of rooted gene tree input, which can be difficult or impossible to obtain. Our new local search algorithm allows for the first time, estimation of species trees under the gene duplication and duplication and loss models from large sets of unrooted gene trees. Applying our fast local search algorithm, we found that the species tree resulting from the 13-taxon, 1356 gene data set is consistent with recent large-scale analyses of the nuclear and mitochondrial data sets [24,8]. These analyses together provide intriguing evidence that *Malpighiales* may not be placed near *Fabales* as traditionally thought. Greater taxon sampling and careful evaluation of support for alternative hypotheses may help to resolve this issue; however, the result demonstrates the importance of a genomic perspective on phylogenetics, which is enabled by our GTP approach.

The current version of *fasturec* includes a full implementation of our new local search algorithm, and a greedy strategy based on cluster frequency counts to build the species tree to initiate the heuristic search. Based on many experiments with other local search heuristics for tree space exploration, the generation of a good initial tree is crucial for the successful inference of species trees with a low cost (i.e., a locally optimal tree). Our early tests suggest that using initial trees based on cluster frequency counts perform well for small sets of taxa (e.g., up to 30), where the optimal tree can be quickly located by a few NNI moves in the neigbourhood of the initial tree. However, for larger number of taxa the strategy of an initial tree inference based on iterative leaf adding interleaved with local search is likely the most powerful and efficient. Therefore, we are currently implementing these extensions in *fasturec*.

Our results also provide a basic graph theoretic properties of rooting unrooted gene trees that could be beneficial for the development of efficient local search algorithms for other GTP problems in their unrooted setting.

Acknowledgements. The reviewers have provided several valuable comments that have improved the presentation. This work was conducted in parts with support from the Gene Tree Reconciliation Working Group at NIMBioS through NSF award #EF-0832858. PG was supported by the grant of MNiSW #N N301 065236. OE was supported in parts by NSF awards #0830012 and #106029. PG and OE was supported by the grant of NCN #2011/01/B/ST6/02777.

References

1. Bansal, M.S., Burleigh, J.G., Eulenstein, O., Wehe, A.: Heuristics for the Gene-Duplication Problem: A $\Theta(n)$ Speed-Up for the Local Search. In: Speed, T., Huang, H. (eds.) RECOMB 2007. LNCS (LNBI), vol. 4453, pp. 238–252. Springer, Heidelberg (2007)
2. Bansal, M.S., Eulenstein, O.: An $\Omega(n^2/\log n)$ speed-up of TBR heuristics for the gene-duplication problem. IEEE/ACM TCBB 5(4), 514–524 (2008)
3. Bansal, M.S., Eulenstein, O., Wehe, A.: The gene-duplication problem: Near-linear time algorithms for NNI-based local searches. IEEE/ACM TCBB 6(2), 221–231 (2009)
4. Beiko, R.G., Doolittle, W.F., Charlebois, R.L.: The Impact of Reticulate Evolution on Genome Phylogeny. Systematic Biology 57(6), 844–856 (2008)
5. Bender, M.A., Farach-Colton, M.: The lca Problem Revisited. In: Gonnet, G.H., Viola, A. (eds.) LATIN 2000. LNCS, vol. 1776, pp. 88–94. Springer, Heidelberg (2000)
6. Bininda-Emonds, O.R.P.: Phylogenetic supertrees: combining information to reveal the tree of life (2004)
7. Bouchenak-Khelladi, Y., Salamin, N., Savolainen, V., Forest, F., Bank, M., Chase, M.W., Hodkinson, T.R.: Large multi-gene phylogenetic trees of the grasses (poaceae): progress towards complete tribal and generic level sampling. Mol. Phyl. Evol. 47(2), 488–505 (2008)
8. Burleigh, J.G., Bansal, M.S., Eulenstein, O., Hartmann, S., Wehe, A., Vision, T.J.: Genome-scale phylogenetics: inferring the plant tree of life from 18,896 discordant gene trees. Systematic Biology 60, 117–125 (2011)
9. Delsuc, F., Brinkmann, H., Philippe, H.: Phylogenomics and the reconstruction of the tree of life. Nature Reviews Genetics 6(5), 361–375 (2005)
10. Edgar, R.C.: MUSCLE: multiple sequence alignment with high accuracy and high throughput. Nucleic Acids Research 32, 1792–1797 (2004)
11. Eulenstein, O., Huzurbazar, S., Liberles, D.A.: Reconciling phylogenetic trees. In: Dittmar, Liberles (eds.) Evolution After Gene Duplication. Wiley (2010)
12. Goodman, M., Czelusniak, J., Moore, G.W., Romero-Herrera, A.E., Matsuda, G.: Fitting the gene lineage into its species lineage, a parsimony strategy illustrated by cladograms constructed from globin sequences. Systematic Zoology 28(2), 132–163 (1979)

13. Górecki, P., Tiuryn, J.: Inferring phylogeny from whole genomes. Bioinformatics 23(2), e116–e222 (2007)
14. Guigó, R., Muchnik, I., Smith, T.F.: Reconstruction of ancient molecular phylogeny. Molecular Phylogenetics and Evolution 6(2), 189–213 (1996)
15. Holland, B.R., Penny, D., Hendy, M.D.: Outgroup misplacement and phylogenetic inaccuracy under a molecular clock a simulation study. Syst. Biol. 52, 229–238 (2003)
16. Huelsenbeck, J.P., Bollback, J.P., Levine, A.M.: Inferring the Root of a Phylogenetic Tree. Systematic Biology 51(1), 32–43 (2002)
17. Jones, D.T., Taylor, W.R., Thornton, J.M.: The rapid generation of mutation data matrices from protein sequences. Computer Applications in the Biosciences 8, 275–282 (1992)
18. Kubatko, L.S., Degnan, J.H.: Inconsistency of Phylogenetic Estimates from Concatenated Data under Coalescence. Syst. Biol. 56(1), 17–24 (2007)
19. Ma, B., Li, M., Zhang, L.: From gene trees to species trees. SIAM Journal on Computing 30(3), 729–752 (2000)
20. Maddison, W.P.: Gene trees in species trees. Systematic Biology 46, 523–536 (1997)
21. Moore, M.J., Soltis, P.S., Bell, C.D., Burleigh, J.G., Soltis, D.E.: Phylogenetic analysis of 83 plastid genes further resolves the early diversification of eudicots. Proceedings of the National Academy of Sciences 107(10), 4623–4628 (2010)
22. Mossel, E., Vigoda, E.: Phylogenetic MCMC algorithms are misleading on mixtures of trees. Science 309(5744), 2207–2209 (2005)
23. Page, R.D.M.: Maps between trees and cladistic analysis of historical associations among genes, organisms, and areas. Systematic Biology 43(1), 58–77 (1994)
24. Qiu, Y., Li, L., Wang, B., Xue, J., Hendry, T.A., Li, R., Brown, J.W., Liu, Y., Hudson, G.T., Chen, Z.: Angiosperm phylogeny inferred from sequences of four mitochondrial genes. Journal of Systematics and Evolution 48(6), 391–425 (2010)
25. Rouard, M., Guignon, V., Aluome, C., Laporte, M., Droc, G., Walde, C., Zmasek, C.M., Périn, C., Conte, M.G.: Greenphyldb v2.0: comparative and functional genomics in plants. Nucleic Acids Research 39, D1095–D1102 (2010)
26. Sanderson, M., Michelle, M.: Inferring angiosperm phylogeny from est data with widespread gene duplication. BMC Evolutionary Biology 7(suppl.1) (2007)
27. Soltis, D.E., Smith, S.A., Cellinese, N., Wurdack, K.J., Tank, D.C., Brockington, S.F., Refulio-Rodriguez, N.F., Walker, J.B., Moore, M.J., Carlsward, B.S., Bell, C.D., Latvis, M., Crawley, S., Black, C., Diouf, D., Xi, Z., Rushworth, C.A., Gitzendanner, M.A., Sytsma, K.J., Qiu, Y., Hilu, K.W., Davis, C.C., Sanderson, M.J., Beaman, R.S., Olmstead, R.G., Judd, W.S., Donoghue, M.J., Soltis, P.S.: Angiosperm phylogeny: 17 genes, 640 taxa. American Journal of Botany 98(4), 704–730 (2011)
28. Stamatakis, A.: RAxML-VI-HPC: maximum likelihood-based phylogenetic analyses with thousands of taxa and mixed models. Bioinformatics 22(21), 2688–2690 (2006)
29. Yu, Y., Warnow, T., Nakhleh, L.: Algorithms for MDC-Based Multi-locus Phylogeny Inference. In: Bafna, V., Sahinalp, S.C. (eds.) RECOMB 2011. LNCS, vol. 6577, pp. 531–545. Springer, Heidelberg (2011)
30. Zhang, L.: From gene trees to species trees ii: Species tree inference by minimizing deep coalescence events. IEEE/ACM TCBB 8, 1685–1691 (2011)

A Robinson-Foulds Measure to Compare Unrooted Trees with Rooted Trees[★]

Paweł Górecki[1] and Oliver Eulenstein[2]

[1] Institute of Informatics, University of Warsaw, Poland
gorecki@mimuw.edu.pl
[2] Dept. of Computer Science, Iowa State University, USA
oeulenst@cs.iastate.edu

Abstract. The Robinson-Foulds (RF) distance is a well-established measure for the comparative analysis of phylogenetic trees, and comparing unrooted gene trees with rooted species trees is a standard application in phylogenetics. However, the RF distance is not defined to compare unrooted trees with rooted trees. Here we propose urRF a new measure based on the standard RF distance to compare an unrooted tree with a rooted tree, and describe a linear time algorithm to compute urRF. Further, we show that our measure relates to several intriguing properties that can be crucial for the in-depth comparative study and the credible rooting of trees. Finally, we present detailed empirical studies based on real data sets demonstrating the performance of our novel RF measure, the efficiency of our linear time algorithm and the quality of determining optimal rootings of unrooted gene trees.

1 Introduction

In phylogenetics comparison measures are the elementary tools for estimating quantitative dissimilarities between phylogenetic trees [25,13]. While a wide variety of phylogenetic measures have been proposed [6,27,9,3], the *Robinson-Foulds (RF) metric* [4,22] is the most widespread. The RF metric between two unrooted trees is the count of the symmetric difference of their *split-representation*; that is, the set of all bi-partitions, or *splits*, of its taxon set induced by an edge removal [25,18]. Analogously, for two rooted trees the RF distance is defined as the count of the symmetric difference of their *cluster-presentation*; that is, the set of all taxon sets of its full subtrees [25]. The RF metric for two trees, both either unrooted or rooted, over a taxon set of size n can be computed in time $O(n)$ [10]. In fact, a randomized approximation scheme was introduced in [21] that computes in sublinear time and with high probability, a $(1 + \epsilon)$ approximation of the exact RF score. Furthermore, the distribution of the RF distance relative to a fixed tree can be computed in linear time [5].

However, unlike other comparison measures [12,14,19,20,23,15], the RF metric has not been adapted for comparing an unrooted tree with a rooted tree. Such comparisons

---

[★] Support was provided to PG by the grant of MNiSW #N N301 065236, to OE by the NSF (#0830012 and #106029), and to PG and OE by NCN #2011/01/B/ST6/02777 and the NIM-BioS Working Group: Gene Tree Reconciliation through NSF #EF-0832858.

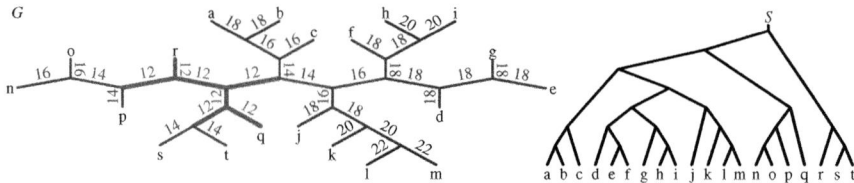

Fig. 1. An unrooted tree G and a rooted tree S. Each edge of G is labeled by the RF score between S and the rooting of G placed on this edge. There are 7 optimal rootings forming a "valley" with the RF score 12.

are often necessary when comparing gene trees with species trees. Gene trees are typically unrooted, since the most commonly used methods to build gene trees, such as maximum parsimony and maximum likelihood, produce only unrooted gene trees, and correctly rooting gene trees can be difficult if not impossible in many cases [26,28]. In contrast, species trees often encompass a trusted root, which, for example, can be based on the NCBI taxonomy [24]. Here, for the first time, we define the RF measure for determining the dissimilarity of an unrooted tree with a rooted tree as a natural extension of the standard RF measure, by rooting the unrooted tree 'optimally'. Our new measure is defined to be the minimum of all RF scores that result from comparing each rooting of the unrooted tree with the rooted tree. While the applicability of some RF methods is severely hampered in practice by limiting the RF metric to only binary trees (e.g., [1]), our measure is generally defined for multifurcated trees. The naive algorithm to compute our RF measure from an unrooted tree to a rooted tree for the same n taxa requires $\Theta(n^2)$ time. However, this runtime becomes prohibitive when a smaller number of larger trees need to be compared (e.g., when gene trees in large tree database need to be identified that support a given trusted species tree). Our contribution is a linear time algorithm for computing our new RF measure, which makes computing our RF measure asymptotically as efficient as computing the standard RF metric.

We further show that our RF measure implies several viable properties that allow for a more comprehensive comparison of phylogenetic trees. For example, for an in-depth comparative study of two trees their unique set of shared clusters or splits have to be carefully analyzed. Our RF measure could make such an analysis difficult, since there could be several optimal rootings of the unrooted tree resulting in distinct rooted trees that share different sets of clusters with the rooted tree. Such non-unique shared cluster sets can then severely complicate the detailed comparison of trees. Surprisingly, as we show in Corollary 1, all cluster sets that are shared between every optimal rooting of the unrooted input tree and the given rooted tree are identical (see Fig. 3 for an example). Another intriguing property, which is shown in Theorem 1, is that all optimal places where the unrooted tree can be rooted induce a subtree in the unrooted tree. This subtree forms a unique *RF valley* in the unrooted tree, as RF scores of rootings along paths from this subtree are monotonically increasing. An example of an RF valley is depicted in Fig. 1 and Fig. 3. In fact, as we show in Theorem 2, the RF valley formed by an unrooted tree and a rooted tree contains a valley that is formed under a seemingly unrelated measurement, the weighted duplication and loss (DL) measure [16]. Consequently, all optimal rootings of an unrooted tree under the DL measure are optimal rootings under our RF measure.

In an empirical study we demonstrate the performance and benefits of our new measure and demonstrate the efficiency of our linear time algorithm in practice. Finally, based on the the theory presented in this paper, we propose a novel approach to the problem of identifying a credible root for unrooted gene trees. In particular, we show that credible rootings in unrooted gene trees (in the context of given species trees) can be determined by analyzing the optimal valleys and their internal properties.

Our work is in parts building on the mathematical framework developed in [16]. In difference, we not only show results for our new measurement, but also extent the previous theory to multifurcated trees.

2 Basic Definitions and Preliminaries

Let $\mathcal{I}$ be the set of taxa. An *unrooted tree* T is an acyclic, connected, and undirected graph that has no degree-two vertices, and every degree-one vertex is labeled with an element from $\mathcal{I}$. The degree-one vertices are called *leaves*; and the remaining vertices are called *internal* vertices. A *rooted* tree $T = \langle V_T, E_T \rangle$ is defined similar to an unrooted tree, with the difference that it has a distinguished vertex, called *root*, which can have a degree of two. By $L(T)$, we denote the set of all leaf labels in T (rooted or unrooted).

Any roooted tree S can be viewed as an upper semilattice with the least upper bound operation $+$, and the top element denoted by $\top$; that is, the root of S. In other words, for vertices $a, b \in V_S$, $a + b$ is the least common ancestor of a and b in S. We also use the binary order relation $a \leq b$ if b is a vertex on a the path between a and the root of S. Let T be a rooted tree such that $L(T) \subseteq L(S)$. We define $M \colon V_T \to V_S$ to be the *least common ancestor (lca) mapping* from T into S that preserves the labeling of the leaves. Note, that if $a_1, a_2, \ldots, a_k \in V_G$ are the children of v, then $M(v) = M(a_1) + M(a_2) + \ldots + M(a_k)$.

Let T be a rooted tree. We denote by $T(v)$ the maximum subtree of T that is rooted at v. For each vertex $v \in V_T$ we define the *cluster* at v as $c(v) := L(T(v))$. A cluster of a root or a leaf in T is called *trivial*. We shall denote the *cluster representation* of T by $\mathcal{H}(T) := \{c(v) \colon v \in V(T)\}$. For example, if $S = ((a, (b, c, d)), e)$ then $\mathcal{H}(S) = \{\{a\}, \{b\}, \{c\}, \{d\}, \{e\}, \{b, c, d\}, \{a, b, c, d\}, \{a, b, c, d, e\}\}$, where $\{b, c, d\}$ and $\{a, b, c, d\}$ are the only non-trivial clusters. The labeling of the leaves of a tree T (rooted or unrooted) will be called *unique* if each species from $\mathcal{I}$ occurs exactly once in T as label of a leaf.

2.1 New RF Measure for Comparing Unrooted with Rooted Trees

For the remainder of this work we assume that any pair of trees compared by our RF measure has unique leaf labelling, that is, for trees T and S, there is a bijection between the leaves of T and the leaves of S that preserves leaf labels. In other words, $L(T) = L(S)$ and each species occurs exactly once as leaf label in a tree.

Definition 1. *The* Robinson-Foulds (RF) *distance between two rooted trees S and T, both with unique leaf labelling, is defined as follows:* $RF(S,T) := |(\mathcal{H}(S) \setminus \mathcal{H}(T)) \cup (\mathcal{H}(T) \setminus \mathcal{H}(S))|$.

This formula can be equivalently expressed in the following way [7]: $RF(S,T) = |I(S)| + |I(T)| - 2|F(S,T)|$, where $I(T)$ is the set of internal non-root nodes of T and $F(S,T)$ is the set of non-trivial clusters from T that are present in S. Formally, $F(S,T)$ can be defined as $\{c(g) : c(g) = c(M(g)), g \in I(S)\}$, where $M \colon S \to T$ is the lca mapping. The RF formula using lca-mappings will be later of use to show properties of the RF cost.

Let $G = \langle V_G, E_G \rangle$ be an unrooted tree. A rooting in G is defined by selecting an edge $e \in E_G$ on which the root is to be placed. Such a rooted tree will be denoted by G_e. To distinguish between rootings of G, all defined symbols for a rooted trees will be extended by inserting index e. For example, M_e is the mapping from G_e to S, etc.

Definition 2. *The unrooted* Robinson-Foulds (urRF) *measure between an unrooted tree G and a rooted tree S, both with unique leaf labeling, is defined as* $urRF(G,S) := \min\{RF(G_e, S) \colon e \in E_G\}$.

The edges $e \in E_G$, for which G_e has the minimal RF score are called *optimal*.

3 Methods

Here we present the properties of urRF and the linear time algorithm for computing the urRF score. For the remainder of this work we assume that G is an unrooted gene tree and S is a species tree, both with unique leaf labelling.

3.1 Transforming Gene Tree

First, we transform G into a directed graph $\widehat{G}$, by replacing each undirected edge $\{v, w\}$ from G by a pair of directed edges $\langle v, w \rangle$ and $\langle w, v \rangle$. Formally, $\widehat{G} = \langle V_G, \widehat{E_G} \rangle$, where $\widehat{E_G} = \{\langle v, w \rangle, \langle w, v \rangle \| \{v, w\} \in E_G\}$.

The edges of $\widehat{G}$ are labeled by the nodes of S as follows. If $v \in V_G$ is a leaf labeled by a, then the edge $\langle v, w \rangle \in \widehat{E_G}$ is labeled by a. For an internal node $v \in V_G$ with exactly k siblings $w_1, w_2, \ldots w_k$, assume that $\langle w_1, v \rangle, \langle w_2, v \rangle, \ldots, \langle w_{k-1}, v \rangle$ are labeled by $s_1, s_2, \ldots, s_{k-1}$, respectively. Then the edge $\langle v, w_k \rangle \in \widehat{E_G}$ is labeled by $\sum_{i=1}^{k-1} s_i$ (+ in this sum is the lca operator).

Each internal node $v \in V_G$ defines a *star* with the center v as indicated in Fig. 2a. An edge that starts in v is called *outgoing* (in the star), while the remaining edges will be called *incoming*. We will refer to the undirected edge $\{v, w_i\}$ as e_i, for all $i = 1, 2, \ldots k$.

Lemma 1. *Under the notation introduced for stars. We have*

(A) *If $b_i \neq \top$, for all $i = 1, 2, \ldots, k$. Then for at most one i, we have $a_i \neq \top$.*
(B) *If $b_1 = \top$, then for $i > 1$, $a_i = \top$. Moreover, if $a_1 \neq \top$ then $b_i \neq \top$ for $i > 1$.*

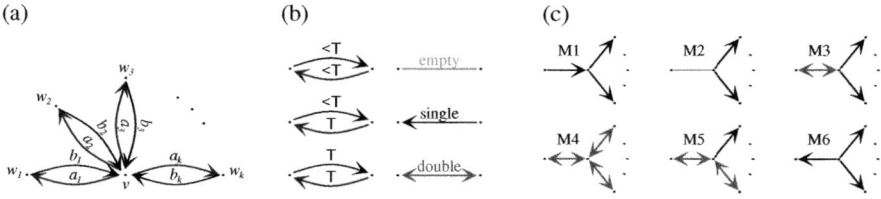

Fig. 2. (a) A star with the center v in $\widehat{G}$ and $k \geq 3$ edges. Here $e_i = \{v, w_i\}$ for $i = 1, 2, \ldots, k$; note that $a_i = \sum_{j=1, j \neq i}^{k} b_j$, etc. (b) A simplified representation of edges (empty, single and double) that will be used through the rest of this work. The notation $< \top$ denotes that the label is a non-root node from S. (c) Star topologies. On the right side of stars there are at least 2 edges. M5 has at least two double edges and at least one single edge.

Proof. (A) It should be clear that for every edge $\{v, w\}$ of G, if the label of $\langle v, w \rangle$ is a and the label of $\langle w, v \rangle$ is b, then $a + b = \top$. Assume that $a_1 \neq \top$ and $a_2 \neq \top$. Let $r = \sum_{i=3}^{k} b_i$. From the star topology, $a_1 = b_2 + r$ and $a_2 = b_1 + r$. From the second equality, we have $r \leq a_2$. On the other hand, $a_2 < \top$ and $b_2 < \top$ are incomparable and $a_2 + b_2 = \top$. Thus, $a_1 = r + b_2 = a_2 + b_2 = \top$, a contradiction. (B) Similar to the proof of (A). □

We use the following classification of stars (Fig. 2). A star with k outgoing edges is said to be of type: **M1** if it has exactly one incoming edge labeled $\top$ and $k - 1$ outgoing edges labeled $\top$ and all these edges are connected to the k siblings of the center; **M2** if it has exactly $k - 1$ edges labeled $\top$ (from Lemma 1 these edges must be outgoing); **M3** if it has all outgoing edges labeled $\top$ and exactly one incoming edge labeled by $\top$; **M4** if it has all edges labeled $\top$; **M5** if it has all outgoing edges labeled $\top$; at least two incoming edges labeled by $\top$ and at least one incoming that is not labeled $\top$; and **M6** if it has no incoming edge labeled $\top$ and all outgoing edges labeled $\top$. In general, it is easy to show, that M6 cannot be present in $\widehat{G}$ if both trees are binary (see Section 3.4).

The proof of the next lemma follows from the star topologies.

Lemma 2. *For every gene tree G we have the following mutually exclusive situations.*

(i) $\widehat{G}$ has exactly one star of type M2 and all other stars are of type M1.
(ii) $\widehat{G}$ has exactly two stars of type M2. They share a common edge. All other stars are of type M1.
(iii) $\widehat{G}$ has exactly one star of type M6. All other stars are of type M1.
(iv) $\widehat{G}$ has an occurrence of a star of type M3, M4 or M5. All other stars are of type M3, M4, M5 or M1.

3.2 RF Cost and the Star Topologies

Here we analyse how the cost changes when we move a position of the root in G. We will use throughout this section the notation of a star introduced in Fig. 2. For an edge e in G, let $\lambda_e = |F(G_e, S)|$. In other words, λ_e is the number of non-trivial clusters from S that are present in the rooted tree G_e.

Proposition 1. *Let $a_1 \neq \top$ (stars M1 and M2). Then, $\lambda_{e_1} \geq \lambda_{e_2} = \cdots = \lambda_{e_k}$. Moreover, $\lambda_{e_1} - \lambda_{e_2} = \theta(c(a_1), L(G_{e_1}(v))$, where, for two clusters x and y, $\theta(x,y) = 1$ if x and y are equal and non-trivial clusters in S, and $\theta(x,y) = 0$ otherwise.*

Proof. It should be clear that for every edge $\{v, w\}$ of G, $\lambda_e = \lambda'_v + \lambda'_w$, where λ'_x is the number of non-trivial clusters from S that are present in the subtree of G_e rooted in x. The above formula can be extended to any star in the following way:

$$\lambda_{e_i} = \sum_i \lambda'_{w_i} + \theta(c(a_i), L(G_e(v))). \tag{1}$$

Please note that, for each $i > 1$, $a_i = \top$ (see stars M1 and M2). Thus, the cluster of a_i is trivial (except $i = 1$), and therefore, $\lambda_{e_i} = \sum_i \lambda'_{w_i}$, for $i > 1$. □

The edge e_i present in stars M1 and M2, such that $a_i \neq \top$ will be called *leading*.

Proposition 2. *Let $a_i = \top$ for all i (stars M3-M6). Then, $\lambda_{e_1} = \lambda_{e_2} = \cdots = \lambda_{e_k}$.*

Proof. It follows from (1) that $\lambda_{e_i} = \sum_i \lambda'_{w_i}$ for all i. □

3.3 Algorithm and the Properties of urRF

Theorem 1. *Let G be an unrooted tree and S be a rooted species tree both with unique leaf labelling. Let Min_G be the set of optimal edges in G for the urRF score computed for G and S.*

  (i) If G has a star of type M3, M4 or M5 then Min_G contains edges from all stars of type M3, M4 and M5 present in G.
  (ii) If G has a star of type M6 then all edges from this star are optimal.
  (iii) If $|\mathrm{Min}_G| = 1$ then G has a star of type M2.
  (iv) If G has a star of type M2 then the symmetric edge in M2 is optimal.
  (v) If Min_G contains at least two edges from a star then all edges from that star are optimal.
  (vi) Min_G is a full subtree of G.
  (vii) An optimal edge in Min_G can be found by a greedy method of gradient descent: leave a star in the reverse direction of the outgoing edge labeled $\top$. Terminate when there is no such edge (M6) or a symmetric edge, that is, a member of M2, M3, M4 or M5 is found (see Algorithm 1).

Proof. It follows easily from the definition of urRF (Def. 2) and λ_e that $\mathrm{Min}_G = \{e \in E_G \mid \lambda_e$ is maximal$\}$. (i) It follows from case (iv) of Lemma 2 that G has only M3-M5 and M1 stars. It is easy to show that all M3-M5 stars in G form a subtree. Hence from Prop. 2 and all edges present in M3-M5 stars have the same cost. Other stars in G are of type M1. Hence, their cost is not larger than the cost of M3-M5 edges by Prop. 1. Similarly, condition (ii) follows from case (iii) of Lemma 2, Prop. 1 (M1 stars) and Prop. 2 (M6 star). (iii) It follows from Prop. 1 that only the leading edge present in M1 or M2 stars can be optimal in G. Assume that this edge is an element of M1 star. Then, the species tree contains at least two nodes, since the two mappings of the leading edge in $\widehat{G}$ are different. Hence, there exists another star in G sharing this edge. Clearly, this star is of type M1 or M3-M5 and the optimal set contains also edges from this star by (i) and (ii), a contradiction. This completes case (iii). The proof of (iv) is similar to (i) and (ii). (v) follows easily from Prop. 2 and Prop. 1. (vi) Let A be the minimal subgraph of G that contains all optimal edges that are present in M2-M6 stars of G. It follows from (i)-(v) that A is a full subtree of G. We show by induction that A can be transformed into A' such that $A'_E = \mathbf{Min}_G$ and A' is a full subtree of G. Assume that A contains an asymmetric edge e. If both nodes of e are internal in G, then there exists a star of s type M1, such that e is a leading edge in this star. If all edges s share the same cost then these edges are optimal in G and A should be extended by s. Clearly, this extension will form a subtree of G. This defines the step of our induction. The induction terminates when there is no asymmetric edge that can extend A. It should be clear, from the construction and star properties, that after the termination $A_E = \mathbf{Min}_G$. (vii) This follows immediately from (i), (ii) and (iv). □

Then, we have to following lemma, that characterizes the properties of optimal rootings (the proof is omitted for brevity):

Lemma 3. $F(G_e, S)$ *is independent of* e *if* e *is an optimal edge.*

The set of clusters $F(G_e, S)$ is shared among optimal rootings. Thus, from now we will call this set *a kernel* of G and S. Finally, we have the result that shows interesting properties of optimal rootings.

Corollary 1. *Let S and G be species and gene trees (S rooted and G unrooted) with the same set of leaf labels. Let T be the strict consensus tree for the set that contains S and all optimal rootings of G. Then the set of all non-trivial clusters of T is the kernel of G and S.*

An example is shown in Fig. 3.

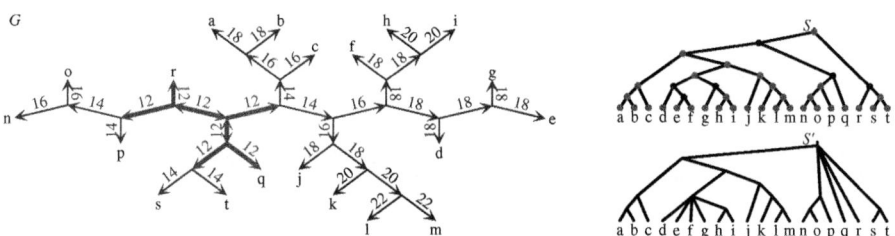

Fig. 3. Gene tree G from Fig. 1 converted into the star-like graph. S is the species tree with marked clusters present in optimal rootings of G. The kernel of G and S contains all non-trivial clusters of nodes of S marked with red circles: ab, abc, $defghi$, hi, lm, klm, $defghijklm$, $abcdefghijklm$, no, nop and st. S' is the strict consensus tree of all optimal rootings of G. It is easy to notice the relation between clusters of S' and the red nodes of S (see Thm. 1).

Our greedy method of gradient descent together with the urRF cost computation is depicted in Alg. 1. The proof of correctness of the optimal edge search (lines 1-6) follows from Thm 1. The formula for urRF score (line 7) follows from Def. 1, the formula for the rooted RF score based on clusters and the following lemma (proof omitted):

Lemma 4. *Under the assumptions and the notation from Alg. 1. Let $e = \langle v, w \rangle$ be the optimal edge found by Alg. 1. Then, $|F(G_e, S)| = \lambda'_{v,w} + \lambda'_{w,v}$. Moreover, for each non-root node $x \in G_e$ we have: $c_e(x) = L(G_e(x))$ is the cluster of x in G_e and*

$$M_e(x) = \begin{cases} m_{v,w} & \text{if } x = v, \\ m_{w,v} & \text{if } x = w, \\ m_{x,y} & \text{if } v \neq x \neq w \text{ and } y \text{ is the parent of } x \text{ in } G_e. \end{cases}$$

Now we summarize the time complexity of Alg. 1. It follows from [2] that a single least common ancestor query (that, is $a + b$ for nodes a and b in S) can be computed in constant time after an initial preprocessing step requiring $O(|S|)$ time. The structures like $\widehat{G}$ with the labeling can be computed in $O(|G|)$ time. The same complexity has the procedure of finding an optimal edge in G. Computing the urRF cost (line 7 of Alg. 1) can be done in $O(|G|)$. It follows from the property, that if x is internal non-root node in G_e then $\theta(c_e(v), c(M_e(v))) = 1$ if and only if $|c_e(v)| = |c(M_e(x))|$. Thus, it is sufficient to assign the size of clusters to each node of G_e and S in order to compute θ. This approach can be easily generalized to compute all costs of rootings in G in $O(|G|)$ time. In summary an optimal edge and the urRF cost can be computed in linear time.

Computer program and examples are freely available at the following website: http://bioputer.mimuw.edu.pl/gorecki/urrf.

Algorithm 1. Optimal edge search with urRF cost computation.

1: **Input** A species tree S, an unrooted gene tree G with at least three leaves and unique leaf labelling.
2: **Output** $urRF(G, S)$.
3: Compute $\widehat{G}$ with the labelling of edges as follows. For each directed edge $\langle v, w \rangle$ in $\widehat{G}$, let $m_{v,w} \in V_S$ be the label of a directed $\langle v, w \rangle$ from $\widehat{G}$:
$$m_{v,w} := \begin{cases} s & v \in G \text{ and } s \in S \text{ are leaves with the same label}, \\ \sum m_{x_i, v} & \text{if } \{x_1, x_2, \ldots, x_k, w\} \text{ is the set of all neighbours of } v \text{ in } G. \end{cases}$$
4: **let** v be a node from V_G.
5: **while** there exists a node w adjacent with v such that $m_{w,v} = \top \neq m_{v,w}$:
    set $v := w$ (continue search, star M1).
6: **if** v is incident with a symmetric edge $\langle v, w \rangle$, that is, $m_{v,w} = \top = m_{w,v}$ or $m_{v,w} \neq \top \neq m_{w,v}$
    then $e := \{v, w\}$ (optimal edge found, stars M2-M5)
    else $e := \{v, w\}$, where w is a neighbour of v (optimal star M6).
7: **return** $|I(G_e)| + |I(S)| - 2 * (\lambda'_{v,w} + \lambda'_{w,v})$, where ($\theta$ defined in Prop. 1)
$$\lambda'_{v,w} := \begin{cases} 0 & v \text{ is a leaf}, \\ \sum \lambda'_{x_i, v} + \theta(c_e(v), c(M_e(v))) & x_1, \ldots, x_k \text{ are the children of } v \text{ in } G_e, k > 1. \end{cases}$$

3.4 Relations between urRF and DL

In this section we present relations between the duplication loss (DL) cost and the urRF measure. Fig. 4 illustrates an example of the DL cost for the trees from previous figures. For more details on DL cost, please refer to [16].

Theorem 2. *Let S be a binary species tree and G a binary unrooted gene tree. Then each optimal rooting for the duplication-loss cost is optimal for the urRF measure.*

Proof. It follows from the star topologies for binary case (see Fig.4) that the star of type M1 has type S1, M2 has type S2 and so on. Please note that the star of type M6 is not present if both trees are binary. From Theorem 7 [16], the optimal rootings for the weighted duplication-cost are determined by all edges from stars of type S4-S5 and by the symmetric edge from stars of type S2-S3. By Thm. 1, these edges are also optimal for the urRF distance. □

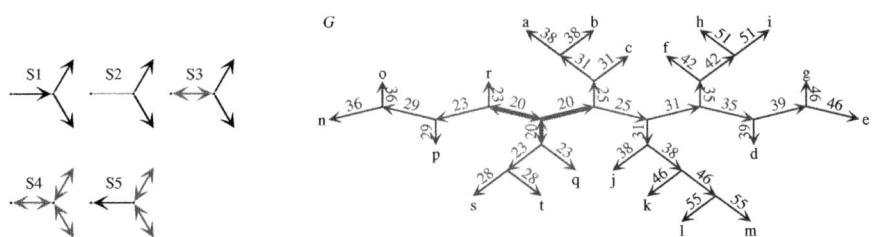

Fig. 4. Left: stars for the binary case (gene and species trees are binary) [16]. Right: duplication-loss cost example with the star-like topology for the trees from Fig. 3. There are 3 optimal edges with cost 20. Please observe that these edges are also optimal for the urRF measure.

4 Experiment

In our experimental study we inferred 3150 unrooted, and mostly multifurcated, gene family trees from the OrthoMCL-DB v5.0 database [8]. For the first time, our new urRF measure allowed us to directly compare each of these unrooted gene trees with the rooted species tree from the NCBI taxonomy [24] based on the cluster presentations of the trees involved.

To construct the gene trees, we first extracted 3150 orthologous groups of protein sequences from the OrthoMCL database with at least 11 elements. We aligned the protein sequences of each group using the program Muscle [11] by applying the default parameter setting. Maximum likelihood gene trees were inferred from each of the alignments using PhyML [17]. In the resulting unrooted and binary gene trees we collapsed uncertain edges; that were all edges with branch lengths shorter than 0.06.

Using an implementation of our algorithm for the urRF measure we compared each of the 3150 unrooted gene trees with the rooted species tree. The overall computation time of the scores took less than a minute on a standard workstation. As an example, the comparison of one of these gene trees is depicted in Fig. 5. The minimum urRF score for this gene tree is 19 and is obtained by rooting the tree at one of the three edges in light red that are forming its valley. Observe, that all of these optimal rootings of the gene tree are in agreement with the topology of the species tree. Another example is depicted in Fig. 6, which depicts the gene tree for the OG5_132446 family of quinolinate synthetases. This tree has a valley of 48 edges, and rooting the tree at any of these edges results in the minimum urRF score of 29. This valley is by a factor of 16 larger than

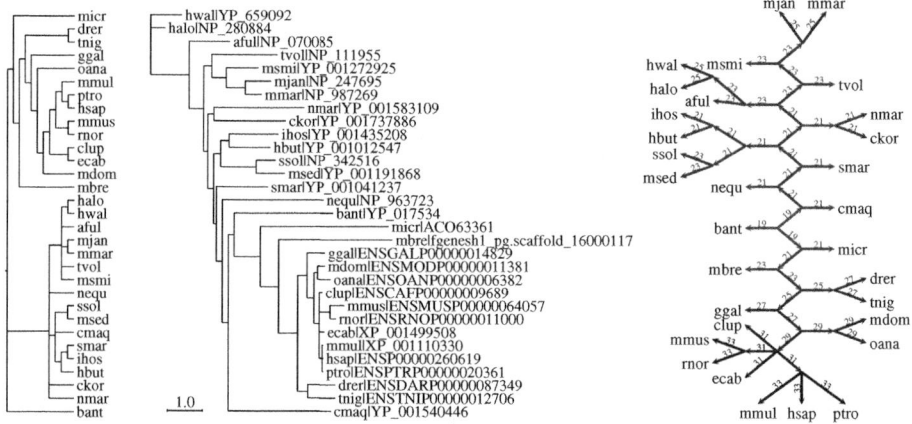

Fig. 5. Middle: an unrooted gene tree of RNA methylases and similar enzymes from the OrthoMCL family OG5_133438[8]. Left: a rooted species tree based on the NCBI taxonomy [24] inferred for the set of species present in the OG5_133438 family. Right: the result of the urRF cost analysis. The minimum score for the valley of the gene tree is 19.

the valley of the previous tree. In our experiments we have observed that larger valleys typically occur when the gene trees have larger urRF distances to the species tree or have many multifurcations. In such a case we propose to choose the rooting placed on one of the double edges of the optimal set. Another option is to choose the center edge[1] of the set of double edges (see the gene tree that is on the bottom in Fig. 6). This observation can be justified by Thm. 2 where the double edges are optimal for both the DL cost and the urRF measure. Observe, that by Thm. 1 the double edges form a subtree in G, and therefore, the operation of choosing the center edge is well defined. It should also be noted that the theory of unrooted reconciliation with the DL-cost is defined for binary trees only.

For brevity a detailed analysis for all of the trees is omitted here, but more results can be obtained at http://bioputer.mimuw.edu.pl/gorecki/urrf.

5 Discussion and Conclusion

Here we addressed the phylogenetic problem of comparing an unrooted tree with a rooted tree. Solving this problem efficiently is, for example, crucial when unrooted gene trees in a large phylogenetic tree database base have to be identified that are supporting the clusters of a credible rooted species tree. We (i) defined a novel RF measure to compare an unrooted tree with a rooted tree that is based on the rooted RF metric, (ii) showed crucial properties that are induced by our measure, (iii) provided a linear time

---

[1] Let $k(e)$ be the maximal length of the path connecting this edge with a leaf. An edge $e \in T$ is called the center (of T) if it has the minimal $k(e)$ among all edges from T.

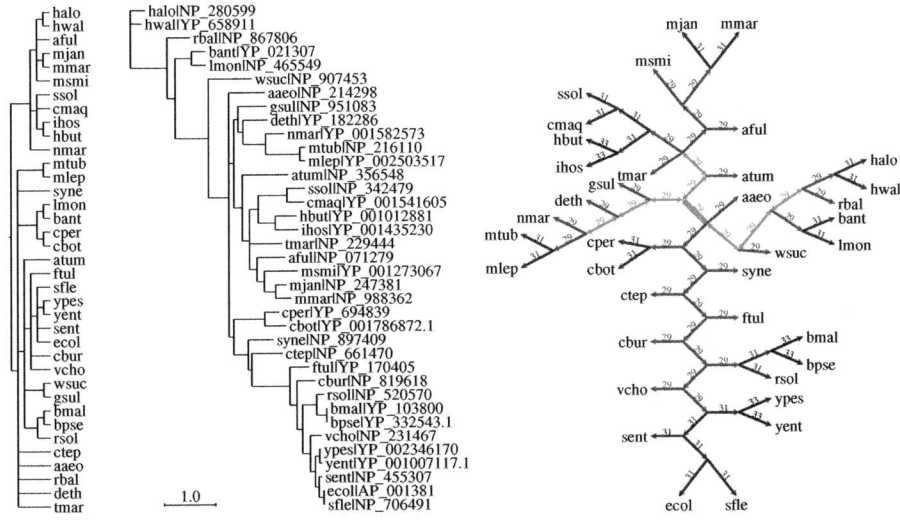

Fig. 6. Middle: an unrooted gene tree for the family of quinolinate synthetases from the OrthoMCL family OG5_132446 [8]. Left: a rooted species tree based on the NCBI taxonomy [24] inferred for the set of species present in the OG5_132446 family. Right: the result of the urRF cost analysis. The minimum score is 29. The double edges are colored green, and the bold green edge is the center edge of the set of double edges. Based on Thm. 1, we advise to root the gene tree at this edge instead of the center edge of the whole tree.

algorithm to compute this measure, and (iv) proposed a novel approach to determine credible roots in a gene trees.

Note, rooting a tree will remove some of the biologically irrelevant clusters/splits from the scoring function. However, it is hard to determine exact relations between our urRF and the RF distance for two unrooted trees, because rooting a tree always removes some of the clusters/splits. In consequence that might increase or decrease the score based on the size of symmetric difference between clusters/splits. In general the rooting can be placed either on the edge, or on a vertex of the unrooted input tree. However, it can be shown that rootings placed on vertices will not improve the optimal score. Moreover, it can be shown that the set of optimal rootings on edges can be extended by internal vertices of the urRF valley.

References

1. Bansal, M.S., Burleigh, J.G., Eulenstein, O., Fernández-Baca, D.: Robinson-foulds supertrees. Algorithms for Molecular Biology 5, 18 (2010)
2. Bender, M.A., Farach-Colton, M.: The lca Problem Revisited. In: Gonnet, G.H., Viola, A. (eds.) LATIN 2000. LNCS, vol. 1776, pp. 88–94. Springer, Heidelberg (2000)
3. Bordewich, M., Semple, C.: On the computational complexity of the rooted subtree prune and regraft distance. Annals of Combinatorics 8, 409–423 (2004)

4. Bourque, M.: Abres de Steiner et reseaux dont varie l'emplagement de certains sommnets. PhD thesis, Department d'Informatique, University Montréal, Montréal (1978)
5. Bryant, D., Steel, M.: Computing the distribution of a tree metric. IEEE/ACM Trans. Comput. Biol. Bioinform. 6(3), 420–426 (2009)
6. Bryant, D., Tsang, J., Kearney, P.E., Li, M.: Computing the quartet distance between evolutionary trees. In: Symposium on Discrete Algorithms, pp. 285–286 (2000)
7. Chaudhary, R., Burleigh, J.G., Fernández-Baca, D.: Fast Local Search for Unrooted Robinson-Foulds Supertrees. In: Chen, J., Wang, J., Zelikovsky, A. (eds.) ISBRA 2011. LNCS, vol. 6674, pp. 184–196. Springer, Heidelberg (2011)
8. Chen, F., Mackey, A.J., Stoeckert, C.J., Roos, D.S.: Orthomcl-db: querying a comprehensive multi-species collection of ortholog groups. Nucleic Acids Research 34(suppl.1), D363–D368
9. DasGupta, B., He, X., Jiang, T., Li, M., Tromp, J., Zhang, L.: On distances between phylogenetic trees. In: SODA, pp. 427–436 (1997)
10. Day, W.H.E.: Optimal algorithms for comparing trees with labeled leaves. Journal of Classification 2(1), 7–28 (1985)
11. Edgar, R.C.: MUSCLE: multiple sequence alignment with high accuracy and high throughput. Nucleic Acids Research 32, 1792–1797 (2004)
12. Eulenstein, O., Mirkin, B., Vingron, M.: Duplication-based measures of difference between gene and species trees. J. Comput. Biol. 5(1), 135–148 (1998)
13. Felsenstein, J.: Inferring phylogenies. Sinauer Associates (2004)
14. Goodman, M., Czelusniak, J., Moore, G.W., Romero-Herrera, A.E., Matsuda, G.: Fitting the gene lineage into its species lineage, a parsimony strategy illustrated by cladograms constructed from globin sequences. Systematic Zoology 28(2), 132–163 (1979)
15. Górecki, P., Tiuryn, J.: Dls-trees: A model of evolutionary scenarios. Theor. Comput. Sci. 359(1-3), 378–399 (2006)
16. Górecki, P., Tiuryn, J.: Inferring phylogeny from whole genomes. Bioinformatics 23(2), e116–e222 (2007)
17. Guindon, S., Delsuc, F., Dufayard, J., Gascuel, O.: Estimating maximum likelihood phylogenies with PhyML. Methods Mol. Biol. 537, 113–137 (2009)
18. Mecham, C.A.: Theoretical and computational considerations of the compatibility of qualitative taxonomic characters. NATO ASI Series, vol. 1, pp. 304–314. Springer, Berlin (1983)
19. Mirkin, B., Muchnik, I.B., Smith, T.F.: A biologically consistent model for comparing molecular phylogenies. J. Comput. Biol. 2(4), 493–507 (1995)
20. Page, R.D.M.: Maps between trees and cladistic analysis of historical associations among genes, organisms, and areas. Systematic Biology 43(1), 58–77 (1994)
21. Pattengale, N.D., Gottlieb, E.J., Moret, B.M.E.: Efficiently computing the robinson-foulds metric. J. Comput. Biol. 14(6), 724–735 (2007)
22. Robinson, D.F., Foulds, L.R.: Comparison of weighted labelled trees. Lecture Notes in Mathematics, vol. 748, pp. 119–126 (1979)
23. Sanderson, M., McMahon, M.: Inferring angiosperm phylogeny from est data with widespread gene duplication. BMC Evolutionary Biology 7(suppl.1) (2007)
24. Sayers, E.W., et al.: Database resources of the national center for biotechnology information. Nucleic Acids Research 37(suppl.1), D5–D15 (2009)
25. Semple, C., Steel, M.A.: Phylogenetics. Oxford University Press (2003)
26. Smith, A.: Rooting molecular trees: problems and strategies. Biol. J. Linn. Soc. 51, 279–292
27. Strimmer, K., von Haeseler, A.: Quartet puzzling: A quartet maximum likelihood method for reconstructing tree topologies. Molecular Biology and Evolution 13, 964–969 (1996)
28. Wheeler, W.: Nucleic acid sequence phylogeny and random outgroups. Cladistics – The International Journal of the Willi Hennig Society 51, 363–368 (1990)

P-Binder: A System for the Protein-Protein Binding Sites Identification

Fei Guo[1,2], Shuai Cheng Li[2], and Lusheng Wang[2,*]

[1] School of Computer Science and Technology, Shandong University,
Jinan 250101, Shandong, China
[2] Department of Computer Science, City University of Hong Kong,
83 Tat Chee Avenue, Kowloon, Hong Kong
cswangl@cityu.edu.hk

Abstract. Determination of binding sites between proteins has a wide range of applications. Understanding energetics and mechanism of complexes remains one of the essential problems in binding site prediction. We develop a system, P-Binder, for identifying binding sites based on structural compatibility, side-chain conformations, amino acid types and contact energies. P-Binder utilizes an enumeration method and side-chain packing program to identify structurally compatible sites. The system reports the sites with the highest ranked configurations, evaluated through a combination of four statistical energy items. We test P-Binder on protein-protein docking Benchmark v4.0. The overall accuracy and coverage are 64.0% and 69.4% for the bound state, and 51.1% and 61.4% for the unbound state. A comparison with some existing techniques shows P-Binder to improve the success rate by at least 12.3%. The system reports improvements in prediction quality, in terms of both accuracy and coverage. The software package is available at http://sites.google.com/site/guofeics/p-binder for non-commercial use.

Keywords: Protein-protein interaction, binding site prediction, free energy function, secondary structure, side-chain conformation.

1 Introduction

Most of the existing methods to identify the binding sites in protein-protein interaction are based on analyzing the differences between interface residues and non-interface residues. The differences in these methods are often the features of the sequence and the structural or physical attributes. Machine learning methods and other statistical approaches are then used on these features to predict binding sites.

Fernández-Recio *et al.* [3] apply protein docking simulations and analysis of the interaction energy landscapes to identify protein-protein binding sites. The rigid-body docking configurations are used to project the docking energy landscapes onto the surfaces. They produce the low-energy regions as the predicted

---
* Corresponding author.

binding sites. ProMate [13] is based on the idea of interface/non-interface circles. Statistics are performed and histograms are created for each feature. Thereafter, the probability for each circle of protein to be an interface is estimated. PPI-Pred [2] generates an interacting/non-interacting patch for each protein. Seven features are extracted for each patch to build an SVM (Support Vector Machine) model, which predicts the interacting patch. Li *et al.* [9] propose another approach (core-SVM). An SVM is built over eight features extracted from the interface residues, and used to compute the probability of whether a residue is a core interface residue.

In addition, algorithms for 3D structures and other methods have also been used to identify interface residues, through investigating protein surface structures. Pro-BS [4] identifies the binding sites by finding two complementary 3D substructures (each from one protein). It uses the rigid transformations to enumerate possible structural configurations to be free from clashes. When the transformations resulting in the largest number of candidate pairs are found, the corresponding interacting residues are reported as binding sites. In PINUP [11], an empirical scoring function is presented to predict binding sites, which is a linear combination of energy score, interface propensity and residue conservation score. Sppider [14] proposes a novel representation for the recognition of binding sites, which integrates enhanced RSA (Relative Solvent Accessibility) prediction-based fingerprints with high resolution structural data. The meta-PPISP program combines three individual servers: cons-PPISP, ProMate and PINUP [15]; The metaPPI program combines five prediction methods: PPI-Pred, PINUP, PPISP, ProMate, and SPPIDER [5].

Our system, P-Binder, is built on some of the earlier ideas, as well as a number of improvements. Many methods assume that protein structures are rigid. In reality, protein-protein binding typically involves structural changes. These changes are modeled in our system with side-chain flexibility. The protein side-chain conformation can be obtained by using the tree decomposition of protein structure, such as SCWRL4 [8] and TreePack [16]. P-Binder first performs rigid transformations, as in Pro-BS, to enumerate the configurations. For each configuration, the side-chain conformation on the interface is re-constructed with SCWRL4. Then, a good approximation to the bound state is predicted. Finally, P-Binder evaluates the approximations obtained from all the configurations, and it reports the highest ranked one. The evaluation of a configuration is based on its energy state, which is estimated using a combination of four energy items (Secondary Structure Energy, Amino acid-Secondary structure Energy, Side-Chain Energy and Atomic Contact Energy). Information of secondary structure and amino acid are considered in two of these energy items, namely SSE (Secondary Structure Energy) and ASE (Amino acid-Secondary structure Energy).

Our comparison of P-Binder to the existing methods, using the protein-protein docking Benchmark v4.0 [6], shows the system to outperform these methods on commonly used assessment measures. P-Binder improves the success rate by at least 12.3%, and it achieves an accuracy and coverage of 64.0% and 69.4% for the bound state, and 51.1% and 61.4% for the unbound state.

2 Methods

Given two protein structures, our task is to find the binding sites between them. Our method consists of the following five steps. The entire process is illustrated in Figure 1.

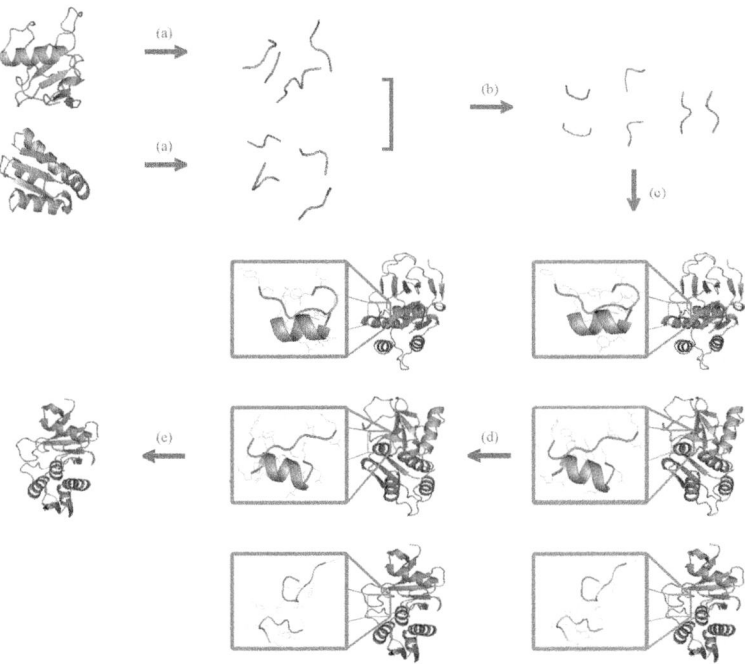

Fig. 1. The process of identifying the binding sites between two proteins. (a) selecting the fragments from each protein using existing methods; (b) identifying surface fragment pairs, one from each protein; (c) enumerating configurations through rigid transformations; (d) using SCWRL4 to construct the side-chain conformations on interface residues; (e) determining the best configuration based on their total energy.

Step 1: We employ a number of existing methods [2,14] to select fragments from the protein structures. These fragments form the basis, from which we will select our candidates for the interface residues.

Step 2: Surface atoms are identified based on an examination of protein structures, as our previous method [4]. Two fragments, one from each protein, are identified as a surface fragment pair, if at least 2/3 of the atoms in both fragments are surface atoms.

Step 3: We perform the rigid transformations to enumerate protein-protein configurations. The begin and end atoms of the surface fragments are used as the axis pair upon which the structure is rotated. A rotation resulting in the minimum overlap size (required to be no more than θ) among the interior points gives us the optimal configuration for the protein structures. The method proposed here is modified from the algorithms for structure comparison [10]. For each pair of proteins, 50 such configurations with minimum overlap sizes are selected for further processing.

Step 4: For each of the configurations obtained in the previous step, we reconstruct the side-chain conformation of the interface residues with an existing method [8]. This gives us a good approximation to the bound state.

Step 5: The configurations are evaluated based on a statistical energy function. The energy is estimated from a linear combination of four energy items. The binding sites of the highest ranked configuration are reported as the results.

2.1 Selecting Fragments on a Single Protein

We first adopt a few existing methods (SPPIDER and PPI-Pred) to select fragments which are potential binding sites. These methods predict whether a residue in a protein structure is an interface residue. If a residue is predicted as an interface by any of the methods, it is labeled as 1; otherwise, it is labeled as 0. We use a sliding window of a length of 3 residues to parse the protein sequence. For each window, if any of the residues is labeled with 1, the fragment of this window is kept for further processing and this fragment is extracted from the sequence. We continue this process on the unextracted portion of the sequence. If all residues in the window are labeled with 0, we continue at the next window.

2.2 Side-Chain Conformation and Side-Chain Energy

One possible way to obtain all the relative configurations of the structure pair is through an enumeration of the possible rigid transformations. However, for a substantial number of proteins, conformational change occurs within the bound state of the protein structures. Therefore, rigid transformation is not sufficiently accurate for predicting binding site. This problem can be addressed through an analysis of the side-chain conformations, since many possible conformations of flexible side-chain include a good approximation to the bound state.

We use SCWRL4 to repack the side-chain for the interface residues, and we also estimate the energy of new conformation. SCWRL4 constructs the rotamers and sub-rotamers of the side-chain for input backbone residues by a new backbone-dependent rotamer library. It uses rigid rotamer model to calculate self energies of a rotamer and pairwise energies between rotamers. Then, rotamer conformations with high self energy are discarded, and disulfide bonds are resolved. SCWRL4 is able to model side-chain in complexes using symmetry operators. Optimization of all parameters determines the interaction graph, via edge decomposition, dead-end elimination and tree decomposition. Finally, an optimization method is employed to produce the minimal energy and yield the corresponding side-chain conformation of interface residues.

Two residues (one from each subunit) are regarded as interacting if they contain a pair of atoms (one from each residue) within a distance d_1 (set to $6\mathring{A}$ in our method). For each obtained configuration, we use SCWRL4 to calculate the side-chain conformations of interface residues. The total energy calculated by SCWRL4 is denoted as the SCE energy item.

2.3 Atomic Contact Energy

The ACE (Atomic Contact Energy) item is an atomic energy measure developed in [17]. It is defined as the free energy of replacing an atom-water contact with an atom-atom contact. The energy for each atom pair has been determined, based on an analysis of atom-pairing frequencies in known proteins. These predetermined scores are given as log likelihood values, thus allowing their summation. For a given configuration, the ACE energy item is a summation of the scores for each atom pair within a threshold distance d_2 (set to 6Å in our method). The ACE energy item gives an estimation of the free energy of the protein interactions. A configuration with a lower ACE value implies a lower free energy, which is considered to be more favorable.

2.4 Secondary Structure Energy

Secondary structure consists of local residues that interact via hydrogen bonds. DSSP (Define Secondary Structure of Proteins) [7] is the standard method for assigning secondary structure type to the amino acids of proteins, given the atomic-resolution coordinates. We use DSSP to determine the types of secondary structure for each residue.

Inspired by the work on the ACE energy, we propose the following method based on the SSE and ASE energy items. We consider eight types of secondary structure and 20 types of amino acid analyzed by DSSP, and one solvent contacting the residues in protein surfaces. The SSE energy item takes 8×8 possible residue pairs, and the ASE energy item takes 20×8 possible residue pairs. Both of them are obtained from the statistical analysis of residue-pairing frequencies in Benchmark v4.0. We define a pair of residues from two subunits as the contact residue pair, if a pair of atoms are within distance d_3 (set to 6Å in our method).

The contact energies in going from unbound proteins to bound structures are correlated to the statistical averages of the observed numbers of pairwise contacts on the interface. The effective contact energy between amino acid type i and type j is defined as

$$S[i,j] = -\ln \frac{N_{i,j}/C_{i,j}}{(N_{i,0}/C_{i,0}) \times (N_{j,0}/C_{j,0})}$$

where type 0 corresponds to the solvent. The number of i-j contact is defined as $N_{i,j} = \sum_p n_{ij,p}$, and the number of i-0 contact is defined as $N_{i,0} = \sum_p n_{i0,p}$. They are the estimation of the actual numbers of contacts, where $n_{ij,p}$ is the contact number between residue type i and residue type j, and $n_{i0,p}$ is the contact number between residue type i and water in each complex. In addition, the expected numbers of contacts are defined as follows:

$$C_{i,j} = \sum_p n_{rr,p} \times \frac{n_{i,p} n_{j,p}}{n_{r,p}^2}$$

and
$$C_{i,0} = \sum_p n_{r0,p} \times \frac{n_{i,p}}{n_{r,p}},$$

where p denotes a complex of protein pair in the data set; $n_{i,p}/n_{r,p}$ is the fraction of residue type i in all residues for each complex; $n_{rr,p}$ and $n_{r0,p}$ are the total numbers of residue-residue contacts and residue-water contacts in each complex, respectively.

For a given configuration, we sum up the scores of all contact residue pairs. Denote the sets of residues from two subunits as S_1 and S_2, respectively, then the SSE and ASE energy items are computed as follows:

$$E_{SSE} = \sum_{s \in S_1, t \in S_2, \|s-t\| \leq d3} S_{SSE}[s,t]$$

and

$$E_{ASE} = \sum_{s \in S_1, t \in S_2, \|s-t\| \leq d3} S_{ASE}[s,t],$$

where $\|s - t\| \leq d3$ means s and t are a pair of contact residues, and $S[s,t]$ is the pre-determined energy item of the residue pair s and t. S_{SSE} is an 8×8 symmetric table for the residue pairs with different types of secondary structure, and S_{ASE} is a 20×8 symmetric table for the residues pairs with different types of amino acid and secondary structure. A configuration with a lower value is more favorable.

2.5 Combination of Energy Scores

We construct 50 configurations for each pair of proteins, and then calculate four energy items for each of these configurations. From these, we select the configuration which optimizes the energies to predict the interface residues. We propose an energy function consisting of a linear combination of four energy items (SSE, ASE, SCE and ACE). The energy function for evaluating the configurations is as following:

$$E_{total} = \alpha E_{ACE} + \beta E_{SSE} + \gamma E_{ASE} + \eta E_{SCE}$$

We optimize all coefficients in above equation by using training data. For a protein pair, five configurations with maximum F-score ($F = 2 \times \frac{Accuracy \times Coverage}{Accuracy + Coverage}$) are labeled as +1, while others are labeled as -1, where +1 denoted a *good* configuration, and -1 denote a *bad* configuration. We calculate the E_{total} for all the 50 configurations, and then report the one with the lowest energy value. If the label of reported result is +1, it is claimed as a success. We try to identify the parameters to maximize the success cases. The grid search method is used [1]. We first identify the best possible combination of values from 0.5 to 10, with a step size of 0.5, for the coefficients. Then, we refine this combination of values, by examining the combinations of values around it with a smaller step size (for example, V_{picked}-0.4, V_{picked}-0.3, ..., V_{picked}+0.4). The optimal values of α, β, γ and η are used for prediction on the other data sets.

For testing data, the predicted configuration is the one with the lowest value of total energy. Two residues of the predicted configuration are reported as the interface residues if they can be connected with two atoms, one from each residue, of distance at most 4.5Å apart.

3 Results

A pair of residues from two proteins are called *interface residues* if any two atoms interact. By interact, we mean the distance between atom pair is less than the sum of the van der Waals radius of two atoms plus 1Å.

Three commonly used measures are utilized to assess the performance of P-Binder. Accuracy (Acc) is the fraction of correctly predicted residues over the total number of predicted interface residues; Coverage (Cov) is the fraction of correctly predicted interface residues over the total number of actual interface residues. A reported result is claimed as a success if the accuracy is no less than 50%. The success rate (Suc) is the fraction of successful predicted cases in the total number of predicted proteins. We also calculate the average predicted size (M_n) and the standard deviation of predicted size (V_n) on the data set.

3.1 Comparison to Existing Methods

We use Benchmark v4.0 to calculate the parameters in the linear combination of total energy function. The values we use for α, β, γ and η in P-Binder are 0.3, 5.5, 4.9 and 5.8, in our comparisons with the existing methods. Each comparison with a method is performed using the test data set used by the compared method in the literature. The predicted results are compared to those reported by the authors of the compared method.

Comparison to Fernández-Recio's method. In this test we compare the performance of P-Binder to Fernández-Recio's method. The test data used by this method consists of 43 complexes [3]. The results are reported in Table 1. The overall accuracy and coverage for P-Binder are 66.1% and 63.5%, respectively. Fernández-Recio method achieves the overall accuracy and coverage of 39.3% and 72.7%, respectively. Furthermore, the success rate for P-Binder is 64.0%, while Fernández-Recio method achieves a success rate of 37.2%. Our method improves the success rate by 26.8%. The average predicted sizes for P-Binder and Fernández-Recio's method are 22.7 residues and 46.3 residues respectively, while the average actual size is 22.2 residues. The standard deviation of predicted size for P-Binder is 20.6, and for Fernández-Recio's method is 40.0. Hence in this test, the binding residues on the protein-protein interface reported by P-Binder are more accurate.

Table 1. Comparison to Fernández-Recio method

	P-Binder					Fernández-Recio				
	Suc[a]	Acc[a]	Cov[a]	M_n	V_n	Suc	Acc	Cov	M_n	V_n
Overall	64.0	66.1	63.5	22.7	20.6	37.2	39.3	72.7	46.3	40.0

[a] *Acc*, *Cov* and *Suc* are the values of percentage (%).

Comparison to metaPPI, meta-PPISP and PPI-Pred. In this group of tests, the test set in metaPPI [5] is used. The data which consists of 41 complexes can be divided into two categories: enzyme-inhibitor (EI) and others. The overall accuracy and coverage from each prediction method on the complexes of both categories are shown in Table 2. P-Binder achieves a success rate of 67.1%; in contrast, metaPPI, meta-PPISP and PPI-Pred have success rates of 54.8%, 43.9% and 43.5%, respectively. P-Binder achieves an accuracy of 65.1%, which improves upon other three methods by at least 16.0%. P-Binder achieves a coverage of 61.3%, which improves upon the others by at least 23.6%. The average predicted sizes for P-Binder, metaPPI, meta-PPISP and PPI-Pred are 21.3 residues, 13.2 residues, 18.2 residues and 27.8 residues respectively, while the average actual size is 22.7 residues. The standard deviation of predicted size for P-Binder is 6.2, and for metaPPI, meta-PPISP and PPI-Pred are 11.8, 13.5 and 17.1, respectively.

Table 2. Comparison to metaPPI, meta-PPISP and PPI-Pred

Type	P-Binder					metaPPI					meta-PPISP					PPI-Pred				
	Suc[b]	Acc[b]	Cov[b]	M_n	V_n	Suc	Acc	Cov	M_n	V_n	Suc	Acc	Cov	M_n	V_n	Suc	Acc	Cov	M_n	V_n
E-I[a]	67.6	76.1	64.3	23.6	6.7	70.5	61.1	36.5	12.9	10.4	55.8	56.4	54.7	24.1	13.5	58.8	45.5	46.9	23.7	15.1
others	66.7	57.3	59.2	22.1	5.8	43.8	40.7	22.2	8.0	10.1	35.6	38.5	25.7	11.8	12.6	32.7	29.3	31.3	19.0	14.7
Overall	67.1	65.1	61.3	21.3	6.2	54.8	49.1	28.1	13.2	11.8	43.9	45.9	37.7	18.2	13.5	43.5	36.0	30.7	27.8	17.1

[a] *E-I* is the type of enzyme-inhibitor.
[b] *Acc*, *Cov* and *Suc* are the values of percentage (%).

Comparison to ProMate and PINUP. In this experiment, P-Binder is compared to ProMate and PINUP. The test data is originally used by ProMate [13], and it consists of 57 non-homologous proteins. The results are reported in Table 3. P-Binder achieves the overall accuracy and coverage of 56.3% and 60.8%, respectively. The overall accuracy and coverage for PINUP are 44.5% and 42.2%, and for ProMate are 52.5% and 13.2%. The overall coverage is improved by at least 18.6%. The average predicted sizes for P-Binder, PINUP and ProMate are 26.5 residues, 19.0 residues and 5.4 residues, while the average actual size is 22.6 residues. The standard deviation of predicted size for P-Binder is 14.5, and for PINUP and ProMate are 8.7 and 16.8, respectively.

Comparison to core-SVM. We compare P-Binder to core-SVM with 50 dimers [9]. Table 4 shows the results. The overall accuracy and coverage for our method are 63.7% and 61.0%, while the values of core-SVM are 53.4% and 60.6%. P-Binder achieves a success rate of 72.0% on 50 protein pairs in those binary

Table 3. Comparison to PINUP and ProMate

	P-Binder					PINUP					ProMate				
	Suc[a]	Acc[a]	Cov[a]	M_n	V_n	Suc	Acc	Cov	M_n	V_n	Suc	Acc	Cov	M_n	V_n
Overall	61.4	56.3	60.8	26.5	14.5	43.8	44.5	42.2	19.0	8.7	63.1	52.5	13.2	5.4	16.8

[a] *Acc*, *Cov* and *Suc* are the values of percentage (%).

complexes. The success rate of core-SVM is not reported in the literature. Our method improves the overall accuracy by at least 10.3%. The average predicted size for P-Binder is 40.5 residues, while the average actual size is 38.3 residues. The standard deviation of predicted size for P-Binder is 25.9.

Table 4. Comparison to core-SVM

	P-Binder					core-SVM				
	Suc[a]	Acc[a]	Cov[a]	M_n	V_n	Suc	Acc	Cov	M_n	V_n
Overall	72.0	63.7	61.0	40.5	25.9	—	53.4	60.6	—	—

[a] *Acc*, *Cov* and *Suc* are the values of percentage (%).

3.2 Evaluation on Benchmark v4.0

To further evaluate our method, we perform test on the protein-protein docking Benchmark v4.0. This benchmark consists of 176 complexes. We use the four groups of data in the previous sections to select the parameters $\alpha=0.2$, $\beta=2.8$, $\gamma=2.5$ and $\eta=5.7$. We test our method on both bound and unbound states.

Results on Bound States. The complexes are classified into broad biochemical categories: Enzyme-Inhibitor (52), Antibody-Antigen (25) and Others (99). For the bound structures, the overall accuracy and coverage of P-Binder are 69.6% and 66.9% for the Enzyme-Inhibitor complexes, 63.5% and 72.6% for the Antibody-Antigen complexes, and 60.9% and 68.8% for Others. A success rate of 76.0% is achieved for the Antibody-Antigen complexes, whereas the numbers for the other two categories are lower. The details are shown in Table 5.

Table 5. Performance of our method for Benchmark v4.0 on bound states

Type[a]	No. of complexes	Suc[b]	Acc[b]	Cov[b]
Enzyme-Inhibitor	52	65.4	69.6	66.9
Antibody-Antigen	25	76.0	63.5	72.6
Others	99	57.6	60.9	68.8
Overall	176	62.5	63.8	68.8

[a] *Type* is based on the broad biochemical categories.
[b] *Acc*, *Cov* and *Suc* are the values of percentage

Results on Unbound States. The unbound protein pairs are classified into three categories: 121 rigid-body (easy) cases, 30 medium difficult cases, and 25 difficult cases, according to the magnitude of conformational change after binding [12]. For the unbound proteins, the overall accuracy and coverage of P-Binder are 55.0% and 66.6% for the rigid-body cases, 49.3% and 51.2% for the medium difficult cases, and 33.9% and 45.0% for the difficult cases. The success rate of P-Binder is 51.3% for the rigid-body cases, which is better than that for the other two categories. In general, the accuracy and coverage decrease as the magnitude of conformational change increases. The details are shown in Table 6.

Table 6. Performance of our method for Benchmark v4.0 on unbound states

Subset[a]	Type[b]	No. of cases	Suc[c]	Acc[c]	Cov[c]
Rigid body	Enzyme-Inhibitor	40	55.0	58.9	66.8
	Antibody-Antigen	22	54.5	52.0	69.7
	Others	59	47.5	53.5	65.3
	Subtotal	121	51.3	55.0	66.6
Medium difficult	Enzyme-Inhibitor	7	42.9	48.6	52.9
	Antibody-Antigen	1	0	46.1	45.7
	Others	22	45.5	49.7	50.9
	Subtotal	30	43.4	49.3	51.2
Difficult	Enzyme-Inhibitor	5	40.0	42.1	46.5
	Antibody-Antigen	2	0	32.0	43.2
	Others	18	22.2	31.9	44.8
	Subtotal	25	24.0	33.9	45.0
Overall		176	46.1	51.0	60.9

[a] Subset is based on the magnitude of conformational change after binding.
[b] Type is based on the broad biochemical categories.
[c] *Acc*, *Cov* and *Suc* are the values of percentage (%).

An example: the configuration discovered by our method for 1ay7(A:B) is shown in Figure 2. The C_α RMSD (Root Mean Square Deviation) between the experimental structure and the predicted complex is 5.8Å. The configurations predicted by P-Binder are fairly accurate for some cases.

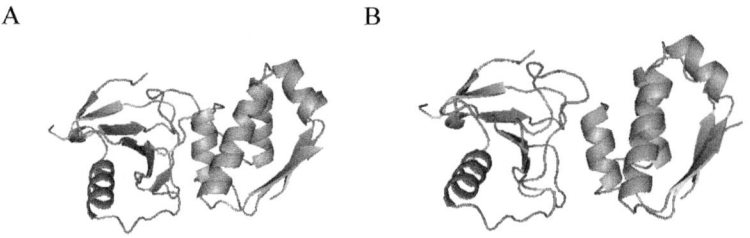

Fig. 2. Configuration discovered by P-Binder for 1ay7(A:B). (A) is the configuration by our method; and (B) is the experimental structure. The C_α RMSD between them is 5.8Å.

3.3 Assessment of the Energy Items

To assess the effectiveness of the energy items, we analyze the performance with Benchmark v4.0. For evaluating the effectiveness of energy items, we re-optimize the coefficients in each case with only three of four items. We re-evaluate the configurations of 176 complexes by leaving one energy item out. The results are shown in Table 7. The overall accuracy and coverage for cases without ACE are 63.9% and 58.3%, for cases without SSE are 61.2% and 51.8%, for cases without ASE are 61.3% and 51.6%, and for cases without SCE are 42.4% and 72.1%. The success rates of all four cases are less than that for the case using all energy items. When the SCE energy item is excluded, the success rates dropped significantly. This confirms our hypothesis that side-chain conformations is an important factor to consider in binding site prediction.

Table 7. Performance of different energy items on Benchmark v4.0

	Suc[a]	Acc[a]	Cov[a]
$E_{SSE+ASE+SCE}$	71.0	63.9	58.3
$E_{ACE+ASE+SCE}$	70.6	61.2	51.8
$E_{ACE+SSE+SCE}$	70.6	61.3	51.6
$E_{ACE+SSE+ASE}$	35.9	42.4	72.1
$E_{ACE+SSE+ASE+SCE}$	71.4	63.8	68.8

[a] Acc, Cov and Suc are the values of percentage (%).

4 Conclusions and Discussion

We developed P-Binder for identifying protein-protein binding sites. The system uses a linear combination of four different energy items to evaluate the configurations between proteins. P-Binder is novel in using of SCWRL4 to construct side-chain conformations for predicting the bound state. Our comparison of the system with some existing methods in predicting binding sites shows P-Binder to perform well.

Although our method shows better overall performance, there are some complexes where its predictions are among the worst. As future work, we are considering to combine the strengths of other programs with P-Binder. In addition, the main chain conformation changes are challenging tasks in the field.

Acknowledgments. This work is supported by a grant from the Research Grants Council of the Hong Kong Special Administrative Region, China [Project No. CityU 121608].

References

1. Al-Khayyal, F.: Jointly constrained bilinear programs and related problems: an overview. Computers and Mathematics with Applications 19(11), 53–62 (1990)
2. Bradford, J.R., Westhead, D.R.: Improved prediction of protein-protein binding sites using a support vector machines approach. Bioinformatics 21(8), 1487–1494 (2005)

3. Fernández-Recio, J., Totrov, M., Abagyan, R.: Identification of protein-protein interaction sites from docking energy landscapes. J. Mol. Biol. 335(3), 843–865 (2004)
4. Guo, F., Li, S.C., Wang, L.: Protein-protein binding sites prediction by 3d structural similarities. J. Chem. Inf. Model. 51(12), 3287–3294 (2011)
5. Huang, B., Schröder, M.: Using protein binding site prediction to improve protein docking. Gene 422, 14–21 (2008)
6. Hwang, H., Vreven, T., Janin, J., Weng, Z.: Protein-protein docking benchmark version 4.0. Proteins 78, 3111–3114 (2010)
7. Kabsch, W., Sander, C.: Dictionary of protein secondary structure: pattern recognition of hydrogen-bonded and geometrical features. Biopolymers 22, 2577–2637 (1983)
8. Krivov, G.G., Shapovalov, M.V., L, D.R.: Improved prediction of protein side-chain conformations with scwrl4. Proteins 77(4), 778–795 (2009)
9. Li, N., Sun, Z., Jiang, F.: Prediction of protein-protein binding site by using core interface residue and support vector machine. BMC Bioinformatics 9, 1–13 (2008)
10. Li, S.C., Bu, D., Xu, J., Li, M.: Finding Largest Well-Predicted Subset of Protein Structure Models. In: Ferragina, P., Landau, G.M. (eds.) CPM 2008. LNCS, vol. 5029, pp. 44–55. Springer, Heidelberg (2008)
11. Liang, S., Zhang, C., Liu, S., Zhou, Y.: Protein binding site prediction using an empirical scoring function. Nucl. Acids Res. 34(13), 3698–3707 (2006)
12. Mintseris, J., Wiehe, K., Pierce, B., Anderson, R., Chen, R., Janin, J., Weng, Z.: Protien-protein docking benchmark 2.0: an update. Proteins 60, 214–216 (2005)
13. Neuvirth, H., Raz, R., Schreiber, G.: Promate: a structure based prediction program to identify the location of protein-protein binding sites. J. Mol. Biol. 338, 181–199 (2004)
14. Porollo, A., Meller, J.: Prediction-based fingerprints of protein-protein interactions. Proteins 66(3), 630–645 (2007)
15. Qin, S., Zhou, H.X.: meta-ppisp: a meta web server for protein-protein interaction site prediction. Bioinformatics 23(24), 3386–3387 (2007)
16. Xu, J., Berger, B.: Fast and accurate algorithms for protein side-chain packing. Journal of the ACM 53(4), 533–557 (2006)
17. Zhang, C.: Extracting contact energies from protein structures: A study using a simplified model. Proteins 31(3), 299–308 (1998)

Non-identifiable Pedigrees and a Bayesian Solution

Bonnie Kirkpatrick

University of British Columbia

Abstract. Some methods aim to correct or test for relationships or to reconstruct the pedigree, or family tree. We show that these methods cannot resolve ties for correct relationships due to identifiability of the pedigree likelihood which is the probability of inheriting the data under the pedigree model. This means that no likelihood-based method can produce a correct pedigree inference with high probability. This lack of reliability is critical both for health and forensics applications.

Pedigree inference methods use a structured machine learning approach where the objective is to find the pedigree graph that maximizes the likelihood. Known pedigrees are useful for both association and linkage analysis which aim to find the regions of the genome that are associated with the presence and absence of a particular disease. This means that errors in pedigree prediction have dramatic effects on downstream analysis.

In this paper we present the first discussion of multiple typed individuals in non-isomorphic pedigrees, $\mathcal{P}$ and $\mathcal{Q}$, where the likelihoods are non-identifiable, $Pr[G \mid \mathcal{P}, \theta] = Pr[G \mid \mathcal{Q}, \theta]$, for all input data G and all recombination rate parameters θ. While there were previously known non-identifiable pairs, we give an example having data for multiple individuals.

Additionally, deeper understanding of the general discrete structures driving these non-identifiability examples has been provided, as well as results to guide algorithms that wish to examine only identifiable pedigrees. This paper introduces a general criteria for establishing whether a pair of pedigrees is non-identifiable and two easy-to-compute criteria guaranteeing identifiability. Finally, we suggest a method for dealing with non-identifiable likelihoods: use Bayes rule to obtain the posterior from the likelihood and prior. We propose a prior guaranteeing that the posterior distinguishes all pairs of pedigrees.

Keywords: Pedigree genetics, discrete probability, identifiability.

1 Introduction

Motivation. Pedigrees are useful for disease association [20], linkage analysis [1], and estimating recombination rates [4]. Most of these calculations involve the pedigree likelihood which is formulated using probabilities for Mendelian inheritance given a graph of the relationships. Since the known algorithms for computing the likelihood are exponential, there have been many attempts to speed up the exact likelihood calculation [6,1,14,7,3,11,8]. Due to the running-time issue, other statistical methods have been introduced which perform genome-wide association studies that use a faster correction for the relationship structure [2,20,21].

Pedigree reconstruction, introduced by Thompson [19], is very similar to methods used for phylogenetic tree reconstruction. The aim is to search the space of pedigree graphs for the graph that maximizes the likelihood, which is the probability of the observed data being inherited on the given pedigree graph. However, the pedigree reconstruction problem differs from the phylogenetic reconstruction problem in several important ways: 1) the pedigree graph is a directed acyclic graph whereas the phylogeny is a tree, 2) while the phylogenetic likelihood is efficiently computed, the only known algorithms for the pedigree likelihood are exponential, either in the number of people or the number of sites [10], and 3) the phylogenetic likelihood is identifiable [17], while we demonstrate that the pedigree likelihood is non-identifiable for the pedigree graph.

Whether the pedigree likelihood is identifiable for the pedigree graph is crucial to forensics where relationship testing is performed using the likelihood on unlinked sites [13]. The scenario is that an unknown person, a, leaves their DNA at the crime scene, and it is a close match to a sample, b, in a database. The relationship between a and b is predicted, and any relatives of b who fit the relationship type are under suspicion. Our results indicate that the number of people who should fall under suspicion might be larger than previously thought. For example, paternity and full-sibling testing are both common and very accurate. However, half-sibling relationships are non-identifiable from avuncular relationships and from grand-parental relationships with unlinked sites. As we will see later, for both unlinked and linked sites, different types of cousins relationships are also non-identifiable, even with the addition of genetic material from a third related person. Due to these non-identifiable relationships, a known relationship between a third person, c and b is not enough information for conviction without also checking whether there is a perfect match between the DNA of c and a and whether there is additional information.

The likelihood is also used to correct existing pedigrees where relationships are mis-specified [12,16,15]. Much of their success comes from changing relationships that result in zero or very low likelihoods. Again, the accuracy of these methods will be effected by the non-identifiable likelihood. For similar reasons, the accuracy of pedigree relationship prediction [15] and reconstruction methods [19,9] is greatly influenced by the likelihood being non-identifiable, since these methods rely on the likelihood or approximations of it to guide relationship prediction.

The kinship coefficient is known to be non-identifiable for the pedigree graph [18]. The kinship coefficient is an expectation over the condensed identity states which describe the distinguishable allelic relationships between a pair of individuals. Pinto et al. [13] showed that there are cousin-type pairs of pedigrees having the same kinship coefficient. However, these results apply only to *unlinked* sites, a special case of the *linked* sites.

This work considers identifiable pedigrees on *linked* sites. Thompson [18] provided an early discussion of this topic. Donnelly [5] discovered that cousin-type relationships are non-identifiable if two pedigrees have the same total number of edges separating the two genotyped cousins from the common ancestor.

In this paper, we make use of a method by Kirkpatrick and Kirkpatrick [8] to collapse the original hidden states of the likelihood HMM into the combinatorially largest partition which is still an HMM. Using this tool-box, we are able to show that two pedigrees are non-identifiable if and only if they have an isomorphism between their collapsed state spaces. We relate this isomorphism to known results on the non-identifiability of the kinship coefficient. We introduce a method of removing edges from a pedigree to obtain a minimal pedigree having the same likelihood. We then show that two pedigrees that have different minimal sizes must be identifiable. We connect this notion of removing edges to the pruning introduced by McPeek [11] which is clearly implementable in polynomial time, and we also introduce a result stating that pedigrees with discrete non-overlapping generations such as those obtained from the diploid Wright-Fisher (dWF) model are always identifiable.

We give several examples of the kinship coefficient and pedigree likelihood being non-identifiable. We give the only known non-identifiability example where there are more than two individuals with data. Finally, we discuss a Bayesian method for integrating over this uncertainty.

2 Background

A *pedigree graph* is a directed acyclic graph $P = (I(P), E(P))$ where the nodes are individuals and edges are parent-child relationships directed from parent to child. All individuals in $I(P)$ must have either zero or two incoming edges. If an individual has zero incoming edges, then that individual is a *founder*. The set of founders for pedigree graph P is $F(P)$.

A *pedigree* is a tuple $\mathcal{P} = (P, s, \chi, \ell)$ where P is the pedigree graph, function $s : I(P) \to \{m, f\}$ are the genders, set $\chi \subseteq I(P)$ is the individuals of interest, and $\ell : \chi \to \mathbb{N}$ are the *names* of the individuals of interest. If $i \in I(P)$ has two incoming edges, $p_0(i)$ and $p_1(i)$, then one parent must be labeled $s(p_j(i)) = m$ and the other $s(p_{1-j}(i)) = f$ for $j \in \{0, 1\}$.

The likelihood, $Pr[G \mid \mathcal{P}, \theta]$, is a function of the genotypes G, the recombination rates θ, and the pedigree $\mathcal{P}$. However, we will abuse notation by referring to a pedigree by its pedigree graph and writing $Pr[G \mid P, \theta]$. In these instances, the set χ will be clear from the context.

Two pedigrees $\mathcal{P}$ and $\mathcal{Q}$ are said to be *identifiable* if and only if $Pr[G \mid \mathcal{P}, \theta] \neq Pr[G \mid \mathcal{Q}, \theta]$ for some values of G and θ. If $\mathcal{P}$ and $\mathcal{Q}$ are not identifiable, we call them *non-identifiable*.

Two pedigree graphs, P and Q are *isomorphic* if there exists a mapping $\phi : I(P) \to I(Q)$ such that $(u, v) \in E(P)$ if and only if $(\phi(u), \phi(v)) \in E(Q)$. This is an isomorphism of the pedigree graph rather than of the pedigree, because the genders are not necessarily preserved by the map ϕ. From now on, we will assume that P and Q are not isomorphic.

Two isomorphic pedigrees might have different gender labels, and they would be identifiable when considering sex-chromosome data. We restrict our discussion to autosomal data, where these two pedigrees would be non-identifiable.

The Hidden Markov Model. Rather than writing out the cumbersome likelihood equation, we will define the likelihood by specifying the HMM. For each pedigree $\mathcal{P} = (P, s, \chi, \ell)$, there is an HMM, and everything in this section is defined relative to a specific pedigree $\mathcal{P}$. To specify the HMM, we need to specify the hidden states, the emission probability, and the transition probabilities. We will begin with the hidden states.

An inheritance vector $x \in \{0,1\}^n$ has length $n = |E(P)|$. Each bit, x_e, in this vector indicates which grand-parental allele, maternal or paternal, was inherited along edge $e \in E(P)$. An *inheritance graph* R_x contains two nodes for each individual in $i \in I(P)$, called i_0 and i_1, and edges $(p_j(i)_{x_e}, i_j)$ for each $(p_j(i), i) \in E(P)$. The sets χ_0 and χ_1 are the paternal and maternal alleles, respectively, of the individuals of interest. We will refer to the collective set $\chi_0 \cup \chi_1$ as the *alleles of interest*. Each node in R_x represents an allele. The inheritance graph is a forest with each root being a founder allele. The inheritance vectors are the *hidden states* of the HMM. Let $\mathcal{H}_P$ be the hypercube of dimension $|E(P)|$; its vertices represent all the inheritance vectors.

This inheritance graph represents identity-by-descent (IBD) in that any pair of individuals of interest $i, i' \in \chi$ are IBD if there exists an inheritance vector x such that one pair of (i_0, i'_0), (i_0, i'_1), (i_1, i'_0) or (i_1, i'_1) are connected. The *identity states*, are the sets of the partition induced on the alleles of interest by the connected components of R_x, namely $D_x = \{y \in \mathcal{H}_P | CC(R_y) = CC(R_x)\}$. The *transition probabilities* are a function of the per-site recombination rates $\theta = (\theta_1, .., \theta_{T-1})$ for T sites. Let X_t be the random variable for the hidden state at site t. The probability of recombining from hidden state x to state y at site t is

$$Pr[X_{t+1} = y \mid X_t = x, \theta] = \theta_t^{H(x,y)}(1-\theta_t)^{n-H(x,y)} \quad (1)$$

where $H(x,y) = |x \oplus y|_1$ is the Hamming distance between the two bit vectors, $\oplus$ indicates the XOR operation, and $|.|_1$ is the L_1-norm. In some instances, we may make the θ implicit, because it is clear from context.

The *emission probability* depends on the data, which is the genotype random variable G. Each individual of interest $i \in \chi$ has two rows in the genotype matrix which encode, for each column t, the alleles that appear in that individual's genome. For example, $\{g_{it}^0, g_{it}^1\}$ from the 0th and 1st rows for individual i at site t is the (unordered) set of alleles that appear in that individual's genome. The data for all the individuals at site t is an n-tuple $g_t = (\{g_{it}^0, g_{it}^1\} | \forall i)$ and $g = (g_1, ..., g_T)$ is the data at all T sites. The pedigree HMM deconvolves these unordered alleles by considering all possible orderings of the genotypes when assigning them to the hidden alleles.

Specifically, let $CC(R_x)$ be the connected components of R_x. Then the emission probability at site t is

$$Pr[G_t = g_t \mid X_t = x, P] \propto \sum_{\tilde{g}_t} \prod_{c \in CC(R_x)} \mathbf{1}\{n(c, \tilde{g}_t) = 1\} Pr[h(c, \tilde{g}_t)]$$

where $\tilde{g}_t$ is the ordered alleles (g_{it}^0, g_{it}^1) that appear in g_t, $n(c, \tilde{g}_t)$ is the number of alleles assigned to c by $\tilde{g}_t$, and $h(c, \tilde{g}_t)$ is the allele of $\tilde{g}_t$ that appears in c. Notice that by definition of the identity states, $\{D_x | \forall x\}$, $Pr[G_t \mid X_t = x_1] = Pr[G_t \mid X_t = x_2]$ for all $x_1, x_2 \in D_x$.

This completes the definition of the HMM and the likelihood. Now, our task is to find pairs of pedigree graphs (P, Q) such that $Pr[G \mid P, \theta] = Pr[G \mid Q, \theta]$ for all G and θ. We can do this by considering multiple equivalent HMMs and finding the "optimal" HMM that describes the likelihood of interest. Given two optimal HMMs, we can easily compare their likelihoods for different values of G and θ.

The Maximum Ensemble Partition. In this paper, we will use a method similar to that discussed by Browning and Browning [3] and improved by Kirkpatrick and Kirkpatrick [8]. This method relies on an algebraic formulation of the hidden states of the Hidden Markov Model (HMM) that is used to compute the pedigree likelihood. Specifically, we can collapse the original hidden states into the combinatorially largest partition which is still an HMM. From the collapsed state space (termed the maximum ensemble partition), we can easily see that certain pairs of pedigrees have isomorphic HMMs and thus identical likelihoods.

For pedigree $\mathcal{P} = (P, s, \chi, \ell)$, consider a new HMM with hidden states Y_t in a state space that is defined by a partition, $m(P) := \{W_1, ..., W_k\}$, of $\mathcal{H}_P$, meaning that for all i, j, $W_i \cap W_j = \emptyset$ and $\cup_{i=1}^k W_i = \mathcal{H}_P$. For the HMM for Y_t to have the same likelihood as the HMM for X_t the *Markov property* and the *emission property*, defined next, must be satisfied.

Let the transition probabilities of Y_t be the expectation of X_t as follows, for all i, j, and for $x \in W_i$

$$Pr[Y_{t+1} = W_j \mid Y_t = W_i] = Pr[X_{t+1} \in W_j \mid X_t = x] \quad (2)$$

$$= \sum_{y \in W_j} Pr[X_{t+1} = y \mid X_t = x]. \quad (3)$$

Conditioning on θ is implicit on both sides of the equation. The *Markov property* is required for Y_t to be Markovian:

$$\sum_{y \in W_j} Pr[X_{t+1} = y \mid X_t = x_1] = \sum_{y \in W_j} Pr[X_{t+1} = y \mid X_t = x_2]$$

for all $x_1, x_2 \in W_i$ for all i and for all W_j. For more details, see [3,8].

The *emission property* states that the emission probabilities of X_t impose a constraint on Y_t. This constraint is that the partition, $\{W_1, ..., W_k\}$ must be a sub-partition of the partition induced on the hidden states by the emission probabilities:

$$E_x(P) = \{y \in \mathcal{H}_P \mid Pr[G_t = g_t \mid X_t = x] = Pr[G_t = g_t \mid X_t = y] \, \forall g_t\}.$$

We call the set $\{E_x(P) | \forall x\}$ the *emission partition* since it partitions the state-space $\mathcal{H}_P$.

It has been shown in [8] that the partition $\{W_1, ..., W_k\}$ which satisfies the Markov property and the emission property and which maximizes the sizes of the sets in the partition—i.e. $\max_{i \in \{1,...,k\}} |W_i|$—can be found in time $O(nk2^n)$ where n is the number of edges, and k is a function of the known symmetries of the pedigree graph $k \leq 2^n$. We call this partition the *maximum ensemble partition*.

It turns out that the maximum ensemble partition is unique, making the derived HMM the unique "optimal" representation for the likelihood. We will exploit this fact to find non-identifiable pairs of pedigrees.

3 Methods

We will define a general criteria under which a pair of non-isomorphic pedigree graphs have identical likelihoods for all input data and recombination rates, as wells as define a uni-directional polynomial-checkable criteria whereby we can determine whether some pairs of pedigrees are identifiable. In the following section, we will apply these results to investigate when pedigrees are identifiable, to give several examples where the pedigrees are non-identifiable, and to suggest a Bayesian solution.

Given two non-isomorphic pedigree graphs P and Q, and their maximum ensemble partitions $m(P)$ and $m(Q)$, respectively. We say that ψ is a *proper isomorphism* if ψ is a bijection $m(P)$ onto $m(Q)$ such that the following hold:

Transition Equality. $Pr[Y_{t+1}^P \mid Y_t^P, \theta] = Pr[\psi(Y_{t+1}^P) \mid \psi(Y_t^P), \theta] \quad \forall t$
Emission Equality. $Pr[G_t \mid Y_t^P, P] = Pr[G_t \mid \psi(Y_t^P), Q] \quad \forall t$

where Y_t^P is the random variable for the hidden state for pedigree P.

Theorem 1. *There exists isomorphism $\psi : m(P) \to m(Q)$ satisfying the transition and emission equalities if and only if the likelihoods for $\mathcal{P}$ and $\mathcal{Q}$ are non-identifiable, $Pr[G \mid \theta, P] = Pr[G \mid \theta, Q]$, for all G and $\theta = (\theta_1, ..., \theta_{T-1})$ where T is the number of sites and $T \geq 2$.*

Proof. ($\Rightarrow$) Given a proper isomorphism $\psi : m(P) \to m(Q)$ that satisfies the transition and emission equalities, the likelihoods are necessarily the same, by definition of the Hidden Markov Model.

($\Leftarrow$) Given that the two pedigrees are identifiable, we will construct ψ. Consider pedigrees P and Q. They both have unique maximum ensemble partitions $m(Q)$ and $m(P)$ [8]. By the definition of $Pr[G \mid \theta, Q]$, this distribution can be represented by an HMM, called $\mathcal{M}(Q)$, over state-space $m(Q)$. By the equality $Pr[G \mid \theta, P] = Pr[G \mid \theta, Q]$, we know that there is an HMM for P, $\mathcal{M}(P)$, with the same transition matrix and emission probabilities as $\mathcal{M}(Q)$. Since $\mathcal{M}(Q)$ has maximum ensemble state-space $m(Q)$, then by uniqueness, there is no other state-space that is as small. By the equality of the two distributions, we know that $\mathcal{M}(P)$ also has maximum ensemble state-space $m(P)$. But since $m(P)$ is the unique maximum ensemble state-space for $\mathcal{M}(P)$, there must be an isomorphism $\psi : m(P) \to m(Q)$ satisfying the transition and emission equalities. □

To apply this method, we need to obtain $m(P)$ and $m(Q)$ and the appropriate proper isomorphism ψ. To obtain $m(P)$ and $m(Q)$ we rely on the maximum ensemble algorithm [8]. The proper isomorphism is obtained by examining the transition probabilities of the respective HMMs.

Corollary 1. *For unlinked sites $\theta_t = 0.5$ for all $1 \leq t \leq T-1$, for any pedigree graphs P and Q with maximum ensemble states $|m(P)| = |m(Q)|$ and identical identity states, the pedigrees are non-identifiable. (proven in Appendix)*

We are now in a position to relate non-identifiability on pedigree HMMs to non-identifiability of an important calculation that relies on independent sites—the kinship coefficient. The *kinship coefficient* for a pair of individuals of interest is defined as the probability of IBD when randomly choosing one allele from each individual of interest. Let the two individuals of interest be $\chi = \{a, b\}$. We write the kinship coefficient for χ as $\Phi_I(P)_\chi = \sum_x \frac{\eta(x,\chi)}{4} \frac{1}{2^n}$ where $\eta(x, \chi)$ is the number of pairs of alleles of interest $\chi_0 \cup \chi_1$ sharing the same connected component in R_x and $\chi_0 \cup \chi_1 = \{\{a_0, b_0\}, \{a_0, b_1\}, \{a_1, b_0\}, \{a_1, b_1\}\}$.

Corollary 2. *For unlinked sites $\theta_t = 0.5$ for all $1 \leq t \leq T-1$, given two non-identifiable pedigree graphs, P and Q, with two individuals of interest $\chi = \{a, b\}$, the kinship coefficient is identical. (proven in Appendix)*

This last corollary is a uni-directional implication. There are some pairs of pedigrees P and Q for which the kinship coefficient is equal but for which the likelihood is identifiable, see Fig 1.

The final set of results we introduce will try to answer the question of when are pedigrees identifiable. Since some algorithms use the likelihood to choose the best pedigree graph or relationship type, these results give some guarantees for when those algorithms will make correct decisions. We wish to show that under some definition of "necessary" edges for some individuals of interest, pedigrees P and Q with different numbers of necessary edges have no proper isomorphism and are, therefore, identifiable. We will relate our definition of a necessary edge to the literature. And, we will establish an even more restricted class of pedigrees for which no pair of pedigrees is identifiable. This is the class of all dWF pedigrees.

For an edge, e, in the pedigree, let σ be the indicator vector with bits $\sigma_f = 0$ for all $f \neq e$ and $\sigma_e = 1$. For pedigree P having states $\{W_1, ..., W_k\}$, we will define an edge $e \in E(P)$ to be *superfluous* if and only if the following two properties hold

1) $Pr[X_{t+1} = y | X_t = x] = Pr[X_{t+1} = \sigma \oplus y | X_t = \sigma \oplus x]$, for every $y \in W_j$ and $x \in W_i$ and for ever i and j, and
2) $Pr[G_t | X_t = x] = Pr[G_t | X_t = \sigma \oplus x]$ for all $x \in \mathcal{H}_P$.

Conversely, an edge e is *necessary* if it is not superfluous. For an example, see the edge adjacent to the grand-father in P' of Fig 2.

Lemma 1. *We say that an edge is removed if its bit is set to a fixed value in all the inheritance vectors. Any superfluous edge can be removed without changing the value of the likelihood. (proven in Appendix)*

Theorem 2. *If pedigrees P and Q have a different number of* necessary *edges, then there is no proper isomorphism and the likelihoods for P and Q are identifiable. (proven in Appendix)*

In order to connect our definition of superfluous edges to the literature we will reiterate McPeek's formulation of *superfluous* individuals [11]. An *individual* $i \in I(P)$ *is superfluous* if for every pair $\{a,b\} \in \chi$ at least one of the following holds:

1. $i \notin A(a) \cup A(b)$ where $A(a)$ is the ancestors of a
2. $A(i) \cap \{a,b\} = \emptyset$ and there exists some $c \in I(P) \setminus \{a,b\}$ and $d \in I(P)$ such that for every $e \in \{i\} \cup A(i)$ for every $l \geq 1$ and every directed path $q = (q_0, ..., q_l)$ of length l with $q_0 = e$ and $q_l \in \{a,b\}$, we have $c = q_m$ and $d = q_{m+1}$ for some $0 \leq m \leq l-1$.

This last condition states that every directed path from i or an ancestor of i to $\{a,b\}$ must pass through directed edge (c,d).

The reason for the definition of superfluous individuals is that it is polynomial-time checkable. If one were to directly check the definition of superfluous edges, one would find it necessary to compute the emission partition and the maximal ensemble state space which requires exponential time. Despite this, from the definition of superfluous edges, it is easy to see the operational consequence: edges can be removed from the pedigree. Superfluous edges and superfluous individuals are related as follows.

Lemma 2. *An individual is* superfluous *if and only if all the edges adjacent to that individuals are* superfluous. *(proven in Appendix)*

Theorem 2 tells us that when two pedigrees have a different number of necessary edges they are certainly identifiable. While this criteria is useful if we are interested in a particular pedigree, it does not allow us to draw broad conclusions about a class of pedigrees. Ideally, if we want to integrate over the space of pedigrees, we would want to integrate only over identifiable pedigrees for efficiency of computation.

The class of diploid Wright-Fisher (dWF) pedigrees are haploid Wright-Fisher genealogies which are two-colorable where there is a color for each gender. These pedigrees have discrete non-overlapping generations, and all the individuals of interest are 'leaves' of the genealogy.

Theorem 3. *Two non-isomorphic, dWF pedigrees P and Q contain only necessary edges and have individuals of interest χ labeling the 'leafs' which are the individuals with no children. Then pedigrees P and Q are identifiable. (proven in Appendix)*

4 Examples

We will consider several examples. The first of which is a trio of pedigrees that are non-identifiable with data from unlinked sites. This fact is well known due to their identical kinship coefficients. However, these three pedigrees are identifiable with data from linked sites. The second example is an extension of the well-known non-identifiable cousin-type relationships. In this example, we extend the relationship from two to three individuals of interest and show that the relationships remain non-identifiable. To the best of our knowledge, this is the first example of non-identifiable pedigrees on more than two individuals of interest.

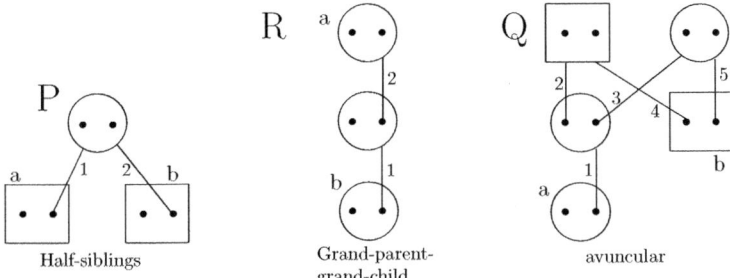

Fig. 1. Half-siblings, grand-parent-grand-child, and avuncular relationships are identifiable. Individuals are drawn as boxes, if male, and circles, if female. The individuals of interest are $\chi = \{a, b\}$. Alleles are drawn as disks with a line between the allele and the parent it was inherited from. For each edge, numbered $e \in \{1, ..., 5\}$, the binary value x_e in the inheritance vector indicates which parental allele was chosen for that hidden state where zero indicates the paternal and the leftmost of the two alleles. The numbers labeling the edges indicate in which order the bits appear in their respective vectors. These three relationships have identical kinship coefficients. The likelihoods of these relationships are identifiable given data on linked sites.

Half-Sibling, Avuncular, and Grandparent-Grandchild Relationships. The first example we will consider is the well-known trio of pedigrees where the kinship coefficient is identical: half-sibling, avuncular, and grandparent-grandchild relationships. There are two individuals of interest, a and b for whom we have data. These three relationships are drawn in Fig 1.

The maximum ensemble partition for each of these three pedigrees are $\{W_1^P = \{00, 11\}, W_2^P = \{01, 10\}\}$ for the half-siblings, $\{W_1^R = \{00, 01\}, W_2^R = \{10, 11\}\}$ for the grand-parent-grand-child, and for the avuncular relationship:

$$W_1^Q = \{00000, 01010, 00101, 01111, 10000, 11010, 10101, 11111\}$$
$$W_2^Q = \{00001, 01011, 00100, 01110, 10010, 11000, 10111, 11101\}$$
$$W_3^Q = \{00010, 00111, 01000, 01101, 10001, 10100, 11011, 11110\}$$
$$W_4^Q = \{00011, 00110, 01001, 01100, 10011, 10110, 11001, 11100\}$$

To get the transition probabilities, we need to sum Equation 1 as in Equation 2. Since for the first two pedigrees, P and R, there are only two states, we need

only compute the transition probability for one state (the others are obtained by observing that the transition probabilities sum to one). For pedigree P, we have
$$Pr[Y^P_{t+1} = W^P_1 \mid Y^P_t = W^P_1] = (1 - \theta_t)^2 + \theta_t^2 = 2\theta_t^2 - 2\theta_t + 1.$$
For pedigree R,
$$Pr[Y^R_{t+1} = W^R_1 \mid Y^R_t = W^R_1] = (1 - \theta_t)^2 + \theta_t(1 - \theta_t) = 1 - \theta_t.$$

It is evident that there is no proper isomorphism that has transition equality for pedigrees P and R. For pairs P, Q and R, Q there is no proper isomorphism, because all three pedigrees contain only necessary edges and both $|m(P)| \neq |m(Q)|$ and $|m(R)| \neq |m(Q)|$. So, we conclude that these pedigrees are identifiable as long as the number of sites $T \geq 2$ and $\theta_t < 0.5$ for all $1 \leq t \leq T - 1$. Despite the well-known fact that these three pedigrees have identical kinship coefficients, these pedigrees *are* identifiable when the data is from multiple linked sites. To the best of our knowledge, this paper is the first to prove this simple fact.

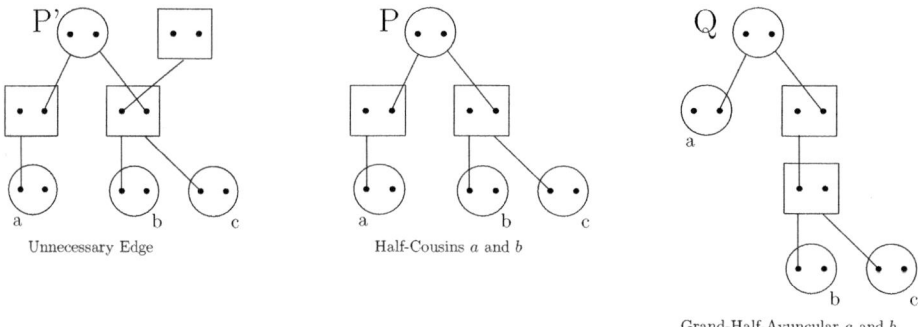

Fig. 2. Half-cousins and grand-half-avuncular relationships are non-identifiable even when there is a third individual of interest. Pedigree P is derived from pedigree P' by removing the superfluous edge. The two pedigree graphs, P and Q are not isomorphic, yet the likelihoods are non-identifiable, meaning that no amount of data on the individual a, b, and c will distinguish these likelihoods.

Half-Cousins and Full-Cousins Relationships. To the best of our knowledge Donnelly [5] was the first to remark that pairs of pedigrees either of the half-cousin or of the full-cousin type and having equal numbers of edges are non-identifiable. Figure 6 of [5] illustrates this situation. Suppose we have two pedigrees P_{d_a, d_b} and $P_{d'_a, d'_b}$ each having two individuals of interest, $\chi = \{a, b\}$ at the leaves, and the most recent common ancestors of χ have the same relationship type in both pedigrees, either half or full relationships. Let d_a and d_b be the number of edges or meioses that separate individuals a and b from their common ancestor(s) in pedigree P_{d_a, d_b}. Then as long as $d_a + d_b = d'_a + d'_b$, the two pedigrees are non-identifiable.

Donnelly remarked this means that no amount of autosomal genetic information can distinguish these two pedigrees, "unless of course information is available on a third person related to both of the individuals in question." Figure 2 shows that for some third individuals these relationships remain non-identifiable. To the best of our knowledge, this is the first example of a pair of non-identifiable pedigrees each having three individuals of interest.

By Theorem 1 and Corollary 2 we can show that *both* the pedigree likelihood and the kinship coefficient are non-identifiable for half-cousin-type relationships, see Figure 2. The isomorphism is omitted for space reasons. We believe that a similar result can be obtained for the full-cousin-type relationship. However, the number of edges is large enough that calculation is difficult due to the exponential algorithm.

These examples mean that the likelihood alone is not a practical tool for testing relationships, for inferring pedigrees, or for correcting pedigrees that have relationship errors since the pedigrees under consideration might be non-identifiable.

5 A Potential Solution

This paper has focused on the likelihood $Pr[G|P, \theta]$, since it is currently the object being used for relationship testing and pedigree reconstruction. However, a common alternative to the likelihood is the posterior distribution obtained via Bayes rule

$$Pr[P|G, \theta] = \frac{Pr[G|P, \theta]Pr[P]}{Pr[G|\theta]} = \frac{Pr[G|P, \theta]Pr[P]}{\sum_Q Pr[G|Q, \theta]Pr[Q]}.$$

The utility of this expression is that the posterior $Pr[P|G, \theta]$ will distinguish between non-identifiable pedigrees provided that the prior has the property that $Pr[P] \neq Pr[Q]$ when P and Q are non-identifiable. Indeed, the uniform distribution over dWF pedigrees is such a prior. Taking care with the zero-probability pedigrees which do not occur under the dWF model, we suggest a refinement. Let W be the set of all dWF pedigrees, and let $\bar{W}$ be the pedigrees which are not dWF. Then, let $Pr[P] = 1/(|W| + 1)$ for $P \in W$, and for an arbitrary ordering $Q_1, ..., Q_{|\bar{W}|}$ with $Q_i \in \bar{W}$, let $Pr[Q_i] = (1/2^i)/(Z(|W| + 1))$ where $Z = \sum_{i=1} 1/2^i$. Since the number of non-diploid WF pedigrees are countably infinite, we can approximate Z using its limit $Z = 1$.

Now that we have a prior, the challenge of using the posterior is that the partition function, the denominator $Pr[G|\theta]$, is most certainly intractable. This is because there are an exponential number of pedigrees and the likelihood algorithm has exponential run-time for each pedigree.

The intractability of the partition function points to the use of sampling methods, in particular, the Metropolis-Hastings Markov Chain Monte Carlo approach might be well suited to this problem. Indeed, MCMC facilitates computing the proposed prior, because we can simply take the Q_i in the order that they are encountered by the Markov chain. If we obtain a sample pedigree P^τ, we can draw a new pedigree $P^{\tau+1}$ by proposing a pedigree Q according to a proposal distribution $q[Q|P^\tau]$ and then choosing to accept, $P^{\tau+1} = Q$ with probability

$$\min\left\{1, \frac{Pr[G|Q,\theta]Pr[Q]}{Pr[G|P^\tau,\theta]Pr[P^\tau]} \frac{q[P^\tau|Q]}{q[Q|P^\tau]}\right\}$$

otherwise $P^{\tau+1} = P^\tau$ remains unchanged. A sequence of $P^1, P^2, ..., P^\tau$ is guaranteed to converge to the stationary distribution $Pr[P^\tau|G,\theta]$. After convergence at time-step τ, take δ pedigree samples $\{P^\tau, P^{k+\tau}, ..., P^{\delta k+\tau}\}$ where k is the number of steps between samples. Those samples can yield information about the posterior distribution, such as the confidence for each edge. One could also take the most probable pedigree that was sampled, and treat that as the estimated pedigree.

The complexity here comes down to three issues, first the likelihood calculation which is exponential, second the prior on the pedigrees which might be tailored to a specific set of pedigrees having positive probabilities, i.e. those containing particular "known" edges, and third calculating the proposal distribution which should be tractable and produce non-zero pedigrees. The latter is critical, because MCMC methods will not converge if they repeatedly propose zero-probability events. This can probably be overcome by using moves inspired by the phylogenetic prune and re-graft method. As yet, all these details are an open problem.

Alternative to integration over the whole space of pedigrees, if we have a single pedigree of which we are fairly confident, we could use this method to integrate over 'nearby' pedigree graphs to get a measure of our confidence in our chosen pedigree. We could use Theorem 2 as a guide to integrate only over a set of pedigrees all having the same number of necessary edges while giving a zero prior to all other pedigrees. Such an approach might even be computationally feasible due to the polynomial-time checkable definition of necessary edges. This would allow us to incorporate into our calculations the uncertainty we have about our chosen pedigree relative to its non-identifiable 'neighbors'.

6 Discussion

This paper reviews the pedigrees that were known to be non-identifiable, namely the half-cousin-type and full-cousin-type relationships. It also introduces a troubling new pair of non-identifiable pedigrees that are also half-cousin-type pedigrees but which contain three individuals of interest. This is the first discussion of non-identifiable pedigrees with genetic data available for more than two individuals, demonstrating that identifiability is not restricted to pedigrees having two individuals with data.

We introduce a general criteria that can be used to detect non-identifiable pedigrees. We show how non-identifiable likelihoods relate to non-identifiable kinship coefficients. An example is given showing that the kinship coefficient can be identical while the likelihood is sufficient to distinguish the pedigrees. Finally, we show that a broad class of pedigree pairs, namely those with different numbers of necessary edges, are identifiable, and the necessary edges can be obtained in polynomial time. We also introduce a class of pedigrees, i.e. diploid Wright-Fisher genealogies, which are provably identifiable.

In order to effectively deal with non-identifiable pedigrees, we can use Bayes rule to obtain the posterior as a function of the likelihood and the prior. Some mild conditions on the prior mean that the posterior will distinguish among the potential pedigrees. The class of dWF pedigrees provides such a prior. Furthermore, we could use Theorem 2 as a guide to integrate over the uncertainty we have about a pedigree structure.

References

1. Abecasis, G.R., Cherny, S.S., Cookson, W.O., et al.: Merlin-rapid analysis of dense genetic maps using sparse gene flow trees. Nature Genetics 30, 97–101 (2002)
2. Bourgain, C., Hoffjan, S., Nicolae, R., et al.: Novel case-control test in a founder population identifies p-selectin as an atopy-susceptibility locus. American Journal of Human Genetics 73(3), 612–626 (2003)
3. Browning, S., Browning, B.L.: On reducing the statespace of hidden Markov models for the identity by descent process. Theoretical Population Biology 62(1), 1–8 (2002)
4. Coop, G., Wen, X., Ober, C., et al.: High-Resolution Mapping of Crossovers Reveals Extensive Variation in Fine-Scale Recombination Patterns Among Humans. Science 319(5868), 1395–1398 (2008)
5. Donnelly, K.P.: The probability that related individuals share some section of genome identical by descent. Theoretical Population Biology 23(1), 34–63 (1983)
6. Fishelson, M., Dovgolevsky, N., Geiger, D.: Maximum likelihood haplotyping for general pedigrees. Human Heredity 59, 41–60 (2005)
7. Geiger, D., Meek, C., Wexler, Y.: Speeding up HMM algorithms for genetic linkage analysis via chain reductions of the state space. Bioinformatics 25(12), i196 (2009)
8. Kirkpatrick, B., Kirkpatrick, K.: Optimal State-Space Reduction for Pedigree Hidden Markov Models. ArXiv e-prints (February 2012)
9. Kirkpatrick, B., Li, S.C., Karp, R.M., Halperin, E.: Pedigree Reconstruction Using Identity by Descent. In: Bafna, V., Sahinalp, S.C. (eds.) RECOMB 2011. LNCS, vol. 6577, pp. 136–152. Springer, Heidelberg (2011)
10. Lauritzen, S.L., Sheehan, N.A.: Graphical models for genetic analysis. Statistical Science 18(4), 489–514 (2003)
11. McPeek, M.S.: Inference on pedigree structure from genome screen data. Statistica Sinica 12(1), 311–336 (2002)
12. McPeek, M.S., Sun, L.: Statistical tests for detection of misspecified relationships by use of genome-screen data. Amer. J. Human Genetics 66, 1076–1094 (2000)
13. Pinto, N., Silva, P.V., Amorim, A.: General derivation of the sets of pedigrees with the same kinship coefficients. Hum. Hered. 70(3), 194–204 (2010)
14. Sobel, E., Lange, K.: Descent graphs in pedigree analysis: Applications to haplotyping, location scores, and marker-sharing statistics. American Journal of Human Genetics 58(6), 1323–1337 (1996)
15. Stankovich, J., Bahlo, M., Rubio, J.P., et al.: Identifying nineteenth century genealogical links from genotypes. Human Genetics 117(2-3), 188–199 (2005)

16. Sun, L., Wilder, K., McPeek, M.S.: Enhanced pedigree error detection. Hum. Hered. 54(2), 99–110 (2002)
17. Thatte, B.D.: Reconstructing pedigrees: some identifiability questions for a recombination-mutation model. ArXiv e-prints (August 2010)
18. Thompson, E.A.: The estimation of pairwise relationships. Annals of Human Genetics 39(2), 173–188 (1975)
19. Thompson, E.A.: Pedigree Analysis in Human Genetics. Johns Hopkins University Press, Baltimore (1985)
20. Thornton, T., McPeek, M.S.: Case-control association testing with related individuals: A more powerful quasi-likelihood score test. American Journal of Human Genetics 81, 321–337 (2007)
21. Thornton, T., McPeek, M.S.: ROADTRIPS: case-control association testing with partially or completely unknown population and pedigree structure. American Journal of Human Genetics 86(2), 172–184 (2010)

Iterative Piecewise Linear Regression to Accurately Assess Statistical Significance in Batch Confounded Differential Expression Analysis

Juntao Li[1,2], Kwok Pui Choi[2], and R. Krishna Murthy Karuturi[1,*]

[1] Computational & Mathematical Biology, Genome Institute of Singapore
[2] Department of Statistics and Applied Probability, National University of Singapore
{lij9,karuturikm}@gis.a-star.edu.sg, stackp@nus.edu.sg

Abstract. Batch dependent variation in microarray experiments may be manifested through systematic shift in expression measurements from batch to batch. Such a systematic shift could be taken care of by using an appropriate model for differential expression analysis. However, it poses greater challenge in the estimation of statistical significance and false discovery rate (FDR), if the batches are confounded (collinear) with the biological groups of interest. Batch confounding problem occurs commonly in the analysis of time-course data or data from different laboratories. We demonstrate that batch confounding may lead to incorrect estimation of the expected statistics. In this paper, we propose an *iterative piecewise linear regression* (iPLR) method, a major extension of our previously published *Stepped Linear Regression* (SLR) method, in the context of SAM to re-estimate the expected statistics and FDR. iPLR can be applied to one-sided or two-sided statistics based tests. We demonstrate the efficacy of iPLR on both simulated and real microarray datasets. iPLR also provides a better interpretation of the linear model parameters.

1 Background

Batch dependent systematic variations or batch effects are commonly observed across multiple batches of microarray experiments. Batch effect influences the expression measurements of all genes in the arrays; the effect on a single gene is random but similar across all arrays in the batch and different from other genes and batches [1]. It has been observed by many researchers that the established normalization and preprocessing methods cannot fully eliminate batch effects [2] and hence developed procedures to account for batch effects at probe level in the differential expression models [3,4].

A few popular methods are SVD (Singular Value Decomposition) / PCA (Principal Component Analysis) [3], DWD (Distance Weighted Discrimination) [4] and empirical Bayes methods [2] which treat batch as a factor assuming that the experimental batches are not confounded with the biological groups of interest i.e. batch and treatment variables are not collinear. In other words, each

---

* Corresponding author.

batch contains arrays of samples from different biological groups. However the problem is not amenable to such analysis if the biological groups are confounded with that of the batches i.e., the arrays in a batch receive samples from one biological group of interest and the arrays in the other batch contain samples from the other biological group of interest. It results in collinearity of batch and treatment variable which means the above methods are not applicable. It is unavoidable in many practical situations as one wants to compare the data from one experiment or laboratory to the data from another experiment or laboratory which essentially means batch confounded biological groups. Time course experiments spread over long time horizons may also result in batch confounding when samples from different time-points are compared for change of expression. Similarly, batch confounding is unavoidable in huge experiments even though all groups were generated in the same laboratory.

Batch confounding has severe influence on differential expression analysis as the biologically differentially expressed genes are mixed up with large number of mere batch affected expression measurements. Even after microarray data preprocessing and normalization, batch confounding still exists in the data. It may lead to gross incorrect estimation of statistical significance, i.e. false discovery rate (FDR), to an intolerable limit as several batch affected biologically irrelevant genes will also have significantly lower p-values. This is true irrespective of whether the statistical significance is assessed using resampling as in SAM (Significance Analysis of Microarrays) [5] or parametric distribution as in LIMMA (Linear Models for Microarray Data) [6]. In the absence of gold standard positive and negative gene sets in genome-wide expression studies, FDR, being an important parameter, needs to be accurately estimated. For example, FDR has been used to estimate the effects of certain treatment or condition on a cell culture via the number of genes passed the FDR cut-off [7]. Hence it is important to estimate FDR as accurately as possible even in the batch confounded data analysis to facilitate correct conclusions on the significantly affected genes.

In [8], we used stepped linear regression (SLR) to address FDR estimation in batch confounded data. However, the statistical significance and π_0 estimation are sensitive to the choice of the predefined cutoff parameter in SLR. Moreover, SLR is not applicable to one-sided tests. To address these issues, we developed a method called *iterative piecewise linear regression* (iPLR) which is a major modification of SLR. iPLR, similar to SLR, re-estimates the expected differential expression statistics under the assumption that the expression difference due to batch variation is smaller than that of the biological variation. FDR is estimated based on the re-estimated expected statistics i.e. the null distribution is re-estimated. After applying iPLR, we can get accurate significance assessment and biologically significant genes. Moreover, our method provides a better interpretation for the linear model in this paper and incorporated procedure to handle one-sided tests.

We present our iPLR in the context of SAM [5]. SAM is a statistical technique for finding significantly differentially expressed genes in microarray experiments. SAM assigns a score called d-score to each gene on the basis of change in gene expression relative to the standard deviation of replicated measurements. The genes with d-scores greater than certain threshold are declared to be differentially expressed. This threshold corresponds to certain false discovery rate (FDR), the percentage of genes identified by chance for the given d-score by permuting the class labels. Those genes will be regarded as significantly biologically relevant genes according to the data. However, in the case of batch confounding, many of them may not be actually relevant to the underlying biology of interest. Our iPLR helps correct this artifact.

Though iPLR is presented in the context of SAM analysis for simplicity, the method is equally applicable to any reasonable statistical procedure based on resampling strategy. Our results show that iPLR is effective in estimating FDR accurately both in simulated as well as real data with batch confounding. We demonstrate how iPLR corrects for the incorrectly magnified assessment of effects of certain conditions or treatments on gene expression.

2 Methods

The iterative piecewise linear regression (iPLR) is based on the following assumptions: (a) for those biologically differentially expressed genes, the biological influence is much greater than those of the batch effects' influence; (b) the batch effect is independent of biological effect; and (c) the proportion of biologically non-differentially expressed genes (π_0) is larger than 0.5.

2.1 Re-estimate the Expected Statistics

For a SAM analysis based on a two-sided test statistics [9], we first obtain the SAM computed statistic d_i for each gene $g_i (i = 1, 2, \ldots, n)$, and without loss of generality we assume $d_1 \leq d_2 \leq \ldots \leq d_n$. To compute FDR, SAM performs permutations to estimate the expected values of these order statistics $\bar{d}_1 \leq \bar{d}_2 \leq \ldots \leq \bar{d}_n$.

When batch effect exists in the data and is confounded with biological effect, we propose the linear model between observed statistics d_i and expected statistics $\bar{d}_i$ as follows:

$$d_i = a\bar{d}_i + b + c_i + e_i, \tag{1}$$

where a and b are batch effect factors, c_i is the biological effect factor and e_i is the model error ($i = 1, 2, .., n$). $c_i = 0$ if gene g_i has no differential expression between different classes of the experiment.

It is difficult to estimate batch effect factors without knowing biological effect factor c_i. Therefore, we simply approximate c_i by a linear function in $\bar{d}_i$ when $c_i \neq 0$. Based on (1), we are led to consider

$$d_i = \begin{cases} a\bar{d}_i + b + e_i & \text{if } c_i = 0 \\ a\bar{d}_i + b + c_{a+}\bar{d}_i + c_{b+} + e_i & \text{if } c_i > 0 \\ a\bar{d}_i + b + c_{a-}\bar{d}_i + c_{b-} + e_i & \text{if } c_i < 0 \end{cases} \tag{2}$$

where c_{a+}, c_{b+} and c_{a-}, c_{b-} are the coefficients of the linearity between c_i and $\bar{d}_i$. From (2), we perform the iterative piecewise linear regression (iPLR) to estimate the batch effect factors a and b. In iPLR, we use iterative approach to identify the regression section with $c_i = 0$ to estimate the batch effect factors a and b. The proportion of non-differentially expressed genes, π_0, will be estimated by $\frac{\#\{c_i=0\}}{n}$.

After estimating the batch effect factors a and b, we can re-estimate the expected statistics to eliminate the batch effect in FDR estimation. The model in (1) is about comparing quantiles or ordered statistics (origin is about the 50th percentile or median which is usually close to 0) for the observed test statistics (a combination of null hypotheses and alternative hypotheses) and the test statistics of null distribution obtained by resampling. If π_0 is very close to 1, then quantiles of both distributions will be very close to each other. So the test statistics and the expected statistics will lie close to the diagonal of the observed statistic quantiles versus the expected statistic quantiles plot. Then we can set $a = 1$ to eliminate the batch effect. However, when π_0 is not very close to 1 (for example, $\pi_0 = 0.7$), we have to consider the fact that the distribution of the test statistics is a mixture of the distribution of expected statistics (null distribution multiplied by π_0) and the alternate distribution. If the null distribution of test statistics is uniform, the slope for the observed statistic quantiles versus the expected statistic quantiles plot for $c_i = 0$ is π_0^{-1}. In typical hypothesis testing in microarray experiments, the null distributions are unimodal. In these typical cases, we can approximate this slope as π_0^{-1} to achieve a better estimate of the FDR (we omit the proof due to space limit). Therefore, we set $\hat{a} = \hat{\pi}_0^{-1}$ to eliminate the batch effect. Then, we define the re-estimated expected order statistics as

$$\bar{d}_i^* = \hat{\pi}_0(\hat{a}\bar{d}_i + \hat{b}), \tag{3}$$

and then (1) will be rewritten as

$$d_i = \hat{\pi}_0^{-1}\bar{d}_i^* + c_i + e_i. \tag{4}$$

This is the linear model after eliminating the batch effect factors a and b.

2.2 Iterative Piecewise Linear Regression (iPLR)

The iPLR takes observed statistics d_i and expected statistics $\bar{d}_i$ as input data, and uses iterative approach to search for the best piecewise linear regression model fit in (2). The batch effect factors and π_0 are estimated by this model, then iPLR re-estimates the expected order statistics $\bar{d}_i^*$ by (3). Finally, iPLR outputs the re-estimated FDR. The work flow for iPLR is illustrated in Figure 1(A).

By assumption (c), there are more than 50% non-differentially expressed genes in the dataset. Therefore, the baseline, regression line for $c_i = 0$ part in iPLR, will include more than half of the data and batch effect factors a and b will be estimated from this portion of the data.

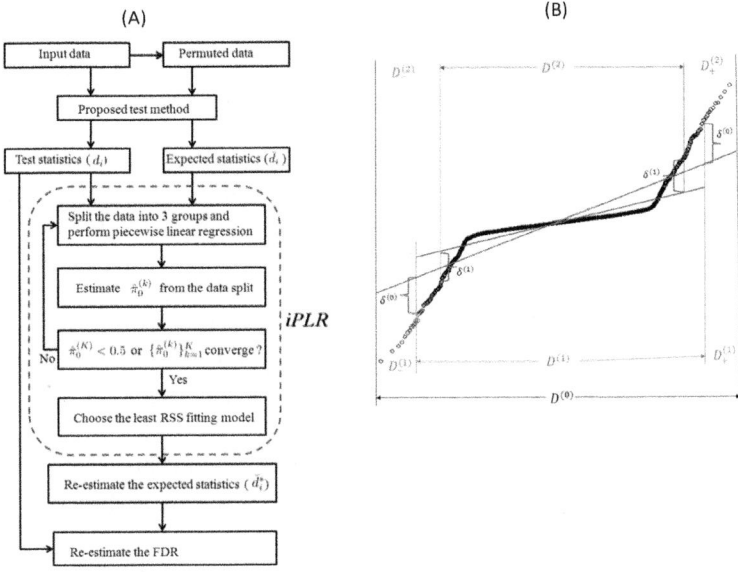

Fig. 1. (A)Work flow for iPLR. (B)Illustration of first two iterations in iPLR. They can be generalized to the following iterations.

At the initialization step, we set $\hat{\pi}_0^{(0)} = 1$, and define the data split

$$S_0 = \{D^{(0)}, D_-^{(0)}, D_+^{(0)}\}, \tag{5}$$

where $D^{(0)} := \{(d_i, \bar{d}_i), i = 1, 2, \ldots, n\}$ is the initial baseline dataset and $D_-^{(0)} = D_+^{(0)} = \phi$ (empty set).

We perform the linear regression in $D^{(0)}$ and this is the baseline in the initial step. Let $\delta^{(0)}$ be the standard deviation of baseline regression errors. The next baseline dataset $D^{(1)}$ is generated by excluding the data points which are far away from the baseline. We define the split $S_1 = \{D^{(1)}, D_-^{(1)}, D_+^{(1)}\}$ as

$$\begin{aligned} D^{(1)} &= \{(d_i, \bar{d}_i)| -z\delta^{(0)} \leq l_i^{(0)} \leq z\delta^{(0)}\}, \quad \hat{\pi}_0^{(1)} = \frac{\#\{D^{(1)}\}}{n} \\ D_-^{(1)} &= \{(d_i, \bar{d}_i)|l_i^{(0)} < -z\delta^{(0)}\}, \quad D_+^{(1)} = \{(d_i, \bar{d}_i)|l_i^{(0)} > z\delta^{(0)}\}, \end{aligned} \tag{6}$$

where $l_i^{(0)}$ is the distance between $(d_i, \bar{d}_i)$ to the regression baseline, z is a pre-defined boundary cutoff. $D_-^{(1)}$ and $D_+^{(1)}$ indicate $c_i < 0$ and $c_i > 0$ and generally distributed at the left and right tails of the data. Then we perform 3-piece linear regression for $D^{(1)}$, $D_-^{(1)}$ and $D_+^{(1)}$ separately. We repeat the above procedure to generate $S_k = \{D^{(k)}, D_-^{(k)}, D_+^{(k)}\}$ and $\hat{\pi}_0^{(k)}$ from $S_{k-1} = \{D^{(k-1)}, D_-^{(k-1)}, D_+^{(k-1)}\}$ and $\hat{\pi}_0^{(k-1)}$ until convergence is reached. The procedure is said to converge at $k = K$ if $\hat{\pi}_0^{(K)} < 0.5$ or the sequence $\{\hat{\pi}_0^{(k)}\}_{k=1}^K$ converges to a constant i.e. $|\hat{\pi}_0^{(K)} - \hat{\pi}_0^{(K-1)}| < 10^{-3}$.

Among the sequence of data-splits S_k for $k = 1, \ldots, K$, we choose the split that gives lowest fitting RSS (residual sum of squares) for 3-piece linear regression, and $\hat{\pi}_0$ is estimated by this split. In iPLR, there are only two free parameters to be pre-selected, boundary cutoff z and stopping cutoff 10^{-3}. We can set $z = 3$ implying the 3 standard deviation boundary. Different choices of z and stopping cutoff influence the search bandwidth and the number of iterations without affecting the outcome significantly. The first two steps (from $D^{(0)}$ to S_1, and S_1 to S_2) for iPLR are illustrated in Figure 1(B).

The following steps detail the computational procedure for iPLR.

1. Set $\hat{\pi}_0^{(0)} = 1$. Perform the linear regression for $D^{(0)}$ to determine the baseline. Compute the distance of each point to the baseline and the standard deviation of baseline regression errors $\delta^{(0)}$.
2. Calculate $\hat{\pi}_0^{(1)}$ using (6) and perform a 3-piece linear regression for $D^{(1)}$, $D_-^{(1)}$ and $D_+^{(1)}$ separately. Compute the standard deviation of baseline regression errors $\delta^{(1)}$ and RSS of the 3-piece regression.
3. Repeat step 2 to obtain S_k and $\hat{\pi}_0^{(k)}$ from $S_{(k-1)}$ until $k = K$ for which $\hat{\pi}_0^{(K)} < 0.5$ or $|\hat{\pi}_0^{(K)} - \hat{\pi}_0^{(K-1)}| < 10^{-3}$.
4. Choose the estimation of $\hat{\pi}_0$ as $\hat{\pi}_0^{(k)}$ with the least RSS fitting for the iPLR, and the batch effect factors a and b are estimated in the baseline regression using this $\hat{\pi}_0$.
5. Re-estimate the expected statistics using (3). Re-estimate the FDR for each gene.

2.3 iPLR for One-Sided Test

The above procedure is designed for 3-piece linear regression for a two-sided test. If the test statistics are from one-sided test, the biological effect $c_i \geq 0, i = 1, 2, \ldots, n$. Therefore, (2) becomes

$$d_i = \begin{cases} a\bar{d}_i + b + e_i & \text{if } c_i = 0 \\ a\bar{d}_i + b + c_{a+}\bar{d}_i + c_{b+} + e_i & \text{if } c_i > 0 \end{cases}. \quad (7)$$

Then iPLR procedure can be modified to take care of this one-sided test. Indeed, we only need to set $D_-^{(k)} = \phi, k = 1, 2, \ldots, K$ in the definition of the split. Subsequently, iPLR performs a 2-piece linear regression by removing one piece for $D_-^{(k)}, k = 1, 2, \ldots, K$ and the rest of iPLR procedure remains the same.

3 Results

We demonstrate that the effects of batch confounding on FDR estimation and the efficacy of iPLR in alleviating it using both simulated data and real data. Using simulated data, we show that iPLR does not introduce any artifacts in FDR estimation for data without batch confounding, and that iPLR corrects the influence of batch confounding if it is present in the data.

3.1 Two-Class Simulations

A two-group data was simulated using the following rule

$$x_{ijk} = \mu_{ik} + \eta_{ik} + \epsilon_{ijk}, \tag{8}$$

where x_{ijk} is an expression measurement of gene $g_i(i = 1, 2, \ldots, n = 10000)$ in sample $S_j(j = 1, 2, \ldots, 10)$ in group $G_k(k = 1, 2)$, and ϵ_{ijk} are standard normal noise. The biological effect μ_{ik} and global batch effect η_{ik} (η_{ik} is the effect of batch on the gene expression x_{ijk} which is different from the effect of batch confounding on the relationship between d_i and $\bar{d}_i$) are defined as follows:

$$\begin{aligned}
&\mu_{i1} = 0 \text{ for } 1 \leq i \leq n, \\
&\eta_{i1} = 0 \text{ for } 1 \leq i \leq n, \\
&\mu_{i2} = \begin{cases} \theta_{i1} \sim N(0, \sigma_\mu^2) & \text{for } 1 \leq i \leq m \\ 0 & \text{for } m < i \leq n \end{cases}, \\
&\eta_{i2} = \theta_{i2} \sim N(0, \sigma_\eta^2) \text{ for } 1 \leq i \leq n,
\end{aligned} \tag{9}$$

where m is the number of differentially expressed genes and n is the total number of genes. The model parameters signify that the batch effect and biological effect are independent and the level of differential expression and batch effect varies from gene to gene. The fraction $1 - (m/n)$ is π_0: the fraction of non-differentially expressed genes or genes not affected by biological treatments.

We simulated two different datasets of $n = 10000$ genes each using two different settings for two choices of $\sigma_\eta = 0$ and 2 while keeping $\sigma_\mu = 4$ and $\pi_0 = 0.7$. Dataset with $\sigma_\eta = 0$ is simulated without batch effects and analyzed with our procedure in order to find out whether our procedure would introduce any artifacts in FDR estimation or not (i.e., FDR estimates before and after re-estimation of $\bar{d}_i$ should be close to each other for non-batch affected data). Dataset with $\sigma_\eta = 2$ is batch effect confounded with reasonably different values of π_0 whose FDR estimates before re-estimation are expected to be far from reality while the FDR estimates after re-estimation are expected to be close to the reality. We used SAM on each of the four datasets to obtain d-statistics for original as well as permuted data. We applied our procedure on each pair of d-statistic sets.

The estimates of π_0 before and after applying iPLR are 0.236 and 0.8599 for batch confounded dataset. This shows that iPLR succeeds in reducing the batch effects in FDR estimation. Both estimates (before and after applying iPLR) of π_0 for dataset without batch confounded are 0.7902 and 0.7833, agree quite well with the true values. Indeed, the estimates after applying iPLR are slightly better than that of not using iPLR. This shows that iPLR does not introduce any bias in the absence batch effect. The other π_0 estimation methods which are only based on the p-value distribution will also be affected by batch confounded data. For example, we used a cross-validatory approach [10] to estimate π_0 for two datasets and the results ($\hat{\pi}_0 = \{0.8037, 0.3192\}$) are similar to SAM estimation. It shows that p-value distribution based procedures cannot solve the bias introduced by batch confounding.

The estimated FDR and true FDR for both before and after iPLR re-estimation procedure are plotted in Figures 2(A) and 2(B). Ideally, the estimated FDR should be close to the real FDR (smooth black curve), the closer the better. Both plots for dataset without batch confounded are similar and are close to the real FDR curve. The estimated FDR after iPLR is slightly higher than real FDR at FDR> 0.6 which is not important for practical purposes. However, FDR plots for original SAM and after iPLR re-estimation are quite different for batch confounded dataset. Plot for SAM is almost 0 showing how severely the FDR was underestimated. But FDR after iPLR re-estimation is closer to the real FDR, even though FDR still underestimated the real FDR due to the influence of batch confounding on the permutation procedure [11].

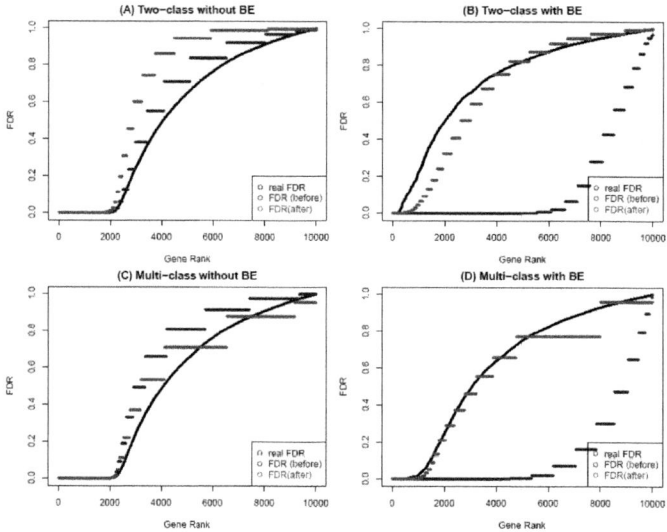

Fig. 2. FDR comparison for simulation data sets two class and multi-class simulations. Black points indicate the real FDR. Blue points indicate estimated FDR before iPLR re-estimation and the red points indicate estimated FDR after iPLR re-estimation. BE indicates batch confounding effects.

3.2 Multi-class Simulations

iPLR is not only applicable to two-class analysis with two batches. We can easily adapt it to analyze multi-class dataset with more than two batches. We do this by considering a 2-piece linear regression instead of a 3-piece linear regression since up-regulated and down-regulated gene groups in two-class analysis will be merged into one single differentially expressed gene group in multi-class analysis.

We simulated three-class datasets using similar procedure as in two-class simulations by adding one more class for each dataset. As in two-class simulations, we simulated 1 dataset without batch effect ($\sigma_\mu = 4, \sigma_{\eta 1} = \sigma_{\eta 2} = 0$) and 1 datasets with batch effect ($\sigma_\mu = 4, \sigma_{\eta 1} = \sigma_{\eta 2} = 1$) for $\pi_0 = 0.7$. We used SAM multi-class method to generate d-score and permuted score, and then performed iPLR (2-piecewise linear regression) to re-estimate the FDR. FDR comparisons are shown in Figure 2(C) and 2(D). Similar to the results of two-class simulations, iPLR can accurately estimate the batch effect factors in datasets and the FDR.

3.3 Fission Yeast Data

Having shown the efficacy of iPLR re-estimation on simulated data, we next demonstrate the utility of our iPLR re-estimation on real gene expression data, mip1 mutant (Δmip1) differential expression in S. pombe (or fission yeast) compared to its wild-type. The data was obtained from [12] containing 28 wt/wt spotted two-color array data and 6 Δmip1/wt data for $\sim$ 5000 open reading frames (ORFs). The purpose is to find the genes influenced by mip1 mutation (Δmip1). The data have been global and local normalized. The wt/wt data contains two batches of equal number of arrays which we call wt1/wt1 (or wt-rep1) and wt2/wt2 (or wt-rep2). SAM estimates π_0 to be 0.255 on wt1/wt1 vs. wt2/wt2 data, while it should be 1 as both wt1 and wt2 samples are the same except that they were hybridized into two different batches of arrays at two different times. After application of our iPLR procedure, the estimate of π_0 is improved to 0.997.

We analyzed wt/wt (combining wt1/wt1 and wt2/wt2) versus Δmip1/wt using SAM to identify differentially expressed genes, the corresponding SAM plot is shown in Figure 3(A1-2). Δmip1/wt was hybridized altogether on a different batch of arrays at completely different time. This resulted in batch effects again and the underestimation of π_0 (0.22). It demonstrates that even using the data which was preprocessed and normalized, the batch confounding effect persists on FDR estimation. We applied our iPLR procedure on this dataset. π_0 estimate is closer to 1 (0.974) than the otherwise unrealistic estimate (0.22) by SAM alone. Results of comparing estimated FDR before and after iPLR are shown in Figures 3(A3). As shown in these figures, FDR estimation before iPLR is extremely low for most genes, but after iPLR procedure, they are closer to what are expected. This results showed that iPLR is a practically useful technique.

3.4 Human Ewing Tumor Data

Another dataset we analyzed is human Ewing tumor data from [13]. It is affymetrix microarray data of A673 cells treated with DMSO vehicle control expression profiled at 24 hours (6 replicates), 3 days (5 replicates), and 5 days (6 replicates). The data have been quantile normalized. Since the experiments were performed on three different days, it is unrealistic to assume no batch confounding effect. In fact, the data from 3 different days are well separated by the day of the sample from the array clustering result. It suggests that biological effect and batch effect are confounded.

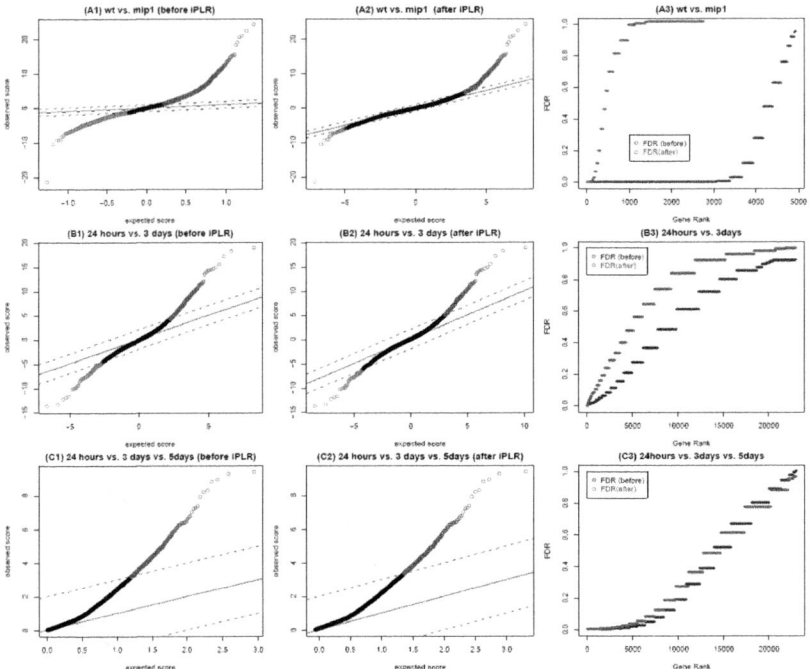

Fig. 3. (A1) The SAM plot before re-estimation for wt/wt vs. Δmip1/wt dataset. (A2) The SAM plot after re-estimation for wt/wt vs. Δmip1/wt dataset. (A3) FDR comparison for wt/wt vs. Δmip1/wt dataset. (B1) The SAM plot before iPLR re-estimation for 24 hours vs. 3 days dataset. (B2) The SAM plot after re-estimation for 24 hours vs. 3 days dataset. (B3) FDR comparison for 24 hours vs. 3 days dataset and 24 hours vs. 5 days dataset. (C1) The SAM plot before iPLR re-estimation for 24 hours vs. 3 days vs. 5 days dataset. (C2) The SAM plot after re-estimation for 24 hours vs. 3 days vs. 5 days dataset. (C3) FDR comparison for simulation 24 hours vs. 3 days vs. 5 days dataset. Blue points indicate estimated FDR before iPLR re-estimation and the red points indicate estimated FDR after iPLR re-estimation.

First we analyzed the datasets from 24 hours and 3 days using SAM alone, and using iPLR combined with SAM. SAM estimated π_0 to be 0.68, while we expect a higher π_0 as the vehicle control data between two days should not be very different. After application of iPLR, the estimation of π_0 is 0.84 closer to what is expected. The SAM plots for theses two comparisons and the comparisons of estimated FDR before and after applying iPLR are shown in Figures 3 (B1-3).

We also compared these three groups: 24 hours, 3 days and 5 days. It is seen that FDR is improved after applying iPLR and estimation of π_0 is closer to real π_0. The SAM plots and the comparisons of estimated FDR before and after applying iPLR are shown in Figure 3 (C1-3). Human tumor data also demonstrates that preprocessing and normalization cannot remove the effect of batch confounding on FDR estimation.

4 Conclusions

We proposed *iterative piecewise linear regression* (iPLR) to correct the bias introduced in the estimation of null distribution when experimental batches are confounded with treatment groups of interest. In FDR estimation, this correction is critical in gene expression studies where one wants to compare data obtained from different laboratories or from the same laboratory but collected at different times. Our results on the real data, which was preprocessed and normalized appropriately, demonstrated that the effect of batch confounding continues to exist in the normalized data also and leads to erroneous FDR estimation. iPLR plays an important role in such a case, it works at the downstream of a resampling based method such as SAM. In iPLR, we assume that batch effects are small and influences all spots on the array in unexpected but definite manner which varies from batch to batch. Under this assumption which was used in the popularly used location/scale model for batch effects [2], the influence is mainly on the estimation of FDR via badly estimated null distribution, underestimated proportion of non-differentially expressed genes and by the inevitable influence of change of mean value on permutation procedure. The SAM manual cites this behavior as one that could be biologically more meaningful to be left to the biologists to decide. When it is reasonable to assume in gene expression studies that π_0 is more than 0.5, and under realistic assumptions of low batch effects, we proposed iPLR method to resolve this problem. iPLR procedure is equally applicable to any differential expression analysis procedure for any number of classes. It is only for the sake of simplicity in describing our methodology and evaluating the results in the context of SAM (a widely used method for differential expression analysis).

Similar problem has been addressed in the evaluation of enrichment of gene sets in a list of genes [14], the GSA (Gene Set Analysis) algorithm. GSA handles the problem by making the mean and standard deviations of the distributions of both observed statistics and permutation statistics to be the same. The idea is simple and effective for GSA because π_0 in GSA is generally close to 1. However, it may not work well in several gene expression studies if π_0 is well below 1. This may lead to severe overestimation of standard deviation and make the idea ineffective for this purpose. Hence, iPLR may play an important contribution.

We have shown the efficacy of our iPLR method on both simulated and real data. These results demonstrate that iPLR combined with SAM is robust to batch confounding effects of treatments. Results in simulation study suggest that iPLR improves the estimate of π_0 to some extent than using SAM alone even in the absence of batch confounding effects. More extensive experiments will be conducted in the future to verify this hypothesis. Furthermore, there is still room to improve iPLR. As shown in Figure 2, re-estimated FDR deviates considerably from real FDR for dataset without batch confounding. However, iPLR in its current form is still useful in making the right choice of differential expression significance threshold in the wake of better and meaningful FDR estimation.

References

1. Li, C., Wong, W.H.: Dna-chip analyzer (dchip). In: The Analysis of Gene Expression Data: Methods and Software, pp. 28–46. Springer, Heidelberg (2003)
2. Johnson, W.E., Li, C., Rabinovic, A.: Adjusting batch effects in microarray expression data using empirical bayes methods. Biostatistics 8, 118–127 (2007)
3. Alter, O., Brown, P.O., Botstein, D.: Singular value decomposition for genome-wide expression data processing and modeling. Proc. Natl. Acad. Sci. USA 97, 10101–10106 (2000)
4. Benito, M., Parker, J., Du, Q., Wu, J., Xiang, D., Perou, C.M., Marron, J.S.: Adjustment of systematic microarray data biases. Bioinformatics 20, 105–114 (2004)
5. Tusher, V.G., Tibshirani, R., Chu, G.: Significance analysis of microarrays applied to the ionizing radiation response. Proc. Natl. Acad. Sci. USA 98, 5116–5121 (2001)
6. Smyth, G.K.: Linear models and empirical bayes methods for assessing differential expression in microarray experiments. Statistical Applications in Genetics and Molecular Biology 3(1) (2004)
7. Storey, J.D., Tibshirani, R.: Statistical significance for genomewide studies. Proc. Natl. Acad. Sci. USA 100, 9440–9445 (2003)
8. Li, J., Liu, J., Karuturi, R.K.M.: Stepped linear regression to accurately assess statistical significance in batch confounded differential expression analysis. Bioinformatics Research and Applications, 481–491 (2008)
9. Chu, G., Narasimhan, B., Tibshirani, R., Tusher, V.: SAM, significance analysis of microarrays. Users guide and technical document
10. Celisse, A., Robin, S.: A cross-validation based estimation of the proportion of true null hypotheses. Journal of Statistical Planning and Inference 140, 3132–3147 (2010)
11. Xie, Y., Pan, W., Khodursky, A.B.: A note on using permutation-based false discovery rate estimates to compare different analysis methods for microarray data. Bioinformatics 21, 4280–4288 (2005)
12. Chu, Z., Li, J., Eshaghi, M., Karuturi, R.K.M., Lin, K., Liu, J.: Adaptive expression responses in the pol-gamma null strain of s. pombe depleted of mitochondrial genome. BMC Genomics 8, 323 (2007)
13. Stegmaier, K., Wong, J.S., Ross, K.N., Chow, K.T., Peck, D., Wright, R.D., Lessnick, S.L., Kung, A.L., Golub, T.R.: Signature-based small molecule screening identifies cytosine arabinoside as an EWS/FLI modulator in ewing sarcoma. PLoS Medicine 4, e122 (2007)
14. Efron, B., Tibshirani, R.: On testing the significance of sets of genes. The Annals of Applied Statistics 1, 107–129 (2007)

Reconstruction of Network Evolutionary History from Extant Network Topology and Duplication History

Si Li[1], Kwok Pui Choi[1,2], Taoyang Wu[1], and Louxin Zhang[1]

[1] Department of Mathematics
[2] Department of Statistics and Applied Probability,
National University of Singapore, Singapore 119076
{g0800874,stackp,matwt,matzlx}@nus.edu.sg

Abstract. Genome-wide protein-protein interaction (PPI) data are readily available thanks to recent breakthroughs in biotechnology. However, PPI networks of extant organisms are only snapshots of the network evolution. How to infer the whole evolution history becomes a challenging problem in computational biology. In this paper, we present a likelihood-based approach to inferring network evolution history from the topology of PPI networks and the duplication relationship among the paralogs. Simulations show that our approach outperforms the existing ones in terms of the accuracy of reconstruction. Moreover, the growth parameters of several real PPI networks estimated by our method are more consistent with the ones predicted in literature.

1 Introduction

Recent progress in experimental systems biology provides us with an unprecedented amount of genome-wide protein-protein interaction (PPI) data [7]. In order to obtain a deeper insight into the molecular machinery behind these interactions, many network models have been proposed to study or model PPI evolution [2, 15, 18]. However, PPI networks of extant organisms are only snapshots of network evolution, and inferring the whole network evolution history remains a challenging problem in computational biology [10].

Unlike many networks studied in technology and sociology, the main growth mechanism of PPI network is gene duplication and divergence [17]: when a new node is added to the network, it copies all the interactions of an existing node designed as the anchor node; subsequently some edges adjacent to one of these two nodes are randomly lost. This mechanism was explicitly converted to a network growth model by Vazquez et al. in [16]. Since then many extensions have been put forth, see for examples, [3, 11, 14]. Here we shall focus on a particular one called duplication-mutation with complementarity (DMC), which is the best model to fit the *D. melanogaster* (fruit fly) PPI network according to a recent study by Middendorf et al. [9].

When a growth model is fixed, the problem of reconstructing the evolutionary history of an observed network is to infer the relative order of the nodes according to which the network evolved (see Section 2.2 for definitions). Better understanding of this problem can provide further insights into not only how PPI networks are formed, but also how they will possibly evolve in the future. Several approaches to address this problem have been proposed in recent years. In order to obtain better ways of predicting protein modules, Dutkowski and Tiuryn introduced a Bayesian network framework to infer the posterior probability of interactions between ancestral nodes based on a duplication and speciation model [4]. A similar approach was used by Pinney [13] to infer ancestral interactions between bZIP proteins. Gibson and Goldberg proposed a merging algorithm to reconstruct the evolutionary history of PPI networks using gene trees [6]. A novel framework for estimating the topology of the ancestral networks based on maximal likelihood was presented by Navlakha and Kingsford in [10]. Recently, Patro et al. [12] used a maximal parsimony approach that appends edges in observed networks to duplication history forest.

Here we introduce a new history inferring framework based on the maximal likelihood principle. In contrast to the model-based methods in [10], our approach incorporates not only the topology of observed networks, but also the duplication history of the proteins contained in the networks. Although the evolution of topology is often determined by some growth mechanisms, the duplication history of the proteins can be inferred independently by phylogenetic studies [12, 13]. After establishing some theoretical results concerning the DMC model, we reduce the problem of finding most probable history of ancient networks to an optimization problem, and propose some efficient heuristic algorithms to solve the latter problem. Simulations show that our method provides better inference than the ones in [10]. Moreover, we also applied our algorithm to the PPI networks of *S. cerevisiae* (budding yeast), *D. melanogaster* and *C. elegans* (worm), and the growth parameters obtained by our approach are more consistent with the ones predicted in [5, 17].

The rest of the paper is organized as follows: Section 2 provides the framework of reconstruction, including the technical background and the inference method. In Section 3 we present the inference results for simulations and real data sets. We conclude in Section 4 with a brief discussion and some possible related research directions.

2 Methods

2.1 Modeling Network Evolution

In the DMC model $\mathcal{M} := \mathcal{M}(p_c, p)$, where p_c and p are the two parameters that specify the model, we start with an initial graph G_0, the so-called seed graph. At each time step t, the graph G_t is obtained from G_{t-1} by the following procedures (see Fig. 1 for an illustration): (1) (Duplication) A node u_t is chosen uniformly at random from the set of nodes in G_{t-1}, and a new node v_t is added and connected to every neighbor of u_t. Here u_t and v_t are often referred to as the anchor node

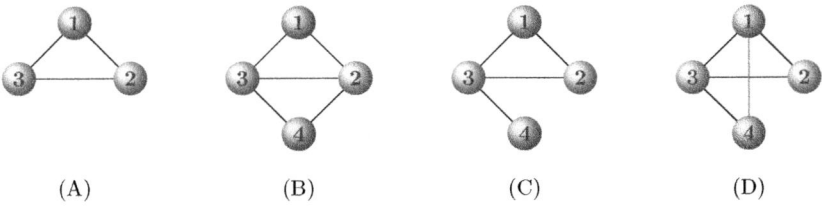

Fig. 1. Illustration of the DMC model. (B) is obtained from (A) by one duplication step, with node 1 (represented in maroon) as the anchor node and node 4 as the duplicate node (represented in purple); the probability that node 1 is chosen as the anchor node is 1/3 because the network in (A) contains three nodes. (C) is obtained from (B) by the mutation step, which occurs with probability $p(1-p)/2$. (D) is obtained from (C) by the complementarity step, which occurs with probability p_c.

and duplicate node at step t, respectively. (2) (Mutation) For each neighbor of u_t, say w, we choose one edge from (u_t, w) and (v_t, w) with equal probability, and this chosen edge is deleted with probability $1 - p$. (3) (Complementarity) The nodes u_t and v_t are connected with probability p_c.

Note that the DMC model is Markovian, that is, the probability of obtaining G_t when G_{t-1} is given depends solely on the parameters of $\mathcal{M}$. For example, denoting the network (A) and (D) in Fig 1 by G_{t-1} and G_t, respectively, then the probability $\mathbb{P}(G_t|G_{t-1}, \mathcal{M})$ that G_t is evolved from G_{t-1} by one step under the model $\mathcal{M}$ is $p(1-p)p_c/2$.

2.2 History Reconstruction

Given an observed network G, a *growth history* $\mathcal{H}$ of G is a graph sequence $(G_0, G_1, \cdots, G_n)$ such that $G_n = G$ and for each index t in $\{1, \cdots, n\}$, the graph G_t can be obtained from G_{t-1} in one step under the DMC model $\mathcal{M}$. The first graph G_0 is referred to as the seed graph of the history. In addition, the number n is called the *span* of the history. Clearly, a history $\mathcal{H}$ induces a unique sequence $\theta := \theta(\mathcal{H})$ of duplicate nodes, that is, $\theta = (v_1, \cdots, v_n)$ such that for all t, node v_t is the unique node in G_t, but not G_{t-1}.

Given a network G, let $\mathcal{H}$ be the growth history we hope to infer. The probability of G being evolved according to history $\mathcal{H}$, when viewed as a function of the unknown history $\mathcal{H}$, is called the *likelihood function* $L(\mathcal{H} \mid G, \mathcal{M})$ that is given by

$$L(\mathcal{H} \mid G, \mathcal{M}) = \prod_{t=1}^{n} \mathbb{P}(G_t|G_{t-1}, \mathcal{M}).$$

We adopt a maximal likelihood approach to infer the history of G as below.

Problem 1. Given a network G together with a natural number n and model $\mathcal{M}$, construct a growth history $\mathcal{H}$ that maximizes the likelihood $L(\mathcal{H} \mid G, \mathcal{M})$ among all histories with span n.

This problem is expected to be difficult since the number of possible histories grows exponentially, and we are not aware of any results concerning whether this problem is polynomial-time solvable. Before introducing a variant of the above problem that is more tractable, we present some necessary tools in the following two subsections.

2.3 Duplication Forest

We begin with duplication history, which is closely related to network history as gene duplication is a major driving force of PPI network evolution [17]. The idea of encoding the duplication history by a forest of binary tree was used in [10, 12]. Patro et al. [12] incorporated duplication history in a parsimony approach to reconstruct network history.

A growth history $\mathcal{H}$ of a PPI network induces a unique duplication forest. Initially, we have a forest Γ_0 consisting of isolated nodes that are identical to the set of nodes in the seed graph. At each step t, the forest Γ_t is obtained from Γ_{t-1} by replacing the anchor node u_t with a cherry $\{u_t, v_t\}$ consisting of u_t and the duplicate node v_t. Here a *cherry* $\{u, v\}$ is referred to a subtree consisting of two leaves u and v and the internal node adjacent to them.

The duplication forest of a PPI network can also be inferred independently without using its growth history. For instance, such a forest can be reconstructed by the phylogenetic relationships between the genes in the network [13]. This observation is key to our investigation.

2.4 Backward Operator

In this subsection, we will introduce a backward operator that is important in our inference framework.

Consider one step in a growth history, that is, a graph G_t obtained from G_{t-1} in one step by using anchor node u_t and duplicate node v_t. Now we want to define a backward operator $\mathcal{R}$ such that G_{t-1} can be determined by this operator and the triplet (G_t, u_t, v_t). To this end, let $\mathcal{R}_{v_t}^{u_t}(G_t)$ be the graph obtained from G_t by merging the two nodes u_t and v_t in G_t, that is, (i) for each neighbor w of v_t such that $w \neq u_t$ and w is not adjacent to u_t, add an edge (w, u_t); (ii) delete the node v_t and all edges incident to it.

Similarly, the backward operator can be applied to the duplication forest, that is, $\mathcal{R}_{v_t}^{u_t}(\Gamma_t)$ is the forest obtained from Γ_t by replacing the cherry $\{u_t, v_t\}$ with the leaf u_t. Note that this definition is consistent with the above one in the following sense: If Γ_t is the duplication forest corresponding to the network G_t, then $\mathcal{R}_{v_t}^{u_t}(\Gamma_t)$ is the duplication forest associated with $\mathcal{R}_{v_t}^{u_t}(G_t)$. When the anchor node u_t is clear from the context, we also write $\mathcal{R}_{v_t}$ for $\mathcal{R}_{v_t}^{u_t}$.

2.5 Growth History with Known Duplication Forest

Using the backward operator introduced above, we shall introduce a scheme to represent a growth history with known duplication forest by a node sequence. Throughout this paper, we use the convention that a node sequence consists of distinct nodes, while a node list may contain repeated nodes.

In general, a node sequence $\theta = (v_1, \cdots, v_n)$ and a duplication forest Γ are said to be *compatible* if there exists a (necessarily unique) sequence $(\Gamma_1^\theta, \cdots, \Gamma_n^\theta)$ of forests such that $\Gamma_n^\theta = \Gamma$, and $\Gamma_{t-1}^\theta = \mathcal{R}_{v_t}(\Gamma_t)$ holds for each $t \in \{1, \cdots, n\}$. Note that a necessary and sufficient condition for θ and Γ being compatible is that v_t belongs to a cherry in Γ_t^θ for each t. Denoting the sibling of v_t in Γ_t^θ, that is, the unique leaf in Γ_t that forms a cherry with v_t, by u_t, we say the list $\pi = (u_1, \cdots, u_n)$ is the *anchor* list determined by Γ and θ.

As mentioned above, a growth history $\mathcal{H} = (G_0, \cdots, G_n)$ specifies a duplicate sequence $\theta = (v_1, \cdots, v_n)$ and a duplication forest Γ. Clearly, the sequence θ and forest Γ must be compatible. On the other hand, given a duplication forest Γ associated with a network G and a sequence θ that is compatible with Γ, then there exists a unique growth history $\mathcal{H}$ such that θ is induced from $\mathcal{H}$. In other words, when the duplication forest Γ is fixed, a growth history $\mathcal{H}$ is uniquely determined by the duplicate sequence θ associated with it. In this context, the likelihood function is defined as

$$L(\theta \mid G, \Gamma, \mathcal{M}) := \prod_{i=1}^{n} \mathbb{P}(G_i^\theta \mid G_{i-1}^\theta, \Gamma, \mathcal{M}),$$

where $\mathbb{P}(G_i^\theta | G_{i-1}^\theta, \Gamma, \mathcal{M})$ is the probability that G_i^θ is evolved from G_{i-1}^θ in one step under the DMC model $\mathcal{M}$ and using the anchor node u_t specified by θ and Γ. Note that in general the probability $\mathbb{P}(G_i^\theta | G_{i-1}^\theta, \Gamma, \mathcal{M})$ is different from $\mathbb{P}(G_i^\theta | G_{i-1}^\theta, \mathcal{M})$. Indeed, the latter can be regarded as an "average" of the former over all possible anchor nodes.

Now, the problem of inferring growth history with given duplication forest, a variant of Problem 1 that will be studied in this paper, can be formally stated as below.

Problem 2. Given a network G together with a duplication forest Γ and a growth model $\mathcal{M}$, construct a duplicate sequence θ such that the likelihood $L(\theta \mid G, \Gamma, \mathcal{M})$ is maximized.

In the above problem, the parameters in the DMC model $\mathcal{M}$ are specifically mentioned. However, as we shall see later, the parameters of $\mathcal{M}$ are not needed for the history inference problem.

2.6 Theoretical Results

Here we present some theoretical results that are crucial to solve Problem 2. Due to space limitations, all proofs are omitted from this extended abstract and they can be found in the version on arXiv (i.e., arXiv:1203.2430).

Lemma 1. *Given a network G with duplication forest Γ, for any two sequences θ_1 and θ_2 that are compatible with Γ, the graph $G_0^{\theta_1}$ is isomorphic to $G_0^{\theta_2}$.*

Given a duplicate sequence $\theta = (v_1, v_2, \cdots, v_n)$, we shall associate it with three families of numbers that are crucial to our analysis. For each duplicate node v_i in θ, let $\delta(v_i)$ be the indicator function that takes value 1 if v_i is connected to its anchor node u_i, and 0 otherwise; $\alpha(v_i)$ the number of the neighbors shared by v_i and u_i; and $\beta(v_i) := \beta(v_i, G_i^\theta)$ the number of nodes adjacent to v_i or u_i in G_i^θ, but not both. Note that $2\delta(v_i) + 2\alpha(v_i) + \beta(v_i)$ is equal to the sum of the degree of v_i and that of u_i in G_i^θ.

The sum $\delta(\theta) := \sum_{i=1}^n \delta(v_i)$ is called the *complementarity number* of history θ, the sum $\alpha(\theta) := \sum_{i=1}^n \alpha(v_i)$ is called the *extension number* of θ, and $\beta(\theta) := \sum_{i=1}^n \beta(v_i)$ is called the *loss number* of θ.

We complete this subsection with the following two key results. The first one shows that the complementarity number and extension number are constants over all compatible duplicate sequences.

Theorem 1. *Given a network G with duplication forest Γ and two compatible duplicate sequences θ_1 and θ_2, we have $\delta(\theta_1) = \delta(\theta_2)$ and $\alpha(\theta_1) = \alpha(\theta_2)$.*

Theorem 2. *Given a network G with duplication history Γ, the ratio of two likelihood functions for two duplicate sequences θ_1 and θ_2 that are compatible with Γ is given by*

$$\frac{L(\theta_1 \mid G, \mathcal{M}, \Gamma)}{L(\theta_2 \mid G, \mathcal{M}, \Gamma)} = \left(\frac{1-p}{2}\right)^{\beta(\theta_1) - \beta(\theta_2)}.$$

2.7 Reconstruction Algorithms

By Theorem 2, solving Problem 2 is equivalent to solving the following problem.

Problem 3. Given a network G and its duplication forest Γ, construct a duplicate sequence θ such that the loss number $\beta(\theta)$ is minimized among all sequences compatible with Γ.

In this section, we propose some heuristic algorithms to solve Problem 3, and hence Problem 2. The first one is a greedy algorithm called minimal loss number (MLN), in which we choose a duplicate node with the smallest value $\beta(v)$ among all candidate ones.

To motivate our main reconstruction algorithm, we introduce some further notation and results. A duplicate sequence $\theta_1 = (v_1, \cdots, v_n)$ is said to be *swapped* from $\theta_2 = (v_1', \cdots, v_n')$ at position m for some index $m \in \{1, \cdots, n-1\}$ if we have $v_m' = v_{m+1}$, $v_{m+1}' = v_m$, and $v_i' = v_i$ for all other indices i.

Lemma 2. *Given a network G with duplication forest Γ, if θ_1 and θ_2 are two compatible duplicate sequences such that θ_1 is swapped from θ_2 at position m, then we have $G_i^{\theta_1} = G_i^{\theta_2}$ for each index $i \in \{0, \cdots, n\}$ with $i \neq m$.*

Let θ_1 and θ_2 be two compatible duplicate sequences as stated in the above lemma. By Lemma 2 and Theorem 2, $L(\theta_1 \mid G, \Gamma, \mathcal{M}) \geq L(\theta_2 \mid G, \Gamma, \mathcal{M})$ if and only if for $G_m = G_m^{\theta_1} = G_m^{\theta_2}$, we have

$$\beta(v_m, G_m) + \beta(v_{m-1}, \mathcal{R}_{v_m}(G_m)) \leq \beta(v_{m-1}, G_m) + \beta(v_m, \mathcal{R}_{v_{m-1}}(G_m)). \quad (1)$$

Motivated by the above observation, for two cherries $\{u, v\}$ and $\{u', v'\}$ in Γ_t, we say $\{u, v\}$ is more *favorable* than $\{u', v'\}$, denoted by $\{u, v\} \succ \{u', v'\}$, if $\beta(v, G_t) + \beta(v', \mathcal{R}_v^u(G_t)) < \beta(v', G_t) + \beta(v, \mathcal{R}_{v'}^u(G_t))$ holds. Note that in general the relation $\succ$ is not transitive, that is, $\{u, v\} \succ \{u', v'\}$ and $\{u', v'\} \succ \{u^*, v^*\}$ does not imply $\{u, v\} \succ \{u^*, v^*\}$.

Now we present our main inference algorithm called cherry greedy (CG), which runs as follows: At every backward reconstruction step, we choose a node from the most favorable cherry C, that is, the number of cherries C' with $C \succ C'$ is maximized. If several cherries are equally favorable, we uniformly choose one of them. More precisely, starting from $G_t := G$ and $\Gamma_t := \Gamma$, we choose a most favorable cherry (u, v) from Γ_t and uniformly choose one node from the cherry, say v_t, as the duplicate node at this step. Then we construct G_{t-1} as $\mathcal{R}_{v_t}(G_t)$ and $\Gamma_{t-1} = \mathcal{R}_{v_t}(\Gamma_t)$. This process continues until G_0 is obtained.

Since the above algorithm is a stochastic one, that is, among a chosen cherry $\{u, v\}$, u and v has the equal probability of being chosen as the duplicate node. Therefore, one natural way of improving its accuracy is to repeat the algorithm for a certain times and report the best output, where the number of repetitions can be regarded as a tuning parameter. When the real duplicate sequence θ_{real} is known, the best one is defined as the output θ such that Kendall's τ between θ_{real} and θ is maximized (see Section 3 for further details on Kendall's τ), otherwise the one with the smallest loss number is chosen. This strengthened version of the CG algorithm with be refereed to as CGR, where 'R' stands for repetition.

2.8 Estimating Parameters

From the results in Section 2.6 and Section 2.7, it is clear that the parameters of the DMC model are not used in our inference framework. Moreover, here we will present a method by which the parameters can be established after a growth history being inferred.

To this end, assume that a growth history $\mathcal{H} = (G_0, \cdots, G_n)$, together with the duplicate sequence $(v_1, \cdots, v_n)$ and anchor list $(u_1, \cdots, u_n)$, is given. Note that for each neighbor w of node u_i in G_{i-1}, the probability that w is adjacent to both u_i and v_i in G_i is p. In other words, the extension number $\alpha(v_i)$ at i-th step, i.e., the number of the common neighbors shared by u_i and v_i in G_i, has the binomial distribution with parameters p and $\beta(u_i) + \alpha(v_i)$, where $\beta(u_i) + \alpha(v_i)$ is the number of neighbors that u_i has in G_{i-1}. On the other hand, the random variable $\delta(v_i)$ has Bernoulli distribution with parameter p_c. Therefore, we are led to propose the estimators $\hat{p} = \frac{\alpha(\theta)}{\beta(\theta) + \alpha(\theta)}$ and $\hat{p}_c = \frac{\delta(\theta)}{n}$ to estimate the parameters p and p_c respectively.

3 Results

Our reconstructing algorithms, minimal loss number (MLN) and cherry greedy (CG), have been implemented in Perl, which is available upon request. Given a network G and duplication forest Γ, each outputs a hypothetical duplicate sequence θ that approximates the one with the minimal loss number.

To assess the performance, we need to measure the difference between the inferred duplicate sequence and the 'real' one. One popular index for this purpose is Kendall's tau K_τ [1, 10]. Formally, for two sequences $\theta_1 = \{v_1, \cdots, v_n\}$ and $\theta_2 = \{v'_1, \cdots, v'_n\}$ that consist of the same set of nodes, $K_\tau(\theta_1, \theta_2)$ is defined as $K_\tau(\theta_1, \theta_2) = \frac{2(n_c - n_d)}{n(n-1)}$, where n_c is the number of concordant pairs, that is, the number of pairs in θ_1 that are in the correct relative order with respect to θ_2, and n_d is the number of discordant pairs. Note that we have $K_\tau(\theta_1, \theta_2) = 1$ if the two sequences are identical, and $K_\tau(\theta_1, \theta_2) = -1$ if they are exactly opposite.

3.1 Simulation Validation

To validate our algorithms, we generated 100 random network using each DMC model $\mathcal{M}$, where the parameters p_c and p ranged from 0.1 to 0.9 at 0.2 intervals. Each network has 100 nodes and is evolved from the same seed graph K_2 (i.e., the graph with two nodes and one edge).

For each simulated network G, its duplication forest Γ and duplicate sequence θ_{real} were recorded. Next, we reconstructed duplicate sequences using our algorithms. The one using MLN is denoted by θ_{MLN}, and the one using CG by θ_{CG}. We also considered the algorithm CGR, which outputs θ_{CGR}, the one with the highest Kendall's τ among ten runs of CG. We ran some of the experiments more than 10 times but found that more runs did not improve the results much, and hence we ran 10 times throughout. For comparison, we also generated a random duplicate sequence θ_{rand}, which can be interpreted as a 'null model'. Finally, we computed $K_\tau(\theta_{\text{real}}, \theta)$ for $\theta \in \{\theta_{\text{rand}}, \theta_{\text{MLN}}, \theta_{\text{CG}}, \theta_{\text{CGR}}\}$.

The results for $K_\tau(\theta_{\text{real}}, \theta_{\text{rand}})$ and $K_\tau(\theta_{\text{real}}, \theta_{\text{MLN}})$ are summarized in Fig. 2. Our results for $K_\tau(\theta_{\text{real}}, \theta_{\text{rand}})$ agree well with the theoretical mean of $K_\tau(\theta_{\text{real}}, \theta_{\text{rand}})$, which is 0. In addition, the results for $K_\tau(\theta_{\text{real}}, \theta_{\text{CG}})$ and $K_\tau(\theta_{\text{real}}, \theta_{\text{CGR}})$ are summarized in Fig. 3. From these results, we can see that compared to random duplicate sequences, our algorithms have improved the values of Kendall's τ substantially. In addition, in general CG has better performance than MLN. Finally, repeating algorithm CG a few times will increase the performance.

3.2 Comparison with Existing Methods

In this subsection, we compare the performance of our algorithm CG with NetArch, the inference method introduced in [10]. Since duplication forest is not incorporated in the framework proposed in [10], it would be expected that CG will outperform NetArch.

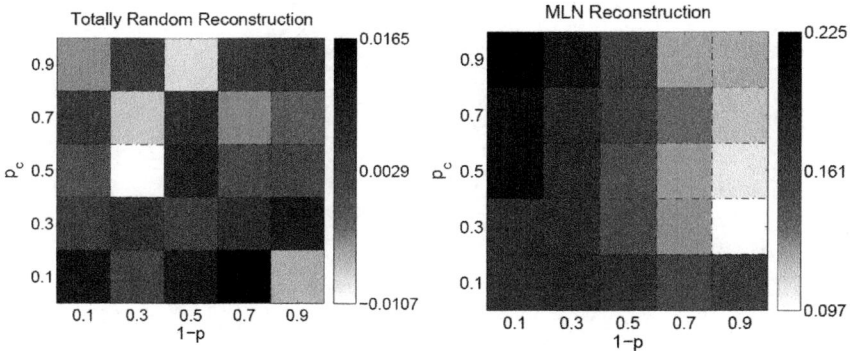

Fig. 2. Results for simulation data sets. The figure in the left is the heat map representing the values of $K_\tau(\theta_{\text{real}}, \theta_{\text{rand}})$, and the one in the right is for $K_\tau(\theta_{\text{real}}, \theta_{\text{MLN}})$. Here the value of Kendall's τ is represented by the intensity of color.

Indeed, Fig. 3 already shows that our algorithm CGR outperforms NetArch because in [10], the authors claimed that the values of Kendall's τ between the real duplicate sequence and the one constructed by their method are between 0.2 and 0 for the same set of combinations of parameters.

Even without using repetition, CG also outperforms NetArch in general. We demonstrate this by comparing the performance of them over 100 simulated random networks. For each simulation, we generated a pair of parameters p and p_c uniformly from the interval $(0, 1)$, and one graph G with 30 nodes from the seed graph K_2 using the DMC model $\mathcal{M}$. As above, the duplication forest Γ and duplicate sequence θ_{real} were recorded. Next, both NetArch and CG were used to reconstruct the duplicate sequence, and their outputs were denoted by θ_{Net} and θ_{CG} history. Finally, the values $\tau_1 := K_\tau(\theta_{\text{real}}, \theta_{\text{CG}})$ and $\tau_2 := K_\tau(\theta_{\text{real}}, \theta_{\text{Net}})$ were computed.

Among the 100 simulated networks, CG outperforms NetArch 87 times, and the distributions of $\tau_2 - \tau_1$ and $\tau_1 - \tau_2$ are summarized in Fig. 4a. Note that for the cases when CG outperforms NetArch, the gains in terms of Kendall's tau is significant, i.e., the average value is 0.2.

Moreover, we also compared the parameters $\hat{p}$ and $\hat{p_c}$ estimated by using CG with the ones p^{best} and p_c^{best} obtained by the method in [10]. Fig.4b are the box plots for the errors of these four estimations, in which the data are calculated as $|p - \hat{p}|$, etc. Note that the closer to 0, the better the estimation is. We can see that our method has smaller means of errors and smaller length of confidence intervals for both p and p_c.

3.3 Application to Real PPI Networks

We downloaded 460 gene trees reconciled in [4]. The gene trees contain genes from *S. cerevisiae* (budding yeast), *D. melanogaster* (fruit fly) and *C. elegans* (worm). For each gene tree, we used the genes of one species and deleted all

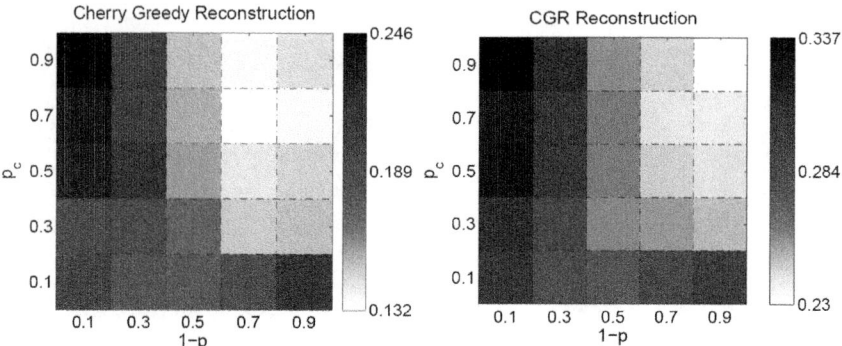

Fig. 3. Results for the algorithm CG and CGR. The figure in the left is the heat map for $K_\tau(\theta_{\text{real}}, \theta_{\text{CG}})$ and the one in the right for $K_\tau(\theta_{\text{real}}, \theta_{\text{CGR}})$. In CGR we run CG for 10 times and report the output with the highest Kendall's τ.

the genes from the other two species to create a gene duplication forest for each species. In addition, we downloaded corresponding PPI networks from the database DIP (http://dip.doe-mbi.ucla.edu/dip/Main.cgi). Since the gene trees obtained in this way are timed, we can infer from them a duplicate sequence θ^*_{real} that approximates the real duplicate sequence.

When we checked the gene trees, we found that some of them, especially the large ones, are very asymmetric about the root, which are not common for the duplication trees associated to networks generated by the DMC model. To handle this asymmetry, we modified our inference algorithm CG by taking account the depth of leaves (i.e., the number of edges between the leave and the root). More precisely, in each backward step we choose the most favorable cherry among the cherries whose depth is larger than a threshold. The output of this modified CG algorithm will be denoted by θ^*_{CG}.

The values of $\tau = K_\tau(\theta^*_{\text{real}}, \theta^*_{\text{CG}})$ for the three networks are listed in Table 1. In addition, the corresponding estimated parameters $\hat{p}$ and $\hat{p}_c$ are also listed. Note that the rate p_c of S. cerevisiae estimated by our approach is 0.05315 while the corresponding one obtained in [10] is 0.7. This indicates that our estimations are more consistent with the ones predicted in [5, 17], in which the authors asserted $p_c \leq 0.1$.

Table 1. The Kendall's τ and estimated parameters for three PPI networks

	S. cerevisiae	C. elegans	D. melanogaster
$\hat{p}$	0.061142	0.020976	0.025953
$\hat{p}_c$	0.053215	0.048443	0.024182
τ	0.378	0.316	0.473

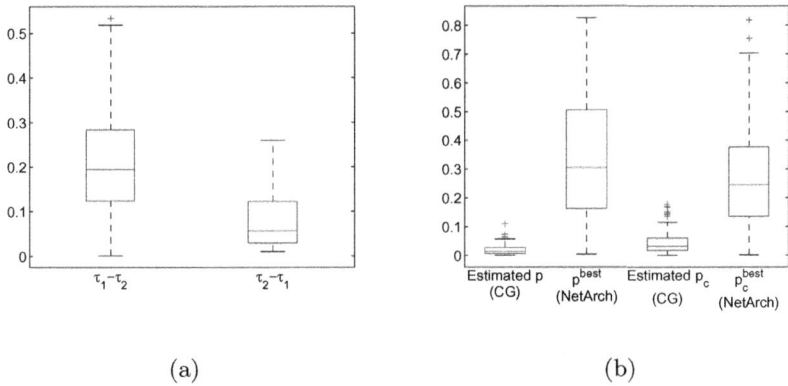

Fig. 4. (a) Box plot for differences between two methods. τ_1 is the Kendall τ for CG and τ_2 for NetArch. For $\tau_1 - \tau_2$, we only consider the cases $\tau_1 > \tau_2$, and likewise for $\tau_2 - \tau_1$. (b) Box plot for errors of estimations of parameters. Here parameters are uniformly generated from the interval $(0, 1)$.

4 Discussion

Assuming the observed network is the result of a growing mechanism as depicted in the DMC model, we have presented a likelihood-based algorithm for recovering the most probable network evolutionary history by exploiting the known duplication history trees of paralogs in the observed network. Through a series of reduction of the search space of all histories to (i) compatible duplicate sequences and (ii) the set of favored duplicate nodes, we have provided a computationally efficient algorithm. Our approach successfully re-traces the network evolution especially in the scenario that the labels of ancestor nodes are not necessarily to be one of the duplicates. As a useful by-product of our reconstruction, we propose natural estimators for the model parameters which are of independent interest. Our approach can be applied to infer the order of duplication events and to trace the topological characteristics of networks as they evolve. Our method, though described in the context of the DMC model, can be adapted to other network growing models. In addition, it can potentially be extended to predict the emergence of interactions and modules during the network evolution, and hence to provide comparison of the evolution history across different species.

Acknowledgments. This work is supported from the Singapore MOE grant R-146-000-134-112. We are grateful to Dr. Navlakha and Kingsford for providing the code in [10].

References

[1] Bar-Ilan, J., Mat-Hassan, M., Levene, M.: Methods for comparing rankings of search engine results. Comput. Netw. 50, 1448–1463 (2006)
[2] Barabasi, A., Oltvai, Z.: Network biology: understanding the cell's functional organization. Nat. Rev. Genet. 5, 101–113 (2004)
[3] Bhan, A., Galas, D., Dewey, T.: A duplication growth model of gene expression networks. Bioinformatics 18, 1486–1493 (2002)
[4] Dutkowski, J., Tiuryn, J.: Identification of functional modules from conserved ancestral protein-protein interactions. Bioinformatics 23, i149–i158 (2007)
[5] Farid, N., Christensen, K.: Evolving networks through deletion and duplication. New J. Phys. 8, 212–229 (2006)
[6] Gibson, T., Goldberg, D.: Reverse engineering the evolution of protein interaction networks. In: Pac. Symp. Biocomp., pp. 190–202 (2009)
[7] Hakes, L., Pinney, J., Robertson, D., Lovell, S.: Protein-protein interaction networks and biology–what's the connection. Nat. Biotech. 26, 69–72 (2008)
[8] Ispolatov, I., Krapivsky, P., Yuryev, A.: Duplication-divergence model of protein interaction network. Phys. Rev. E 71, 061911 (2005)
[9] Middendorf, M., Ziv, E., Wiggins, C.: Inferring network mechanisms: The drosophila melanogaster protein interaction network. Proc. Natl. Acad. Sci. 109, 3192–3197 (2005)
[10] Navlakha, S., Kingsford, C.: Network archaeology: Uncovering ancient networks from present-day interactions. PLoS Comput. Biol. 7, e1001119 (2011)
[11] Pastor-Satorras, R., Smith, E., Sole, R.: Evolving protein interaction networks through gene duplication. J. Theor. Biol. 222, 199–210 (2003)
[12] Patro, R., Sefer, E., Malin, J., Marcais, G., Navlakha, S., Kingsford, C.: Parsimonious Reconstruction of Network Evolution. In: Przytycka, T.M., Sagot, M.-F. (eds.) WABI 2011. LNCS, vol. 6833, pp. 237–249. Springer, Heidelberg (2011)
[13] Pinney, J., Amoutzias, G., Rattray, M., Robertson, D.: Reconstruction of ancestral protein interaction networks for the bZIP transcription factors. Proc. Natl. Acad. Sci. 104, 20449–20453 (2007)
[14] Sole, R., Smith, E., Pastor-Satorras, R., Kepler, T.: A model of large-scale proteome evolutions. Adv. Complex Syst. 5, 43–54 (2002)
[15] Stumpf, M., Kelly, W., Thorne, T., Wiuf, C.: Evolution at the system level: the natural history of protein interaction networks. Trends Ecol. Evol. 22, 366–373 (2007)
[16] Vazquez, A., Flammini, A., Maritan, A., Vespignani, A.: Modeling of protein interaction networks. ComPlexUs 1, 38–44 (2003)
[17] Wagner, A.: The yeast protein interaction network evolves rapidly and contains few redundent duplicate genes. Mol. Biol. Evol. 18, 1283–1292 (2001)
[18] Yamada, T., Bork, P.: Evolution of biomolecular networks–lessons from metabolic and protein interactions. Nat. Rev. Mol. Cell Biol. 10, 791–803 (2009)

POPE: Pipeline of Parentally-Biased Expression

Victor Missirian[1], Isabelle Henry[2], Luca Comai[2], and Vladimir Filkov[1]

[1] University of California at Davis, Department of Computer Science,
Davis CA 95616, USA
vmissirian@ucdavis.edu, filkov@cs.ucdavis.edu
[2] University of California at Davis, Department of Plant Biology and Genome Center, Davis CA 95616, USA

Abstract. While one might expect the phenotypes of progeny to be an additive combination of the parents, Mendelian analysis reveals that this is not always the case. Deviations from additive expectation are observable even at the level of gene expression, and identifying such instances is a prerequisite to the understanding of gene regulation and networks. Many biological studies employ mRNA-seq to identify instances where the overall and allelic expression in hybrids deviates from expectation. We describe a pipeline, POPE (Pipeline of Parentally-biased Expression), that is capable of detecting these instances, building off of a linear model of gene expression in terms of regulatory sequence strength and concentration of synergistic transcriptional regulators. We illustrate the performance of POPE on an existing mRNA-seq data set. POPE is implemented entirely in shell, python, and R, and it is designed for unix-based platforms. The code can be found at http://www.cs.ucdavis.edu/~filkov/POPE/.

Keywords: Computational biology, mRNA-seq, pipeline, additive, non-additive, *trans* effect, *cis* effect.

1 Introduction

A wide array of analysis tools exists to aid in the analysis of mRNA-seq data [19], through fast read mapping [14,15,12] as well as normalization and identification of differentially expressed loci given single [3] or multiple [21,1,9] biological replicates per condition. Two recently published pipelines performs all steps from read mapping to the identification of differentially expressed loci [11,5]. Recent research on hybrid organisms, especially plants, has made apparent a growing need for analysis tools that address a wider range of issues related to allele/isoform-specific expression [24].

Hybrids result from crosses between two parents that are at some evolutionary distance from each other. Their research importance is very broad, and they are practically invaluable to agriculture. The expression of a gene in a hybrid is expected to depend on its level of expression in the two parents, as well as on the effect of potential interactions between the two genomes, either on that gene specifically or on upstream regulators. As such, in a hybrid, a gene is said to exhibit an *additive expression level* if it meets both of the following criteria: i) its expression is not significantly different from the mean expression of its two parents and ii) the ratio of transcripts coming from the two parents is not significantly different from the ratio of expression of that gene in the two parents (an example is shown in Fig. 2c). Departures from additivity can occur due to the presence of *cis-* and *trans-* effects. A *trans* effect is the difference in regulator concentration between two genotypes, either between the two parental strains (*parental trans* effect) or between the hybrid and the mean of the two parents (*hybrid trans* effect). Similarly, we define a *cis* effect as the difference in strength of the sequence of regulatory elements between two parental alleles.

Here, we present a pipeline, POPE (Pipeline of Parentally-biased Expression), that can identify both additively and non-additively expressed genes, as well as detect and distinguish between *cis* and *trans* effects, through the integration of allelic and expression data from mRNA-seq of a hybrid and its two parents. A number of recent studies use mRNA-seq to identify the presence of parental *trans* [17], hybrid *trans* [10,8], and *cis* [10,17] effects. Many other studies have used the presence of *cis* effects to detect parental *trans* effects [6]. In addition, a study on microarrays has identified each of these three effect types and studied the relationships between them [22]. Compared to these POPE is a pipeline for rRNA-seq data, providing a convenient tool for detecting the presence of each of the three effect types. Unlike the others, POPE is based on a unified modeling framework with common statistical justification for all three effects.

In constructing and validating POPE, we addressed a number of specific technical and methodological issues. The first was constructing a pipeline of existing computational tools in order to process raw mRNA-seq data into a workable form. The next was filtering out mapping bias due to mapping of reads from one parental genome onto another, which if unfiltered would result in higher false positives. We used a linear model of gene expression in terms of the strength of *cis* regions and the concentration of synergistic transcriptional regulators to guide the construction of a method to detect both the presence and absence of *cis* and *trans* effects with high confidence. To validate our method, we ran unmodified and control experiments on a real mRNA-seq data set [10], and we provided a case study illustrating how POPE can be used to explore biological hypotheses.

In the next section we describe the structure of the POPE pipeline, including the existing tools it uses, and our approach to filtering out genome mapping bias. Then we describe the theoretical model used by POPE to detect the presence and absence of *cis* and *trans* effects, and by extension, additivity. Finally, we show our pipeline's predictions on real data and provide a case study of how they may be used.

2 POPE Pipeline Structure

POPE incorporates a set of state-of-the-art tools for read alignment, SNP detection, and differential-expression. The structure of POPE is described below and illustrated by a flowchart in Fig. 1. TopHat [23] is used to align reads to a single genomic reference and across specified exon junctions, before tabulating read counts for each locus, keeping only uniquely-mapping reads that match one or more gene models of exactly one locus. We associate aligned reads with sequence features from a user-provided genome annotation using a custom python script and with specific parental alleles using SNP data, which is either extracted from known genomic references using SAMtools v0.1.11 [13] or provided separately. The read count information for each locus and each allele is then passed to the core POPE methods, which apply a fold change based significance test to detect the presence and absence of *cis* and *trans* effects. In parallel, POPE runs a standard differential expression detection method to provide a general picture of the differences between experimental conditions: read counts are 75th-percentile normalized before testing for differential expression using Fisher's Exact Test [3]. Finally, we apply a filtering method to exclude those loci with high genome mapping bias towards one of the two parental alleles, as described in the following subsection.

2.1 Filtering Out Loci with High Genome Mapping Bias

Hybrids, by definition, carry two parental genomes that are more or less evolutionarily distant from each other. While using a single reference genome is more straightforward, we have found that if one of the two parents is more evolutionarily distant from the reference genome than the other, it may be difficult to map its sequences to the reference genome, potentially affecting the evaluation of parental allele contribution and differential expression between parents. To address these concerns, we have provided a pre-processing method that can identify a trusted set of loci, if provided with references of both parents that contain annotations of all relevant sequence features, even if both parents differ from the reference used for read alignment. To obtain this trusted set of loci, we first identify orthologs present across the diverged genomes mentioned above. Considering the instance with three genomes (parent 1, parent 2, and an annotated reference genome that differs from the two parents), we use Tophat to map $n = 100$ randomly selected reads from each coding sequence of each annotated genome against each of the other two genomes individually, and identify orthologs across all three genomes by applying a cutoff of $o = 70$ uniquely mapping reads for each of the genome to genome mappings described above. Selecting a fixed number of reads per coding sequence per genome, rather than generating reads to a particular depth, ensures that the accuracy of our computations is independent of locus length. Finally, for each coding sequence on each parent, we randomly select from that coding sequence the number of reads that we would expect given a sequencing depth of $d = 10$, and map those reads against the reference genome. For each locus, we compute the number of reads from each parental allele that map uniquely to the reference allele. If the largest of these two numbers is greater than or equal to $t = 20$ and the ratio of the smaller number to the larger number is greater than or equal to $r = 0.8$, that locus is retained. The values for parameters n, o, d, t, and r listed above were our best

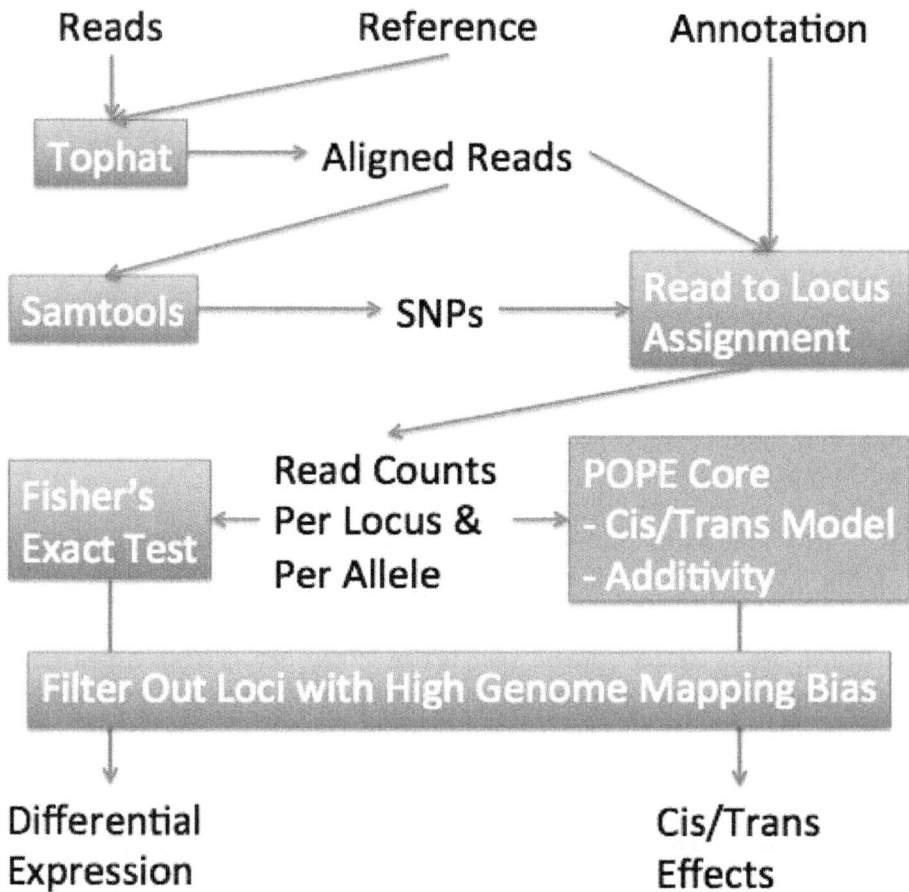

Fig. 1. A flowchart illustrating the main components of the pipeline and how they process mRNA-seq reads and other data in order to predict the presence and absence of *cis* and *trans* effects

attempt at obtaining sufficient results while minimizing the genome mapping bias (data not shown). These parameter values are customizable within POPE so that users can fine-tune their own analyses.

3 Theory of POPE

3.1 Modeling

To identify the presence and absence of *cis* effects, parental *trans* effects, and hybrid *trans* effects from allelic and expression data, we develop a framework that models the interactions between *cis* elements, transcription factors, and the expression level of the target locus. In general, we can assume that the expression of a target locus is a function of the strength of each *cis* module in its regulatory regions and the concentration of the associated transcription factors, i.e. $E = f(c_1, ..., t_1, ...)$. Although the nature of this function is specific to each locus, we must make some theoretical assumptions to provide a basis for predicting the presence and absence of *cis* and *trans* effects.

To facilitate the detection of the presence and absence of *cis* and *trans* effects, we use a very simplified model of the expression function. It has been observed that transcriptional activators may operate synergistically on a given promoter region when each is present in sufficient abundance to saturate its binding sites [16,4]. We propose to simplify our model to contain only two terms: the combined expression level of all transcriptional activators binding to the regulatory sequence of a given locus (T) and the overall binding strength of that regulatory sequence (C). The same authors have also observed cases where there is an approximately linear increase in transcription with respect to activator concentration, altered by a saturation effect at high activator concentrations [4]. To facilitate our analysis, we chose to use a linear model without saturation thresholds. Under the chosen model, the expression of the target locus can be modeled as $E = CT$. We realize that this is a first step in terms of modeling, and we are currently investigating the use of more realistic models.

We can apply the above model to represent the expression E_g^p of each parental allele p under each genotype g as the product of the regulatory sequence strength C_p of the parent and the regulator concentration T_g of the genotype. In a data set with two parents a and b and their hybrid h, we have the following equations

$$\begin{aligned} E_a^a &= C_a * T_a \\ E_b^b &= C_b * T_b \\ E_h^a &= C_a * T_h \\ E_h^b &= C_b * T_h \end{aligned} \quad (1)$$

which we can use to detect the presence and absence of *cis* and *trans* effects.

3.2 Detecting the Presence and Absence of *cis* and *trans* Effects

Following the definition from the Introduction section above, a *cis* effect corresponds to a departure from the condition $\frac{C_a}{C_b} = 1$. Under our regulatory model, we can construct an equivalent statement in terms of the allelic expression levels in the hybrid $\frac{E_h^a}{E_h^b} = 1$. Similarly, a *trans* effect between parents corresponds to a departure from the condition $T_a = T_b$, while a *trans* effect in the hybrid corresponds to a departure from the condition $T_h = \frac{T_a+T_b}{2}$. These two conditions can be described in terms of expression levels as $\frac{E_h^a E_b^b}{E_h^b E_a^a} = 1$ and $\frac{E_h^a+E_h^b}{E_a^a+E_b^b} = 1$, respectively. These null hypotheses are essentially the same as those that have been used in previous studies [10,17]. Under the simplifying assumption that these are the only three types of effect, we can visually identify the presence and absence of *cis* and *trans* effects from accurate estimates of the four expression

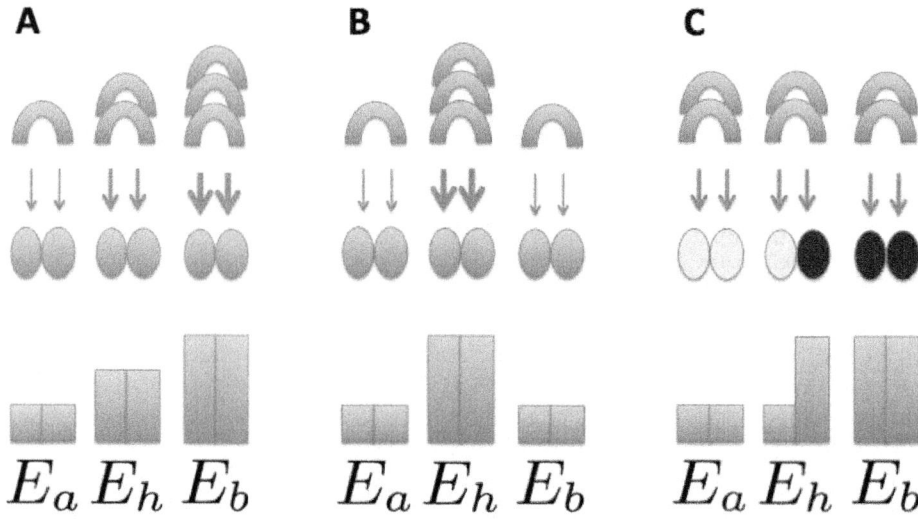

Fig. 2. Illustrations of each of the three basic effects detected by POPE. In each figure, a diagram on the top shows the binding of regulators to *cis* regions, and the bottom of the figure shows the resulting expression levels in each allele of the hybrid (E_h) and parents (E_a, E_b).
(A) In a parental *trans* effect, a difference of regulator concentration between parents results in an intermediate level in the hybrid, and the expression level of each condition is directly proportional to the regulator concentration.
(B) In a hybrid *trans* effect, there is a difference of regulator concentration between the hybrid and the two parents, and the expression level of each condition is directly proportional to the regulator concentration.
(C) In a *cis* effect, there is a difference in *cis* region strength between the two parental alleles, and the expression level of each allele is directly proportional to the strength of the corresponding *cis* region. The observed expression in the hybrid corresponds to our definition of additivity, both in terms of overall expression and allelic bias.

values. Figure 2 illustrates one example for each of these three effects, showing the corresponding regulatory sequence strengths, regulator strengths, and resulting expression pattern. In practice, we also expect there to be integrations of two or more basic effects, which could affect the validity of our predictions, as described in the discussion section.

To account for sampling error, we have developed methods to predict the presence and absence of these *cis* and *trans* effects with high confidence, according to a user-specified minimum fold change threshold t. In the POPE pipeline, when detecting a *cis* or *trans* effect, instead of evaluating the null hypothesis $X = 1$, we evaluate the two related null hypotheses $X = t$ and $X = \frac{1}{t}$. This allows us to place each locus in one of three high confidence categories with respect to a given effect, as shown in Fig. 3, (a) down, (b) up, and (c) no effect. Multiple testing correction is performed using the Benjamini-Hochberg method [2],

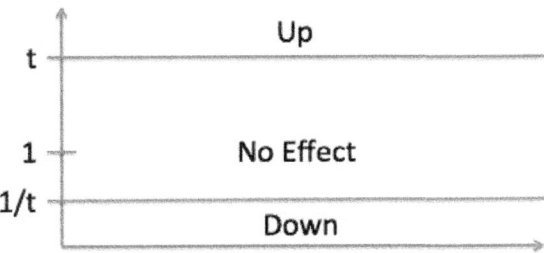

Fig. 3. Diagram illustrating the three high-confidence categories into which each locus can be classified, when detecting each type of effect. Each locus can be visualized as a point on the y-axis that corresponds to the underlying value of the effect-specific variable X. We may be able to assign a given locus to its correct category with high confidence, depending on the amount of available sequencing information and the proximity to a category boundary.

and loci that do not fall into any category with high confidence are considered uncategorizable. One can vary the detection threshold to return more or fewer predictions, as is demonstrated in the results section.

Hypothesis testing for each of the three effects is performed using the same framework. Each null hypothesis takes the form of $X = t$, where X is some function of the expression of each allele in the parents and hybrid. Given such a hypothesis, POPE first computes $obs(X)$, the observed value of X. Then a Monte Carlo simulation is run to generate $sim(X, t)$, a set of observed values of X from repeated simulations of the sequencing process with an underlying fold change of t. The p-value for $X = t$ is estimated as two times the fraction of simulated values $sim(X, t)$ that are greater than or equal to $obs(X)$ or two times the fraction of simulated values that are less than or equal to $obs(X)$, whichever is smaller. We use similar methods to test the complementary null hypothesis $X = \frac{1}{t}$. The particulars of how the observed values are computed and simulations performed depends on the null hypothesis that we are testing, as described in the following paragraphs.

When testing the null hypothesis $\frac{E_h^a}{E_h^b} = t$, $obs(X)$ is just the ratio of read counts assigned to the two alleles. To get the set of values $sim(X, t)$, we simulate allelic read counts under the assumption that $E_h^a = tE_h^b$, using a binomial distribution where the total number of reads remains the same and the probability that a given read is assigned to allele a is $\frac{t}{1+t}$.

When testing the null hypothesis $\frac{E_h^a + E_h^b}{E_a^a + E_b^b} = 1$, $obs(X)$ is the observed normalized expression in the hybrid divided by the mean normalized expression in the two parents. For the simulation, we use the observed data to estimate the mean normalized expression between the hybrid and the mean of the two parents. We then compute simulation-specific underlying values for the normalized expression in the hybrid and the mean of the two parents based on our estimated mean and a fold change of t. We compute simulation-specific underlying values for the normalized expression of each parent using the simulation-specific underlying mean normalized expression and the observed normalized expression ratio between parents. We can simulate read counts for the target and upper quartile loci in each parent, using a binomial distribution where the total number of reads remains the same and the probability that a given read is assigned to the target locus is computed from the simulation-specific underlying normalized expression.

When testing the null hypothesis $\frac{E_h^a E_b^b}{E_h^b E_a^a} = 1$, $obs(X)$ is the observed allelic expression ratio in the hybrid divided by the corresponding normalized expression ratio between the parents. As in Bullard et al. [3], we compute the normalized expression for a locus on a condition to be the number of read counts for that locus divided by the number of read counts for the upper-quartile locus of that condition. For the simulation, we use the observed data to estimate the mean of the underlying allelic expression ratio in the hybrid and the underlying normalized expression ratio between parents. We compute simulation-specific underlying values for the allelic and normalized expression ratios based on our estimated mean and a fold change of t. We compute simulation-specific underlying values for the normalized expression of each parent using the observed mean normalized expression between the two parents and the simulation-specific underlying normalized expression ratio. We can generate read counts from the simulation-specific underlying allelic and normalized expression ratios using the binomial distribution, as described previously.

3.3 Applying the Definition of Additivity

Following our definition of additivity from the Introduction section, to determine if a locus is non-additively expressed in a given hybrid, we must i) compare its expression in the hybrid to the mean expression in the two parents and ii) compare the ratio of transcripts coming from the two parents in the hybrid to the ratio of expression between the two parents themselves. These comparisons correspond to the detection of hybrid *trans* effects and parental *trans* effects. A locus is considered to be additive if and only if it has no significant *trans* effects of either type.

4 Illustration of the Pipeline

4.1 Detecting Additivity from the Absence of *cis* and *trans* Effects

We have evaluated the ability of POPE to detect additivity from the presence and absence of *cis* and *trans* effects using mRNA-seq data from a previously published mRNA-seq dataset [10] on seedling shoots of two Oryza sativa subspecies, Japonica (cv. Nipponbare) and Indica (cv. 93-11), and their reciprocal hybrids. The authors have provided one reads file in fastq format per condition, yielding 19.1, 17.3, 25.1, and 28.4 million mapped reads for Nipponbare, 93-11, Nip x 93-11, and 93-11 x Nip, respectively. Our pre-processing identified 18,327 loci with low genome mapping bias.

We ran POPE on the Nip x 93-11 hybrid and its two parents, using a range of values of t and an adjusted p-value threshold of 0.05. Results are summarized in Table 1. At a higher fold change threshold, POPE predicts fewer loci as having a *cis* or *trans* effect and more loci as having no *cis* or *trans* effect. In addition, we observe that the number of unclassified loci decreases at a higher fold change threshold, which might be due to the vast majority of loci falling well within the "no effect" category boundaries described in Figure 3. At $t = 2$, we were able to detect 8,884 loci for which at least one of the three effects was weak or absent, of which 598 were labeled as additive, since they could be identified as having no parental or hybrid *trans* effect.

To get a better understanding of the accuracy of our predictions, we ran POPE, using $t = 2$, on the same two parents and a generated in-silico hybrid, that is, an average of the two parents. No hybrid *trans* effect was identified out of the 18,327 low genome mapping bias loci surveyed by POPE, confirming a very low False Positive rate in hybrid *trans* effect detection. POPE was even able to confidently confirm the absence of a hybrid *trans* effect in 11,478 of these loci.

Table 1. High confidence categories of loci for each of three types of basic effects output by POPE on a range of fold change threshold t and an adjusted p-val=0.05 on the Nipponbare x 93-11 hybrid from He et al. We consider 18,327 loci with low mapping bias, and exclude loci for which no SNPs were detected for *cis* and parental *trans*. "Up" for *cis* means an allelic bias towards the reference allele (Nipponbare); for parental *trans*, a *trans* effect favoring the reference parent (Nipponbare); for hybrid *trans*, a *trans* effect favoring the hybrid.

	cis			Parental *trans*			Hybrid *trans*		
	$t=1.5$	$t=2$	$t=5$	$t=1.5$	$t=2$	$t=5$	$t=1.5$	$t=2$	$t=5$
Up	195	128	27	38	22	6	455	251	83
Down	32	22	7	611	430	158	216	50	2
None	24	426	2700	4	641	6220	4000	8826	12629
Excluded	11711	11711	11711	11711	11711	11711	0	0	0
Unclear	6365	6040	3882	5963	5523	232	13656	9200	5613
Total	18327	18327	18327	18327	18327	18327	18327	18327	18327

Our results exhibit similar trends to those from He et al. [10], with some differences. We confirm the observed results that more hybrid *trans* effects are up versus down and that most loci with allelic bias favor the reference allele (though, in the latter case, He et al. report the intersection of the results from their reciprocal hybrids). One difference is that POPE predicts many fewer genes with a hybrid *trans* effect for each value of t, which is expected since a value of $t > 1$ implies greater stringency. Discrepancies may be due to different preprocessing techniques, significance thresholds, and POPE's use of a fold change threshold.

4.2 Case Study

We have also provided a case study below to illustrate how our method can be used to explore biological hypotheses. In Arabidopsis, the altered expression of circadian clock genes, including the oscillating regulator CCA1, has been linked to hybrid vigor [18]. Our POPE analysis above finds that a putative CCA1 ortholog in rice [7] is additively expressed with high confidence in the Nip x 93-11 hybrid, suggesting that the hybrid vigor observed in that cross [25] may have a different cause. It is still possible that CCA1's oscillation pattern had an altered amplitude but that mRNA was sampled too close to the inflection point to observe a sufficient fold change in expression. Even so, we believe that POPE presents a powerful tool for testing similar hypotheses.

5 Discussion

The level of divergence between genotypes is crucial to the functioning of our pipeline. The lower the level of divergence, the fewer SNPs there will be per gene, decreasing the ability to detect the presence and absence of *cis* and *trans* effects between the parents, as described above. To an extent, this can be counteracted by increasing the level of coverage. To make a connection to our results on the He et al. data, we observe that only 19% of all sequence features from the MSU 6.1 rice genome annotation [20] have at least one SNP between Nipponbare and 93-11. Since we are using a single reference genome, too much divergence could also cause problems, as it could lead to increased genome mapping bias. Even if we could minimize this effect by choosing two parental genotypes that are equally distant from the available reference, we would expect the divergence between the parents and the reference to result in a low read mapping frequency.

One limitation of our model is that we do not explicitly consider integrated effects, which are necessary to explain the regulator profiles that lead to hybrid *trans* effects. One example is a parental *trans* effect and *cis* effect that operate on the same binding site / regulator pair and that favor opposite parental alleles [25]. Under our model, such an integrated effect would produce an expression profile similar to that of the simultaneous occurrence of all three basic effects. We suggest that it may be possible to estimate the posterior probabilities of specific integrated effects from the frequencies of each type of expression profile.

6 Conclusion

We have introduced POPE, a publicly-available pipeline for the detection of the presence and absence of *cis* and *trans* effects as well as additivity and non-additivity from mRNA-seq of hybrids and their parents. In contrast to previous studies, our pipeline allows the detection of the absence of these effects. We hope that POPE will provide a useful tool for authors conducting mRNA-seq studies on hybrid expression patterns, reducing the need for experimenters to implement their own computational analyses. We are currently extending our pipeline to implement the handling of multiple replicates. Another planned improvement is to explore how different models of gene regulation would affect our pipeline's performance in detecting the presence and absence of *cis* and *trans* effects.

Funding

This work was supported by NSF PGRP grant DBI-0924025.

References

1. Anders, S., Huber, W.: Differential expression analysis for sequence count data. Genome Biol. 11(10), R106 (2010)
2. Benjamini, Y., Hochberg, Y.: Controlling the false discovery rate: A practical and powerful approach to multiple testing. J. R. Stat. Soc. 57(1), 289–300 (1995)
3. Bullard, J.H., Purdom, E., Hansen, K.D., Dudoit, S.: Evaluation of statistical methods for normalization and differential expression in mrna-seq experiments. BMC Bioinformatics 11, 94 (2010)
4. Carey, M., Lin, Y.S., Green, M.R., Ptashne, M.: A mechanism for synergistic activation of a mammalian gene by gal4 derivatives. Letters to Nature 345, 361–364 (1990)
5. Cumbie, J.S., et al.: Gene-counter: A computational pipeline for the analysis of rna-seq data for gene expression differences. PLoS ONE 6(10), e25279 (2011)
6. Emerson, J.J., Li, W.H.: The genetic basis of evolutionary change in gene expression levels. Phil. Trans. R. Soc. B 365(1552), 2581–2590 (2010)
7. Filichkin, S.A., et al.: Global profiling of rice and poplar transcriptomes highlights key conserved circadian-controlled pathways and cis-regulatory modules. PLoS ONE 6(6), e16907 (2011)
8. Groszmann, M., et al.: Changes in 24-nt sirna levels in arabidopsis hybrids suggest an epigenetic contribution to hybrid vigor. Proc. Natl. Acad. Sci. USA 108(6), 2617–2622 (2011)
9. Hardcastle, T.J., Kelly, K.A.: bayseq: Empirical bayesian methods for identifying differential expression in sequence count data. BMC Bioinformatics 11, 422 (2010)
10. He, G., et al.: Global epigenetic and transcriptional trends among two rice subspecies and their reciprocal hybrids. Plant Cell 22(1), 17–33 (2010)
11. Langmead, B., Hansen, K.D., Leek, J.T.: Cloud-scale rna-sequencing differential expression analysis with myrna. Genome Biol. 11(8), R83 (2010)
12. Langmead, B., Trapnell, C., Pop, M., Salzberg, S.L.: Ultrafast and memory-efficient alignment of short dna sequences to the human genome. Genome Biol. 10(3), R25 (2009)
13. Li, H.: Improving snp discovery by base alignment quality. Bioinformatics 27(8), 1157–1158 (2011)
14. Li, H., Durbin, R.: Fast and accurate short read alignment with burrows-wheeler transform. Bioinformatics 25(14), 1754–1760 (2009)
15. Li, R., et al.: Soap2: an improved ultrafast tool for short read alignment. Bioinformatics 25(15), 1966–1967 (2009)
16. Lin, Y.S., Carey, M., Ptashne, M., Green, M.R.: How different eukaryotic transcriptional activators can cooperate promiscuously. Letters to Nature 345, 359–361 (1990)

17. McManus, C.J., et al.: Regulatory divergence in drosophila revealed by mrna-seq. Genome Res. 20(6), 816–825 (2010)
18. Ni, Z., et al.: Altered circadian rhythms regulate growth vigour in hybrids and allopolyploids. Nature 457(7227), 327–331 (2009)
19. Oshlack, A., Robinson, M.D., Young, M.D.: From rna-seq reads to differential expression results. Genome Biol. 11(12), 220 (2010)
20. Ouyang, S., et al.: The tigr rice genome annotation resource: improvements and new features. Nucleic Acids Res. 35(suppl.1), D883–D887 (2007)
21. Robinson, M.D., McCarthy, D.J., Smyth, G.K.: edger: A bioconductor package for differential expression analysis of digital gene expression data. Bioinformatics 26(1), 139–140 (2010)
22. Tirosh, I., Reikhav, S., Levy, A.A., Barkai, N.: A yeast hybrid provides insight into the evolution of gene expression regulation. Science 324(5927), 659–662 (2009)
23. Trapnell, C., Pachter, L., Salzberg, S.L.: Tophat: discovering splice junctions with rna-seq. Bioinformatics 25(9), 1105–1111 (2009)
24. Turro, E., et al.: Haplotype and isoform specific expression estimation using multi-mapping rna-seq reads. Genome Biology 12(2), R13 (2011)
25. Zhang, H.Y., et al.: A genome-wide transcription analysis reveals a close correlation of promoter indel polymorphism and heterotic gene expression in rice hybrids. Molecular Plant 1(5), 720–731 (2008)

On Optimizing the Non-metric Similarity Search in Tandem Mass Spectra by Clustering*

Jiří Novák, David Hoksza, Jakub Lokoč, and Tomáš Skopal

Siret Research Group,
Faculty of Mathematics and Physics, Charles University in Prague,
Malostranské nám. 25, 118 00 Prague, Czech Republic
novak@ksi.mff.cuni.cz

Abstract. Tandem mass spectrometry is a well-known technique for identification of protein sequences from an "in vitro" sample. To identify the sequences from spectra captured by a spectrometer, the similarity search in a database of hypothetical mass spectra is often used. For this purpose, a database of known protein sequences is utilized to generate the hypothetical spectra. Since the number of sequences in the databases grows rapidly over the time, several approaches have been proposed to index the databases of mass spectra. In this paper, we improve an approach based on the non-metric similarity search where the M-tree and the TriGen algorithm are employed for fast and approximative search. We show that preprocessing of mass spectra by clustering speeds up the identification of sequences more than 100× with respect to the sequential scan of the entire database. Moreover, when the protein candidates are refined by sequential scan in the postprocessing step, the whole approach exhibits precision similar to that of sequential scan over the entire database (over 90%).

Keywords: Tandem mass spectrometry, similarity search, non-metric access methods, protein sequences identification, spectral clustering.

1 Introduction

Almost every process on the cell level is secured by proteins whose interactions form the basis of all living organisms. The functions of proteins are determined by their 3D structure, which is derived from protein sequences.

Tandem mass spectrometry (MS/MS) is a widely known method for protein and peptide sequences identification from a sample of proteins "in vitro". Commonly, the sample is analyzed by more runs of a mass spectrometer. A set containing hundreds to thousands of mass spectra is captured in each run. The proteins in the sample are split to many peptide ions where a mass spectrum corresponds to a peptide ion. More peptide ions correspond to a peptide sequence and, similarly, more peptide sequences come from a protein sequence.

---

* This work was supported by Czech Science Foundation (GAČR) projects P202/11/0968, P202/12/P297, 201/09/H057 and by the Grant Agency of Charles University (GAUK) project Nr. 430711.

A mass spectrum is a list of peaks corresponding to peptide fragment ions. The peak is a pair $\left(\frac{m}{z}, I\right)$, where $\frac{m}{z}$ is a mass-to-charge ratio and I is the intensity of a fragment ion occurrence. In a spectrum, there occur several types of fragment ions forming so-called ions series. The most important series for correct peptide sequence identification are y-ions and b-ions. The completeness of these series is crucial for correct spectra interpretation, because the $\frac{m}{z}$ difference between two neighboring peaks in one series, e.g., y_i and y_{i+1}, corresponds to a mass of an amino acid in the peptide. The precursor mass m_p (the mass of the peptide ion before splitting) is also provided as an additional information for each spectrum captured by a spectrometer. The interpretation of spectra is often complicated with post-translational modifications (PTMs) of amino acids, because masses of amino acids are changed in that case, and thus the peaks are shifted [21].

The mass spectrometer does not determine the peptide sequences from mass spectra directly but the spectra must be *interpreted* after they are captured. The successful computational approaches for interpretation of mass spectra (i.e., assigning the peptide sequences to mass spectra) [16] are often based on the similarity search in databases of already known or predicted protein sequences [24]. The databases contain millions of protein sequences [15] and the spectra sets generated by a MS/MS analysis that need to be interpreted (query sets from the database point of view) contain thousands of mass spectra. Thus a sequential scan of entire database for each mass spectrum is time consuming. To speed-up the search, an index over the database of hypothetical mass spectra generated from known peptide sequences (short pieces of proteins) can be constructed.

The simplest way is to index the database of spectra by peptide precursor mass because there is a correlation between the precursor mass of similar peptides [20]. A disadvantage is that indexing of peptides by precursor mass limits the capability of managing spectra with PTMs because mass of peptides with PTMs may differ from the peptides without PTMs from tens to hundreds Daltons. A few approaches were proposed where an inverted file was employed to index the database of protein sequences [11], [14]. Another approach uses a suffix tree [13] and there are also approaches based on the similarity search in metric spaces [22] [4]. The approaches based on the inverted index and on the similarity search in metric spaces commonly do not support the search of spectra with PTMs. We have proposed a fast method based on approximative non-metric similarity search [19], which is able to manage spectra with PTMs.

Despite the search in an indexed database is fast, query sets of mass spectra still contain many noise spectra that should be ignored (on average 90% in the query set [10]). The noise spectra cannot be assigned to peptide sequences because they occur as an artifact of the spectrometer process. A kind of query set preprocessing can be used to eliminate the noise spectra and to speed up the search, because only a small part of the query set needs to be interpreted [25] [23]. Commonly used preprocessing approaches are the spectrum quality filtering [6] [17] and the clustering [8], [7], [5].

The spectrum quality filtering methods analyze many parameters of spectra (the number of peaks and their relative intensity, the precursor mass, the number of complementary y-ions and b-ions, etc.) and assign to each spectrum a score. Since the mass spectrometers from different manufacturers use different physical principles, the significance of parameters may differ from machine to machine. Thus, the score heavily depends on the mass spectrometers which were used to capture the spectra. Only the spectra reaching a specific score are used for further processing while the others are ignored as noisy ones.

On the other hand, the clustering is independent on the properties of different machines because the spectra from different spectrometers are processed the same way, i.e., without the knowledge of significance of particular parameters. The clustering is based on fact that a mass spectrometer generates multiple spectra corresponding to a peptide sequence [28]. Since the spectra corresponding to the peptide sequence are similar, they form a cluster. In the set of spectra obtained from one spectrometer run there are many spectra which are not noise but they correspond to a peptide sequence. A disadvantage in this case is the clustering causes loss of some peptide sequences [28]. Since a mass spectrometry task often consists of multiple spectrometer runs (each run generates a query set), the above disadvantage can be successfully resolved by merging query sets from multiple runs. The number of identifiable peptides that are not clustered decreases with the increasing number of merged query sets, while the noise spectra are successfully eliminated [2].

2 Computational Methods

We briefly introduce the metric access methods, the spectrum similarity employed in MAMs and for clustering of mass spectra, the original idea for spectra interpretation by MAMs, and an extension of this approach using preprocessing and postprocessing.

2.1 Metric Access Methods (MAMs)

A metric is a distance function satisfying the reflexivity, symmetry, non-negativity and triangle inequality. The MAMs were designed for fast search in databases modeled in metric spaces, where the triangle inequality is crucial for organizing objects into metric regions and for pruning irrelevant regions while searching [31]. A distance (partially) violating the triangle inequality is called a semi-metric and the process is denoted as non-metric search. The violation of the triangle inequality is expressed by the triangle error (T-error) tolerance θ [26].

2.2 Spectrum Similarity

The use of MAMs and clustering require a similarity function which says how much two spectra are similar. A commonly employed function is the angle distance (or cosine similarity) [12], [1]; another approach is the sigmoid similarity [7], [25]. In this paper, we have chosen the parameterized Hausdorff distance d_{HP} (Eq. 2), which was successfully applied in non-metric indexes [18].

$$h(\boldsymbol{x}, \boldsymbol{y}) = \frac{\sum_{\boldsymbol{x}_i \in \boldsymbol{x}} \sqrt[n]{\min_{\boldsymbol{y}_j \in \boldsymbol{y}} \{\max(0, |\boldsymbol{x}_i - \boldsymbol{y}_j| - \xi)\}}}{\dim(\boldsymbol{x})} \quad (1)$$

$$d_{HP}(\boldsymbol{x}, \boldsymbol{y}) = \max(h(\boldsymbol{x}, \boldsymbol{y}), h(\boldsymbol{y}, \boldsymbol{x})) \quad (2)$$

where $\boldsymbol{x}$ and $\boldsymbol{y}$ are the vectors of $\frac{m}{z}$ ratios, $\dim(\boldsymbol{x})$ is the length of $\boldsymbol{x}$ and ξ is a mass error tolerance. The angle distance can be computed a little bit faster than d_{HP} but the number of identified peptides and the efficiency of MAMs is lower when the angle distance is utilized. Since lists of peaks in mass spectra are implicitly sorted, both distances can be computed with the linear time complexity $O(p)$, where p is the number of peaks in a spectrum [18].

2.3 Original Method

We briefly describe the previously proposed approach [19], which employs the M-tree [3] and the Trigen algorithm [26]. First, protein sequences from a database are split to peptide sequences and the hypothetical mass spectra are generated from the peptide sequences. Second, the hypothetical mass spectra are indexed by the M-tree (or by another MAM) under the d_{HP} while the TriGen algorithm is utilized to control the T-error tolerance θ (i.e., the efficiency of the M-tree). The search is faster with increasing θ but the number of identified peptides is lower. Finally, a k-nearest neighbor (kNN) query is performed for each query spectrum. For many spectra in the query set, a hypothetical mass spectrum among the k-nearest neighbors corresponds to a peptide sequence that we are looking for. Since the d_{HP} is a coarse function, an additional re-ranking is assumed to determine the correct peptide sequence from the k-nearest neighbors.

2.4 Improvements

We propose an extension of the original approach, where the clustering is employed in preprocessing step to filter out the noise spectra thus speeding the search, and where the sequential scan over the candidates is used in the postprocessing step to increase the number of identified peptide sequences (Fig. 1).

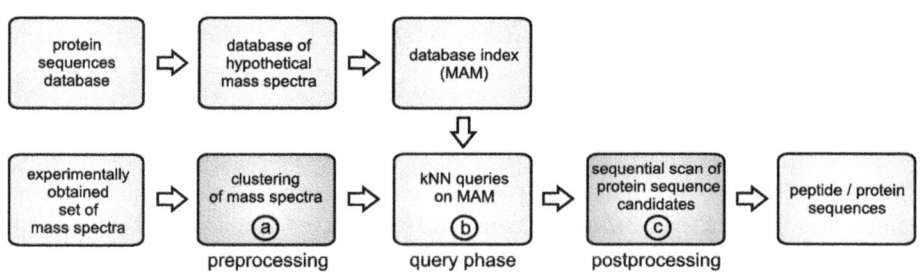

Fig. 1. Sequences identification (original method is yellow, improvements are blue)

Preprocessing. The preprocessing is realized by the clustering (Fig. 1a). A major premise for the clustering is that query sets from more spectrometer runs are merged. Hence, many interpretable spectra that are captured only once per a spectrometer run have a "twin" in the query set so they are not eliminated by the clustering. On the other hand, the noise spectra are successfully cleared away thus many spectra are not searched in the query phase (Fig. 1b), making the search significantly faster.

──────────────── Alg. 1. Clustering of query mass spectra ────────────────

```
1   Clustering(a set of clusters C, a threshold t, a number of cycles w) {
2     let C be initialized with one mass spectrum per cluster;
3     for w cycles {
4       MergeClusters(C,t);
5       SelectCentroids(C);
6       RearrangeClusters(C,t);
7       SelectCentroids(C); }}
8
9   MergeClusters(a set of clusters C, a threshold t) {
10    for all clusters c_i ∈ C {
11      select the spectrum c_{i,0} { // a centroid is stored at c_{i,0}
12      for all clusters c_j ∈ C {
13        if all spectra c_{j,k} have d_{HP}(c_{i,0}, c_{j,k}) ≤ t {
14          store the position p of the cluster c_j with the minimal d_{HP}(c_{i,0}, c_{j,0}); }}}
15      merge the clusters c_i and c_p; } }
16
17  SelectCentroids(a set of clusters C) {
18    for all clusters c_i ∈ C {
19      for all spectra c_{i,j} {
20        P = ∅;
21        for all spectra c_{i,k} {
22          store the maximal distance d_{HP}(c_{i,j}, c_{i,k}) and
23          the position k of the spectrum c_{i,k} in the maximal d_{HP} into P; }
24      select the position p with the minimal d_{HP} from P; }
25      switch the spectra c_{i,0} and c_{i,p}; } // a new centroid has been moved to c_{i,0} }
26
27  RearrangeClusters(a set of clusters C, a threshold t) {
28    for all clusters c_i ∈ C {
29      for all spectra c_{i,m} {
30        P = ∅;
31        for all clusters c_j ∈ C {
32          for all spectra c_{j,n} {
33            if all d_{HP}(c_{i,m}, c_{j,n}) ≤ t { store the distance d_{HP}(c_{i,m}, c_{j,0}) and
34            the position j of the cluster c_j into P; }}}
35        select the position p of the cluster with minimal d_{HP} from P;
36        move the spectrum c_{i,m} to the cluster c_p; }}}
```

One of the best-known algorithms for the clustering is the K-means [30], which is not suitable for clustering of mass spectra because we cannot predict the number of clusters K before the clustering [8]. Moreover, its time complexity is $O(NKd)$, where N is the number of spectra in the query set and d is the dimensionality. The K-means is not suitable for large query sets and high-dimensional data which is exactly the case of mass spectra (usually containing many peaks/dimensions).

A better clustering algorithm for mass spectra is the hierarchical clustering [8][30]. A disadvantage for large query sets of spectra is the time complexity $O(N^2)$. Since we analyze the impact of the clustering on the number of identified peptide sequences, we employ a simple hierarchical-like clustering (Alg. 1).

More efficient clustering algorithms [30] may be used for large query sets, e.g., an approach based on the density clustering (DENCLUE) [9] with the time complexity $O(N \log N)$, which is capable of tackling high-dimensional data and which is robust when dealing with noise data.

The clustering algorithm (Alg. 1) requires a set of clusters C initialized with one mass spectrum per cluster. Then two phases are repeated in w cycles. First, pairs of clusters with the minimal $d_{HP}(c_{i,0}, c_{j,0})$ that $d_{HP}(c_{i,0}, c_{j,0}) \leq t$ are merged, where t is a threshold of the d_{HP} and $c_{i,0}$, $c_{j,0}$ are the centroids of clusters c_i, c_j. The threshold t determines whether the spectra in a cluster are similar or not. Moreover, it determines the number of clusters. If t is too low, each spectrum forms a singleton cluster. If t is too high, all spectra form one cluster. Second, the spectra are rearranged among the clusters. A spectrum is moved to another cluster, if the d_{HP} among the spectrum and all spectra in the target cluster is less or equal t. In case that more clusters are selected, the cluster is picked where the d_{HP} between its centroid and the moved object is minimal.

New centroids of clusters are selected after each phase. Finally, the centroids of clusters containing at least two spectra form the queries, which will be processed by MAM. Another way consists in putting all peaks from all the spectra in the cluster into a representative spectrum [2]. The intensities of the closest peaks are counted up and their $\frac{m}{z}$ values are averaged. Since the increasing number of peaks in a spectrum worsens the efficiency of MAMs because of high intrinsic dimensionality [19], this approach needs a bit improvement for purposes of mass spectra interpretation by the non-metric similarity search. For example, a specified number of peaks with the highest intensity can be selected from the representative spectrum.

Query Phase. The query phase corresponds to the original idea presented in Sec. 2.3, where a kNN query is performed by a MAM for each spectrum selected in the preprocessing (Fig. 1b). The k nearest neighbor peptide sequences to each query spectrum are assigned to the protein sequences of their origin. The protein sequences containing at least one "good" peptide sequence hit (e.g., $d_{HP} \leq 0.65$) are the protein sequence candidates.

The MAM we have chosen for the query phase is the non-metric tree (NM-tree) [27] because it combines the M-tree with the TriGen algorithm in a way that allows to dynamically control the retrieval precision at query time. The NM-tree could be replaced by another MAM since our approach is independent on a specific method.

Postprocessing. The postprocessing is a sequential scan of protein sequence candidates (Fig. 1c), which significantly improves the number of identified peptide sequences because more peptide sequences in a protein sequence correspond to the mass spectra in the query set [29] [25]. The protein sequence candidates (i.e., a small subset of the database sequences selected in the query phase) are split to peptide sequences and their hypothetical mass spectra are compared to the entire set of input spectra (as it was before the clustering phase). The spectra previously missed during the preprocessing or during the query phase are assigned to peptide sequences. The newly identified peptide sequences are assigned to the protein sequences of their origin. Finally, the peptide (or protein, respectively) sequences identified in the query phase and refined in the postprocessing phase form the result. Note that some peptide sequences are lost during the clustering because their spectra are present only once in the query set. Some peptide sequences are lost during the query phase because the search is only approximative (non-metric). The sequential scan of protein sequence candidates helps to reveal a peptide sequence in case it forms a part of a candidate protein sequence which was hit by another peptide sequence.

3 Experiments

We used a dataset containing MS/MS spectra from 2 protein mixtures A and B [10]. Spectra corresponding to peptide sequences were manually annotated. 14 mass spectrometer runs were performed on the mixture A and 8 runs on the mixture B. We show the results for the spectra from the first 6 runs on mixture A and from all runs on mixture B. The spectra were searched in the database of hypothetical mass spectra generated from the database of protein sequences. We used a part of the MSDB [15] database containing 100,000 protein sequences (i.e., 5.6 millions of peptide sequences or hypothetical mass spectra) including the reference protein sequences for mixtures A and B.

All the experiments were carried out on a machine with 2 processors Intel Xeon X5660 (24 cores, 2.8 GHz) with 24 GB RAM and 64-bit OS Windows Server 2008 R2. Even though our implementation supports parallel processing of mass spectra, the stated times of clustering and peptide sequences identification are measured at one core. If not otherwise stated, the following settings were used – protein sequences splitting enzyme: trypsin; maximum number of missed cleavage sites: 1; mass range of peptide ions generated from the database: 500-5,000 Da; fragment ions generated in hypothetical mass spectra: y-ions and b-ions; mass range of generated fragment ions: 300-2,000; $\frac{m}{z}$ error tolerance (ξ): 0.4 Da; number of peaks with highest intensity used in a query: 50; distance measure: d_{HP} (with $n = 30$); clustering threshold (t): 0.65 (values returned by d_{HP} are normalized to $\langle 0, 1 \rangle$), T-error tolerance θ: 0.1.

The number of clusters is the number of those containing at least 2 spectra. Since we perform one kNN query per cluster, the number of clusters determines the number of kNN queries processed on the NM-tree. The number of missed spectra is counted after the clustering phase and before query phase. It is the number of annotated spectra in clusters with single objects and thus missed by clustering. *Independent runs* means that query sets of spectra from more spectrometer runs were processed separately and the results were summed (the number of clusters, number of missed spectra, time of clustering and ratio of identified spectra to annotated spectra) or averaged (time of identification per spectrum). *Merged runs* means that query sets of spectra from more spectrometer runs were processed together.

3.1 Clustering of Spectra from Two Spectrometer Runs

We have verified that clusters formed from merged query sets of spectra from two spectrometer runs contain many more annotated spectra than clusters formed from the query sets which are processed separately (Tab. 1). On average, the clusters formed from spectra from two spectrometer runs contain about 40.7% more annotated spectra than clusters formed from single spectrometer run. Since we perform one kNN query per cluster containing at least 2 spectra, up to 79% of all kNN queries are not performed for the clusters formed from the spectra merged from two runs. For clusters formed from the spectra from single runs, up to 87% of all kNN queries are not performed but there are many missed annotated spectra.

Table 1. Clustering of spectra from single runs and from two merged runs

Dataset	Num. of all spectra	Num. of annotated spectra	Independent runs			Merged runs		
			Num. of clusters	Spectra missed	Clustering time [s]	Num. of clusters	Spectra missed	Clustering time [s]
A1-2	2213	215	321	92	7.9	397	16	16.3
A3-4	1858	158	304	69	5.3	400	25	10.4
A5-6	2021	164	306	49	6.5	385	18	13.7
B1-2	618	121	49	87	0.5	113	33	1.0
B3-4	902	155	86	104	1.1	203	9	2.2
B5-6	1418	208	185	122	3.0	313	23	5.8
B7-8	1661	223	212	123	4.2	365	13	8.0

3.2 Effectiveness and Efficiency of Peptide Sequences Identification

We have tested the impact of the query spectra clustering on the number of finally identified peptide sequences (i.e., after the postprocessing) and on the average time of identification per spectrum. We have compared the sequential scan of entire database and the NM-tree in 3 different ways – without the clustering, with the clustering of two query sets processed independently, and together. When the clustering and/or the NM-tree were employed, the postprocessing was used.

The most peptide sequences (on average 94.6%) were identified when the sequential scan was performed without the clustering (Tab. 2). On average 93.8% peptide sequences were identified when the NM-tree was employed without clustering. The ratio of identified peptides was noticeably worse when the clustering was applied on the query sets from single runs – about 75.3% for the sequential scan and only 65.4% for the NM-tree. When the clustering was applied on the query sets merged from two spectrometer runs, the ratio of identified peptides was almost the same like when no clustering was employed. On average, it was about 93.6% for the sequential scan and 90.1% for the NM-tree. The clustering of query sets merged from 2 runs worsens the ratio of identified peptides about 1% when the sequential scan is performed over entire database and about 3.7% when the NM-tree is employed.

The slowest method was the sequential scan without clustering, where the average time of identification per spectrum was 7.04 s (Tab. 3). The NM-tree without clustering took 0.28 s, thus the speed-up was 25.1×. When clustering was applied on the query sets from single runs, the average time was 0.98 s (speed-up 7.2×) for the sequential scan and 0.04 s (speed-up 176.0×) for the NM-tree. When query sets from two spectrometer runs were merged and the clustering was applied, the average time was 1.59 s for the sequential scan (speed-up 4.4×) and 0.07 s for the NM-tree (speed-up 100.6×). When the NM-tree was employed with clustering, the average speed-up was 4× wrt. NM-tree without clustering.

Table 2. The ratio of identified spectra to annotated spectra [%]

Dataset	Without clustering		With clustering			
			Independent runs		Merged runs	
	Seq. scan	NM-tree	Seq. scan	NM-tree	Seq. scan	NM-tree
A1-2	96.7	96.7	74.0	72.6	95.8	95.8
A3-4	91.1	90.5	69.6	59.5	88.6	81.6
A5-6	93.3	92.7	81.1	78.7	93.3	87.2
B1-2	98.3	95.9	59.5	28.9	97.5	87.6
B3-4	97.4	96.8	81.3	71.0	97.4	96.8
B5-6	91.8	90.9	87.0	78.8	90.9	90.9
B7-8	93.3	93.3	74.4	68.2	91.9	91.0

Table 3. Time of identification per spectrum [s]

Dataset	Without clustering		With clustering			
			Independent runs		Merged runs	
	Seq. scan	NM-tree	Seq. scan	NM-tree	Seq. scan	NM-tree
A1-2	7.36	0.33	1.13	0.05	1.42	0.08
A3-4	7.09	0.30	1.27	0.05	1.63	0.08
A5-6	7.28	0.30	1.17	0.05	1.53	0.07
B1-2	6.75	0.23	0.59	0.02	1.38	0.06
B3-4	6.92	0.24	0.73	0.03	1.75	0.07
B5-6	6.94	0.27	0.99	0.04	1.70	0.08
B7-8	7.20	0.30	0.97	0.04	1.73	0.08

3.3 Clustering of Spectra Merged from More Spectrometer Runs

We have tested the impact of the increasing number of spectra from more spectrometer runs in a query set on the number of annotated mass spectra missed by clustering and on the time of clustering (Tab. 4). We can observe that the number of missed annotated spectra is almost the same when spectra from two or more spectrometer runs are merged, thus merging spectra from more than two spectrometer runs does not significantly improve the effectiveness of peptide sequences identification. Since we employ a simple clustering algorithm (Alg. 1), a disadvantage of merging spectra from too many spectrometer runs is that the time of clustering increases with the quadratic time complexity.

We also measured the ratio of identified to annotated spectra and the average time of identification per spectrum on the NM-tree. The ratio of identified spectra is almost the same when spectra from two or more spectrometer runs are merged (on average 95%). The time of identification a bit increases with increasing number of spectra because of the quadratic complexity of clustering.

We can observe that the ratio of the number of clusters to the number of all spectra in a query set is lower with the increasing number of spectra. This could be an advantage for large query sets of mass spectra because only a small number of the spectra is queried and thus the search is significantly faster. When spectra from 14 spectrometer runs on mixture A were merged, 14365 spectra formed 1188 clusters with more than one spectrum. Thus only 8.3% of all queries were performed on the NM-tree. When spectra from 8 spectrometer runs on mixture B were merged, 4599 spectra formed 711 clusters thus only 15.5% of all queries were performed.

Table 4. Clustering of spectra merged from more spectrometer runs

Dataset	Num. of all spectra	Num. of annotated spectra	Num. of clusters	Ratio of clust. to all spectra [%]	Spectra missed	Clustering time [s]	Ratio of ident. spectra [%]	Time of ident. [s]
A1	1122	119	157	14.0	49	4.0	76.5	0.05
A1-2	2213	215	397	17.9	16	16.3	95.8	0.08
A1-3	3038	274	540	17.8	15	30.6	96.0	0.08
A1-4	4071	373	706	17.3	15	61.2	94.4	0.09
A1-5	5071	451	839	16.5	16	106.4	94.9	0.09
A1-6	6092	537	943	15.5	17	148.3	94.0	0.09

3.4 Impact of Distance Threshold on Clustering

We have tested the impact of the threshold t of d_{HP} on the number of clusters, number of spectra missed by the clustering and on the time of clustering (Tab. 5). We used the dataset A1-2 with 2213 spectra merged from two spectrometer runs. The number of clusters increases with increasing t while the number of spectra missed by clustering decreases. The optimal t seems to be about 0.65 when the number of clusters (or kNN queries performed, respectively) is only 17.9% wrt. the number of kNN queries which must be performed when the clustering is not employed. Moreover, there are only 16 missed spectra. For $t < 0.65$, the number of spectra missed by clustering grows because there are less hits among the hypothetical and the query spectra. The ratio of identified to annotated spectra is still more than 95% because the sequential scan of protein sequence candidates is employed. For $t > 0.65$, the number of clusters increases (up to $t = 0.75$) and the number of missed spectra is almost zero. A disadvantage is that high t may form clusters of spectra not coming from the same peptide. In

Table 5. Impact of distance threshold t on clustering

t	Num. of clusters	Spectra missed	Clustering time [s]	Ratio of ident. spectra [%]	Time of ident. [s]
0.3	162	93	16.6	93.0	0.04
0.4	242	69	16.3	95.3	0.05
0.5	312	49	16.9	95.3	0.06
0.6	368	31	16.0	95.8	0.07
0.65	397	16	16.3	95.8	0.08
0.7	562	4	17.0	95.8	0.11
0.75	633	0	18.0	95.8	0.12
0.8	318	0	24.8	49.8	0.07

practice, the optimal t depends on the number of peaks in query spectra. The optimal t may be higher than 0.65 when a support of PTMs is implemented as described in [19]. The time of identification increases a bit with the increasing t – this corresponds to the increasing number of clusters.

4 Conclusions

We have shown that the clustering of tandem mass spectra significantly improves the efficiency of the method for protein and peptide sequences identification based on the non-metric similarity search in databases of protein sequences. When the NM-tree was employed with clustering, the search was more than 100× faster than the sequential scan without clustering, while the ratio of identified peptides was more than 90% in both cases. The first major premise for successful identification of peptide sequences with clustering is that query sets from at least two spectrometer runs are merged. The second major premise is that the sequential scan of protein sequence candidates is performed because the search using the NM-tree is fast but approximative. The fulfillment of both premises increases the number of identified peptide sequences and speeds up the search.

An important advantage of mass spectra preprocessing by clustering is its independence on the mass spectrometer, which is used to capture the spectra. Since the mass spectrometer can generate spectra in many runs, a disadvantage may be the time complexity of the algorithm, which is used to cluster the spectra. We use only a simple clustering algorithm with the quadratic time complexity, thus the implementation of a more sophisticated clustering algorithm with the time complexity, e.g., $O(N \log N)$ is a subject of our future work.

References

1. Alfassi, Z.B.: On the Normalization of a Mass Spectrum for Comparison of Two Spectra. Journal of the Am. Soc. for Mass Spec. 15(3), 385–387 (2004)
2. Beer, I., Barnea, E., Ziv, T., Admon, A.: Improving large-scale proteomics by clustering of mass spectrometry data. Proteomics 4, 950–960 (2004)
3. Ciaccia, P., Patella, M., Zezula, P.: M-tree: An Efficient Access Method for Similarity Search in Metric Spaces. In: VLDB, pp. 426–435 (1997)
4. Dutta, D., Chen, T.: Speeding up Tandem Mass Spectrometry Database Search: Metric Embeddings and Fast Near Neighbor Search. Bioinf. 23(5), 612–618 (2007)
5. Falkner, J.A., Falkner, J.W., Yocum, A.K., Andrews, P.C.: A spectral clustering approach to MS/MS identification of post-translational modifications. Journal of Proteome Research 7(11), 4614–4622 (2008)
6. Flikka, K., et al.: Improving the reliability and throughput of mass spectrometry-based proteomics by spectrum quality filtering. Proteomics 6, 2086–2094 (2006)
7. Flikka, K., et al.: Implementation and application of a versatile clustering tool for tandem mass spectrometry data. Proteomics 7, 3245–3258 (2007)
8. Frank, A.M., et al.: Clustering millions of tandem mass spectra. Journal of Proteome Research 7(1), 113–122 (2008)
9. Hinneburg, A., Keim, D.A.: An Efficient Approach to Clustering in Large Multimedia Databases with Noise. In: Proc. of KDD 1998, pp. 58–65 (1998)

10. Keller, A., et al.: Experimental Protein Mixture for Validating Tandem Mass Spectral Analysis. OMICS: A Journal of Integrative Biology 6(2), 207–212 (2002)
11. Li, Y., et al.: Speeding up tandem mass spectrometry based database searching by peptide and spectrum indexing. Rapid Comm. Mass Spec. 24(6), 807–814 (2010)
12. Liu, J., et al.: Methods for peptide identification by spectral comparison. Proteome Science 5(3) (2007)
13. Lu, B., Chen, T.: A Suffix Tree Approach to the Interpretation of Tandem Mass Spectra: Applications to Peptides of Non-specific Digestion and Post-translational Modifications. Bioinformatics 19(suppl.2), ii113–ii121 (2003)
14. Mao, R., Ramakrishnan, S.R., Nuckolls, G., Miranker, D.P.: An inverted index for mass spectra similarity query and comparison with a metric-space method: case study. In: SISAP 2010, pp. 93–99 (2010)
15. MSDB, http://www.proteomics.leeds.ac.uk/bioinf/
16. Nesvizhskii, A.I.: A survey of computational methods and error rate estimation procedures for peptide and protein identification in shotgun proteomics. Journal of Proteomics 73(11), 2092–2123 (2010)
17. Nesvizhskii, A.I., et al.: Dynamic Spectrum Quality Assessment and Iterative Computational Analysis of Shotgun Proteomic Data. Molecular & Cellular Proteomics 5, 652–670 (2006)
18. Novák, J., Hoksza, D.: Parametrised Hausdorff Distance as a Non-Metric Similarity Model for Tandem Mass Spectrometry. In: CEUR Proc. DATESO, pp. 1–12 (2010)
19. Novák, J., Skopal, T., Hoksza, D., Lokoč, J.: Non-metric Similarity Search of Tandem Mass Spectra Including Posttranslational Modifications. Journal of Discrete Algorithms (2011), http://dx.doi.org/10.1016/j.jda.2011.10.003
20. Park, C.Y., et al.: Rapid and accurate peptide identification from tandem mass spectra. Journal of Proteome research 7(7), 3022–3027 (2008)
21. Pevzner, P.A., Mulyukov, Z., Dančík, V., Tang, C.L.: Efficiency of Database Search for Identification of Mutated and Modified Proteins via Mass Spectrometry. Genome Research 11(2), 290–299 (2001)
22. Ramakrishnan, S.R., et al.: A Fast Coarse Filtering Method for Peptide Identification by Mass Spectrometry. Bioinformatics 22(12), 1524–1531 (2006)
23. Renard, B.Y., et al.: When less can yield more - Computational preprocessing of MS/MS spectra for peptide identification. Proteomics 9, 4978–4984 (2009)
24. Sadygov, R.G., et al.: Large-scale Database Searching Using Tandem Mass Spectra: Looking up the Answer in the Back of the Book. Nature Met. 1(3), 195–202 (2004)
25. Salmi, J., Nyman, T.A., Nevalainen, O.S., Aittokallio, T.: Filtering strategies for improving protein identification in high-throughput MS/MS studies. Proteomics 9, 848–860 (2009)
26. Skopal, T.: Unified Framework for Fast Exact and Approximate Search in Dissimilarity Spaces. ACM Transactions on Database Systems 32(4), 29 (2007)
27. Skopal, T., Lokoč, J.: NM-Tree: Flexible Approximate Similarity Search in Metric and Non-metric Spaces. In: Bhowmick, S.S., Küng, J., Wagner, R. (eds.) DEXA 2008. LNCS, vol. 5181, pp. 312–325. Springer, Heidelberg (2008)
28. Tabb, D.L., et al.: Similarity among Tandem Mass Spectra from Proteomic Experiments: Detection, Significance and Utility. Anal. Chem. 75(10) (2003)
29. Wang, J., et al.: Peptide identification from mixture tandem mass spectra. Molecular & Cellular Proteomics 9(7), 1476–1485 (2010)
30. Xu, R., Wunsch, D.: Survey of Clustering Algorithms. IEEE Transactions on neural networks 16(3), 645–678 (2005)
31. Zezula, P., Amato, G., Dohnal, V., Batko, M.: Similarity Search: The Metric Space Approach. Advances in Database Systems. Springer, USA (2006)

On the Comparison of Sets of Alternative Transcripts

Aïda Ouangraoua[1], Krister M. Swenson[2], and Anne Bergeron[3]

[1] INRIA Lille, LIFL, Université Lille 1, Villeneuve d'Ascq, France
[2] Université de Montréal and McGill University, Canada
[3] Lacim, Université du Québec à Montréal, Montréal, Canada

Abstract. Alternative splicing is pervasive among complex eukaryote species. For some genes shared by numerous species, dozens of alternative transcripts are already annotated in databases. Most recent studies compare and catalog alternate splicing events within or across species, but there is an urgent need to be able to compare sets of whole transcripts both manually and automatically.

In this paper, we propose a general framework to compare sets of transcripts that are transcribed from orthologous loci of several species. The model is based on the construction of a common reference sequence, and on annotations that allow the reconstruction of ancestral sequences, the identification of conserved events, and the inference of gains and losses of donor/acceptors sites, exons, introns and transcripts.

Our representation of sets of transcripts is straightforward, and readable by both humans and computers. On the other hand, the model has a precise, formal specification that insures its coherence, consistency and scalability. We give several examples, among them a comparison of 24 Smox gene transcripts across five species.

1 Introduction

One of the most intriguing and powerful discoveries of the post-genomic era is the revelation of the extent of alternative splicing in eukaryote genomes, where a single gene sequence can produce a multitude of transcripts [3,5]. The "one gene, one protein" dogma of the last century has not merely been shaken, it has been shattered into pieces, and these pieces tell a story in which genome sequences acquire new function not only by mutation, but by being processed differently.

The main inspiration for this work is the recent paper by [12] that describes the variety of splicing events in vertebrates: the authors carefully annotated and validated hundreds of transcripts from over three hundred genes in human, mouse and other genomes, yielding dozens of conserved or species-specific splicing events. The results are given as combined statistics by species or group of species, and cataloged as one of 68 different kinds of splicing events.

However, beyond recognizing that two transcripts are conserved between species, or that a specific alternative splicing event is conserved, there is no formal setting for the comparison of two or more sets of transcripts that are transcribed from homologous genes of different genomes. The most widely used approach is to resort to comparing all pairs of transcripts within a set, or between two sets (see [19] for a review).

There are several hurdles on the way to a good representation. The first comes from the fact that, when alternate transcripts were scarce, much of the focus was directed towards the representation of alternative splicing events: splicing graphs [8] or pictograms [1] are adequate but do not scale easily to genes that can have dozens of transcripts, or to comparison between multiple species. Other representation techniques, such as bit matrices and codes (see [13,16] and references therein), proposed for the identification and the categorization of alternative splicing events are often more appropriate for computers than for human beings. A second problem is the identification of the features to compare. The splicing machinery is entangled with a myriad of bits and pieces that can vary within and between species: transcripts, coding sequences, exons, introns, splicing donor and acceptor sites, start and stop codons, untranslated regions of arbitrary lengths, frame shifts, etc. Ideally, a model would capture as much as is known about transcripts, including the underlying sequences. In that direction, the goal of the Exalign method [14,18] is to integrate the exon-intron structure of transcripts with gene comparison, in order to find "splicing orthology" for pairs of transcripts. What about orthologous sets of transcripts?

Here we propose a switch from the paradigm of comparing single transcripts between species, to comparing all transcripts with respect to a common reference sequence — derived from a multiple alignment when several species are considered — on which splicing events are represented in a consistent manner. We show that this yields very flexible tools to compare sets of transcripts, that can incorporate the various mechanisms that drive transcript evolution.

The paper is organized as follows. Section 2 presents the model in the single species context. This elementary model is developed in Sections 3 and 4 to include coding sequences and multiple species comparison. Section 6 contains the formal details of the common reference sequence construction.

2 Basic Representations of Transcripts

In this section, we give an intuitive presentation of the basic techniques for representing transcripts. It is self-contained, and the definitions should give the reader a basic understanding of the model. Section 6 presents the formal model, with a level of detail that is not necessary for most situations.

We call an RNA molecule that has made it to a sequencing machine a *transcript*. The sequence for a transcript usually matches the genomic sequence, or *locus*, from which it was transcribed in a piecewise way: parts of the genomic sequence are spliced out, called *introns*, and parts are retained, the *exons*. Different transcripts from the same genomic region may exhibit different combinations of introns and exons in what is called *alternative splicing*, as illustrated in Figure 1 (A) for the human Smox gene.

In order to represent a variety of transcripts in a simple way, we first define blocks of consecutive exons that have no internal variation:

Definition 1. *Given a set of transcripts from the same genomic sequence, an exon block is a maximal sequence of adjacent exons or exon fragments that always appear together in the set of transcripts.*

An exon block may contain introns, for example block B of Figure 1 contains two introns. These introns are inconsequential to our comparisons, since they always start and end at the same position of the genomic sequence in all transcripts of the set: they are said to be *identically spliced*. Maximal intron fragments that are common to all transcripts but not identically spliced form *intron blocks*, such as blocks C, G and I of Figure 1.

Labeling the segments delimited by exon and intron blocks on the genome sequence, from left to right, gives the *reference sequence* $\mathcal{R}$ corresponding to a set of transcripts. Each segment of the reference sequence is simply called a *block*. For example, the reference sequence for the Smox gene of Figure 1 is ABCDEFGHIJ.

A *splicing event* is the removal of a consecutive set of blocks in the reference sequence $\mathcal{R}$. Transcripts can be represented by sequences of block labels, such as AB.D.F.J for transcript H001 in Figure 1 (B), where the dots between blocks indicate the position of the splicing events that created the transcript.

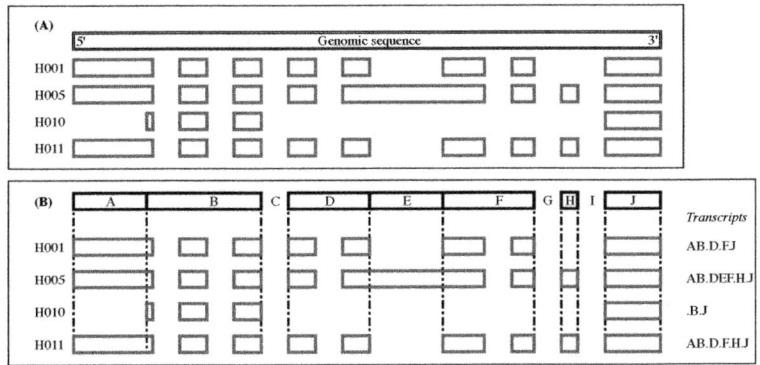

Fig. 1. Four transcripts of the human gene Smox. (A) The transcripts mapped to the genome sequence. Exons are depicted as blue boxes. Transcripts are named with a four symbol code that corresponds to the first letter of the species name, followed by the last three symbols of their transcript identifier in the Ensembl database (see Appendix 1). (B) Blocks of exons or exons fragments that always appear together in the transcripts are depicted by black boxes. Labeling the segments from A to J gives the reference sequence. The four transcripts are represented by the sequences AB.D.F.J, AB.DEF.H.J, .B.J and AB.D.F.H.J.

Introns that are removed by a splicing event are characterized by a *donor site* at their 5' end, known as the exon-intron junction, and by an *acceptor site* at their 3' end, known as the intron-exon junction. The existence of donor or acceptor sites is an attribute of a specific locus or gene sequence. Thus, given a set of transcripts that are transcribed from the same locus, it is natural to annotate a reference sequence with their donor/acceptor sites. We use the following convention:

Definition 2. *In a reference sequence $\mathcal{R}$, we denote by '<b' the fact that the beginning of block b is a donor site, and by 'b>' the fact that the end of block b is an acceptor site.*

For example, the annotated reference sequence for the Smox gene of Figure 1 is:

$$A{>}B{<}C{>}D{<}E{>}F{<}G{>}H{<}I{>}J$$

Since splicing events are consecutive along a transcript, it is also possible to use these annotations to indicate the splicing events that constructed transcripts. For example the four transcripts of Figure 1 may be represented by:

H001: AB<C>D<E>F<GHI>J
H005: AB<C>DEF<G>H<I>J
H010: A>B<CDEFGHI>J
H011: AB<C>D<E>F<G>H<I>J

In the representation of individual transcripts, it is understood that, for internal splicing events, each '<' symbol is paired with the next '>' symbol to create a splicing event. Since donor and acceptor annotations are sets, all the usual set operations can be applied to these annotated sequences. For example, the common annotations for the four transcripts of the human Smox gene is:

$$AB{<}CDEFGHI{>}J$$

In the next sections, we will see that the main advantages of this representation is that it can be used immediately to represent sets of coding transcripts, and to compare sets of transcripts from different species.

3 Integrating Coding Information

Many transcripts will eventually be translated into proteins. The formalism of the preceding section can be adapted to indicate coding exons by adding block separators before each start codon, and after each stop codon. We associate to each block its *coding sequence*: for blocks that contain introns, the coding sequence is the concatenation of its exons or exon fragments; for blocks that contain only intron fragments, the coding sequence is empty. A *coding block* is a block whose coding sequence is not empty.

Figure 2 shows an example using four transcripts of the human Crem gene. Shaded rectangles represent translated regions.

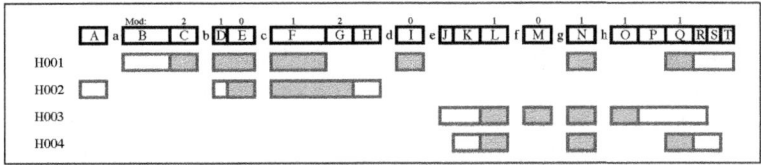

Fig. 2. Four transcripts of the human gene Crem. Shaded rectangles represent translated regions.

The length of a block coding sequence is not necessarily a multiple of three, although the total length of all coding sequences of a transcript must be. When a coding block is skipped, this event may introduce a *frameshift* if the length of the skipped block is not a multiple of three. In order to keep track of the possibility of frameshift, we associate to each block a Mod value which is the remainder of the length of its coding sequence divided by 3. When studying transcripts from a single species, these values can help determine alternative early stop codons, as illustrated by H003 compared to H001 and H004 in Figure 2. However, we will see in Section 6 that these values also have an impact on block construction.

Trimming the blocks that precede the first start codon (in the set), and those that succeed the last stop codon (in the set) does not significantly alter the information content. In the following table, the untranslated blocks of the transcripts of Figure 2 have been painted in blue, and the second column give the shortened versions of the transcripts, where the blocks preceding block C and following block Q have been trimmed.

H001	.BC.DE.F.I.N.QRST	C.DE.F.I.N.Q
H002	A.DE.FGH.	.DE.FGH.
H003	.JKL.M.N.OPQR.	.JKL.M.N.OPQ
H004	.KL.N.QRS.	.KL.N.Q

Early start codons, and late stop codons, are still identifiable in the shortened version, and the untranslated regions that are coding in alternate transcripts are visible. Note that skipping the untranslated region before the start codons is often done in practice due to the variability of the transcription initiation site [5], and splicing events that occur in untranslated regions after a stop codon often lead to a condition known as *nonsense mediated decay* that prevents the translation of the transcript [11].

4 Multi-species Comparisons

Here we show how to adapt the representations of the last two sections to sets of transcripts that come from orthologous loci in multiple species.

Our goal is to construct a common reference sequence using a multiple alignment, and cut it into blocks. The examples of this section are straightforward applications of the representation. Exceptions and limiting cases are treated in Section 6.

Given an alignment of two or more genes and a transcript t from one of these genes, it is always possible to color the exons of t in the corresponding gene sequence. For example, Figure 3 shows an alignment of the human and mouse Ensa gene, human transcript H004 has been colored in blue for the human sequence, and mouse transcript M201 has been colored in red. In the single species context, the blocks of the reference sequence were segments of the genomic sequence, here the blocks are defined as intervals of the columns of the alignment. Block junctions correspond to positions where changes of color occur. The common

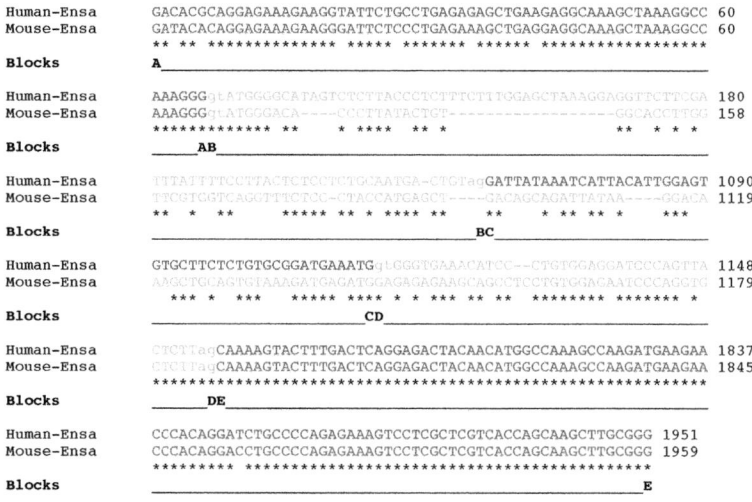

Fig. 3. An alignment of the partial human and mouse Ensa gene, where the exons of human transcript H004 have been colored in blue, and the exons of mouse transcript M201 in red. The blocks of the reference sequence are given below the alignment.

reference sequence for the human and mouse Ensa gene would be the sequence ABCDE as defined by Figure 3. Although block C is an exon in the human transcript, preceded by the canonical 'ag' acceptor site at its 5' junction, and followed by the canonical 'gt' donor site at its 3' junction, these features do not exist in the mouse sequence. However, both transcripts can be represented with the formalism of the preceding section, human as A<B>C<D>E and mouse as A<BCD>E.

This approach can be applied to genes that have a variety of transcripts in many different species. Figure 4 shows 24 transcripts of the Smox gene: 4 from the human gene, 1 from the chimpanzee gene, 2 from the orangutan gene, 10 from the mouse gene and 7 from the rat gene. We chose transcripts that were annotated in the CCDS database [15] for human and mouse, along with transcripts that were annotated as protein coding in the Ensembl database [7].

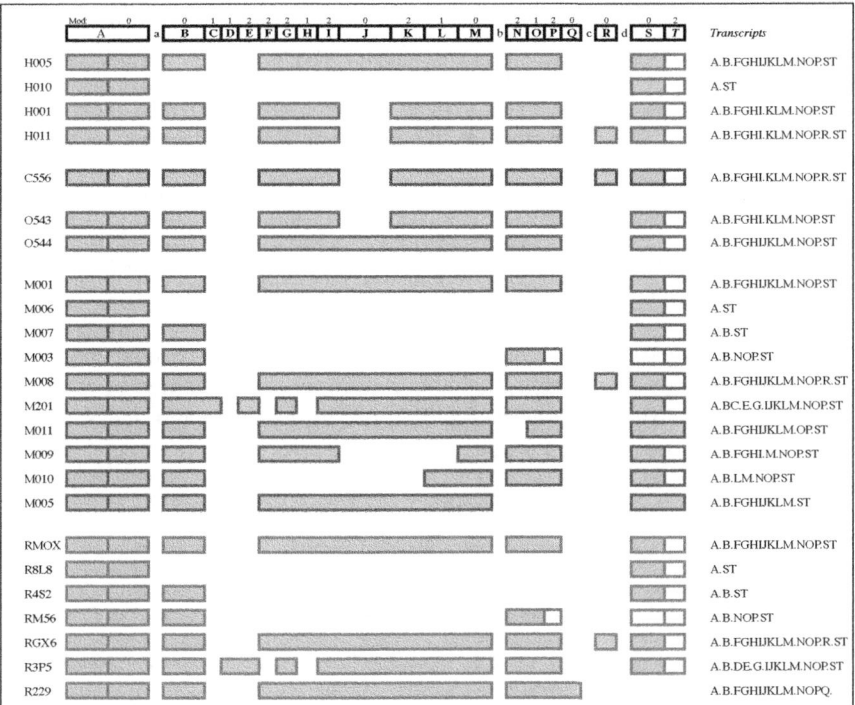

Fig. 4. An alignment of the Smox gene, where the exons of 4 human transcripts have been colored in blue, 1 chimpanzee transcript in dark blue, 2 orangutan transcripts in indigo, 10 mouse transcripts in red, and 7 rat transcripts in green

In the common reference sequence, there are 19 blocks that contain at least one exonic sequence, labeled from 'A' to 'T', and 4 blocks that are common intron fragments, labeled from 'a' to 'd'.

Transcripts that have the same representation are *splicing orthologs* [14]. For example, in Figure 4, the following sets of transcripts are sets of orthologs: {H005, O544, M001, RMOX}, {H010, M006, R8L8}, {H011, C556}, {H001, O543}, {M007, R4S2}, {M003, RM56} and {M008, RGX6}. In order to say more about the relations between these sets, we may compare the reference sequences for each species, constructed using the common reference:

observed in the orangutan. The annotations that are common to all five species reference sequences are: <a>, <C, E>, <b>, <Q, and d>. The common annotations are captured by the sequence:

A<a>B<CDE>FGHIJKLM<b>NOP<QcRd>ST

Annotations can also be used in the context of ancestral reconstruction. Each donor or acceptor site is either present or absent in a locus reference sequence. Given a phylogenetic tree for a set of loci, it is possible to apply a Fitch-like algorithm [6] to each site in order to determine the ancestral states yielding the minimum number of gains/losses of sites. Figure 5 gives an example of such a reconstruction, using the five reference sequences of the Smox gene without their common annotations. Annotations unambiguously assigned to the ancestor of the five species are <J, c> and <d. The rodent ancestor has <F> and <H>, and the primate ancestor has J>.

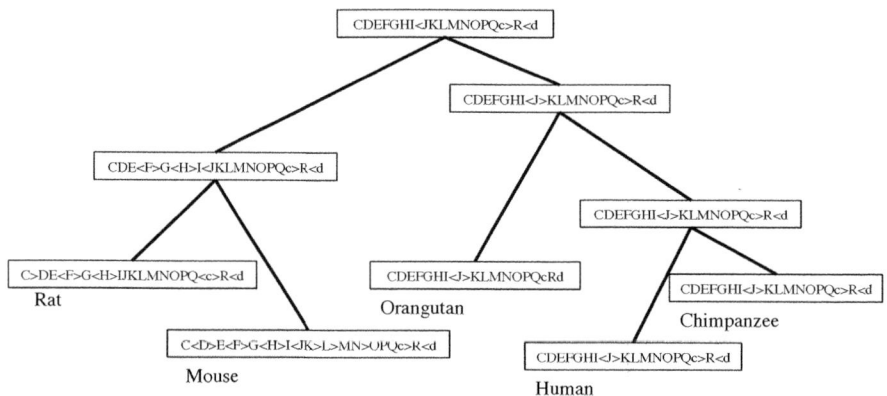

Fig. 5. Ancestor reconstruction. Donor/acceptor sites are assigned to each node with a Fitch-like algorithm to minimize the number of gains/losses of sites.

5 Comparison with Other Formalisms

Most formalisms for the representation of splicing events were developed for the comparative study of transcripts of a single gene [1,8], or the comparison of single events in the same gene or between orthologuous genes [16,12]. In order to be able to assess splicing orthology between two sets of transcripts, as proposed in [18], it is necessary to be able to represent, with the same formalism, exon and intron structure of single transcripts, common and diverging structures of transcripts from the same gene, and common and diverging structures of sets of transcripts from different species.

In the initial phase of this project, we explored some generalizations of splicing graphs [8] and their variants [1,2]. The simplest solution was to construct the graphs using the common blocks of the different species, and compare sets of transcripts using graph comparisons. Assuming that these tasks are simple, there remains a major problem in the fact that two sets of transcripts that have no common elements may still yield the same graph [9]: it is thus necessary to keep track of individual transcripts, together with more general representations that capture their similarities.

The approach of Mudge et al. [12] is adapted to the comparison of splicing events among several species, but focus on the alternative aspect of events: two species may share a transcript ABCDE, but if one has the alternative event A<B>CDE, and the other event ABC<D>E, a local approach would conclude that there is no conserved splicing event. In this case, we think that the fact that the two species have a common transcript should witness a certain degree of similarity. The Bubbles formalism [17] has been proposed to describe the splicing events of a set of transcripts, but can only be applied to the set of common transcripts of a group of species.

Zambelli et al. [18] introduce the concept of "iso-orthology" between transcripts whose less stringent class corresponds roughly to equality of sequences in our proposed representation. However, iso-orthology cannot capture more general similarities between sets of transcripts. For example, in Figure 4, the sets of transcripts of the human, chimpanzee and orangutan are remarkably similar, despite the fact that there are no common iso-orthologs in the three species; the mouse and rat transcripts have five iso-orthologs, but they also each have two transcripts (M201 and R3P5) that are not iso-orthologs, but share a quite distinct conserved feature, exon G, that seems to be unique to the rodents. On the other hand, the similarity, or dissimilarity, of the reference sequences proposed in Section 4 would capture these relations.

6 Formalities of Block Construction

Blocks are constructed using transcript annotations and a multiple alignment. For simplicity, we assume that the transcripts of a species are all transcribed from a contiguous locus, or *gene*, and that the annotations give the position of each exon with respect to the coordinates of the locus.

A multiple alignment of all genes under consideration is obtained using standard software – we used ClustalW [10] for the examples presented in this paper. The quality of the comparison will clearly be influenced by the quality of the alignment: genes that share some highly homologous exons will be easier to compare, even if some exons might be absent in some species.

Let n be the number of columns of the alignment. A transcript t from gene g is represented by a list of n values, 0,1, or '-'. Position i has value 1 if and only if the nucleotide at column i in gene g belongs to an exon of transcript t, or column i belongs to a gap flanked by two nucleotides of the same exon of transcript t. Position i has value 0 if and only if the nucleotide at column i in gene g belongs to an intron of transcript t, or column i belongs to a gap flanked by two nucleotides of the same intron of transcript t. Otherwise the value at position i is the gap character '-'. This yields the *transcript matrix*, which is reminiscent of the bit matrices used in [13].

```
Positions                  1111111111222222222233333333334444444444555555555 6
                  12345678901234567890123456789012345678901234567890123456 7890
Gene x            CATCTGGGTCCGAGGATGCATGCTAGCGGAGGTCCAGCCCTGACCGCT---CCAGCCGGC
Gene y            CATCTGGGTCTGAGGATGCCATGA-------------CTCCTACCCCTAGT----CCGGC

Transcript x.1    00001111111111111111100000000000000000001110011111111111100
Transcript x.2    01110000111111111111100001111110000111111111100111111111111100
Transcript x.3    01111111111111111111100001111110000111111111100111---000011100
Transcript y.1    00001111111000011110000-------------111111001111111111100
Blocks            A_AB__BC__CD__DE__Ea__aF_____FG_GH_____HI_IJ__JK_K
```

Fig. 6. An example of block construction. Values in red correspond to gaps that are within exons.

For example, Figure 6 shows an alignment of two genes, with transcripts $x.1$, $x.2$ and $x.3$ from Gene x, and transcript $y.1$ from Gene y.

A *positive* column in the transcript matrix is a column that contains at least one value 1. Blocks are defined as maximal intervals $[i..j]$ such that 1) column i and j of the transcript matrix are positive, and all the positive columns in the interval are equal, or 2) maximal intervals between two such intervals. For a gene, the *sequence* of a block $[i..j]$ is the subsequence of nucleotides that corresponds to positive columns between i and j. When all the Mod values of the non-empty sequences of a block are equal, then the Mod value of the block can be clearly assigned. Since the block structure should reflect ortholog exons or exon parts, different Mod values require further investigation especially for coding exons.

In the above example, block F is defined by positions [25..37] of the alignment. Of these 13 positions, only 9 positions have a positive column, thus the sequence of block F for Gene x is 'AGCGGACAG', and for Gene y, the empty sequence. For blocks that do not contain any positive column, all the associated sequences are empty. The sequence $\mathcal{R}$ of all blocks of a multiple alignment is called the *reference sequence*. A consecutive sequence of blocks is called an *exon* if the sequence is the longest possible that exactly contains the nucleotide sequence of an exon of one of the transcripts. The sequence of blocks between two exons of the same transcript is called an *intron*. Some 'real' introns may not be representable as sequences of blocks if they are always identically spliced between the same exons. An example is the intron contained in block H in Figure 6.

The reference sequence of the above example is ABCDEaFGHIJK. The four transcripts are represented as follows: $x.1$ as BCDEHIJK, $x.2$ as ACDEFGHIJK, $x.3$ as ABCDEFGHIK, $y.1$ as BCEFGHIJK. Note that the first of these transcripts does not have block F, since it is included in one of its introns. The last has block F, but the corresponding sequence is empty. The second exon of transcript $x.3$ is represented by the sequence FGHI, even though the sequence of block I is empty for Gene x. In this model, we assume that the multiple alignment includes all genes under consideration, even if some genes do not have observed transcripts. Adding new transcripts may split existing blocks.

Proposition 1. *The nucleotide sequence of a transcript from a gene s is the concatenation of the nucleotide sequences of the corresponding blocks $b_1 \ldots b_k$ for gene s.*

As in Section 2, a splicing event removes consecutive blocks from the reference sequence. Any such event defines a donor and/or an acceptor site:

Definition 3. *Block $b = [i..j]$ contains a donor site, denoted '$<b$', if one of the transcripts has an intron starting at position i of the alignment. It contains an acceptor site, denoted '$b>$', if one of the transcripts has an intron ending at position j of the alignment.*

7 Conclusion

We have described a succinct and readable representation for transcripts of alternatively spliced genes. Our representation is amenable to the comparison of sets of transcripts for a single gene, or to sets of transcripts corresponding to orthologous genes across multiple species. To our knowledge, this is the first such representation in the literature; existing studies consider single transcripts, or splicing events across multiple species in isolation. The utility of the representation was demonstrated first on the set of transcripts from the human Smox gene and then on the coding transcripts of the human Crem gene. The Smox gene was then considered in a phylogenetic context where ancestral sets of transcripts were inferred via maximum parsimony. Beyond ancestral inference, we expect that this representation will lead to new tools for phylogeny reconstruction [see [4], for example], transcript discovery, and homologous gene discovery.

References

1. Bollina, D., Lee, B.T., Tan, T.W., Ranganathan, S.: ASGS: an alternative splicing graph web service. Nucleic Acids Res. 34, W444–W447 (2006)
2. Bonizzoni, P., Mauri, G., Pesole, G., Picardi, E., Pirola, Y., Rizzi, R.: Detecting alternative gene structures from spliced ESTs: a computational approach. J. Comput. Biol. 16, 43–66 (2009)
3. Carninci, P., Kasukawa, T., Katayama, S., et al.: The transcriptional landscape of the mammalian genome. Science 309, 1559–1563 (2005)
4. Christinat, Y., Moret, B.M.E.: Inferring transcript phylogenies. In: Proc. of IEEE International Conference on Bioinformatics and Biomedicine, pp. 208–215 (2011)
5. The ENCODE Project Consortium. Identification and analysis of functional elements in 1% of the human genome by the encode pilot project. Nature 447, 799–816 (2007)
6. Fitch, W.M.: Toward defining the course of evolution: minimum change for a specified tree topology. Systematic Zoology 20(4), 406–416 (1971)
7. Flicek, P., Amode, M.R., Barrell, D., et al.: Ensembl 2011. Nucleic Acids Res. 39, D800–D806 (2011)
8. Heber, S., Alekseyev, M., Sze, S.H., Tang, H., Pevzner, P.A.: Splicing graphs and EST assembly problem. Bioinformatics 18(suppl.1), S181–S188 (2002)

9. Lacroix, V., Sammeth, M., Guigo, R., Bergeron, A.: Exact Transcriptome Reconstruction from Short Sequence Reads. In: Crandall, K.A., Lagergren, J. (eds.) WABI 2008. LNCS (LNBI), vol. 5251, pp. 50–63. Springer, Heidelberg (2008)
10. Larkin, M.A., Blackshields, G., Brown, N.P., Chenna, R., McGettigan, P.A., McWilliam, H., Valentin, F., Wallace, A., Wilm, I.M., Lopez, R., Thompson, J.D., Gibson, T.J., Higgins, D.G.: Clustal w and clustal x version 2.0. Bioinformatics 23, 2947–2948 (2007)
11. Mendell, J.T., Sharifi, N.A., Meyers, J.L., Martinez-Murillo, F., Dietz, H.C.: Nonsense surveillance regulates expression of diverse classes of mammalian transcripts and mutes genomic noise. Nature Genetics 36, 1073–1078 (2004)
12. Mudge, J.M., Frankish, A., Fernandez-Banet, J., Alioto, T., Derrien, T., Howald, C., Reymond, A., Guigo, R., Hubbard, T., Harrow, J.: The origins, evolution and functional potential of alternative splicing in vertebrates. Molecular Biology and Evolution 28, 2949–2959 (2011)
13. Nagasaki, H., Arita, M., Nishizawa, T., Suwa, M., Gotoh, O.: Automated classification of alternative splicing and transcriptional initiation and construction of visual database of classified patterns. Bioinformatics 22(10), 1211–1216 (2006)
14. Pavesi, G., Zambelli, F., Caggese, C., Pesole, G.: Exalign: a new method for comparative analysis of exon-intron gene structures. Nucleic Acids Res. 36, e47 (2008)
15. Pruitt, K.D., Harrow, J., Harte, R.A., et al.: The consensus coding sequence (CCDS) project: Identifying a common protein-coding gene set for the human and mouse genomes. Genome Res. 19, 1316–1323 (2009)
16. Sammeth, M., Foissac, S., Guigo, R.: A general definition and nomenclature for alternative splicing events. PLoS Computational Biology 8, e1000147 (2008)
17. Sammeth, M., Valiente, G., Guigo, R.: Bubbles: Alternative Splicing Events of Arbitrary Dimension in Splicing Graphs. In: Vingron, M., Wong, L. (eds.) RECOMB 2008. LNCS (LNBI), vol. 4955, pp. 372–395. Springer, Heidelberg (2008)
18. Zambelli, F., Pavesi, G., Gissi, C., Horner, D.S., Pesole, G.: Assessment of orthologous splicing isoforms in human and mouse orthologous genes. BMC Genomics 11, 534 (2010)
19. Zavolan, M., van Nimwegen, E.: The types and prevalence of alternative splice forms. Curr. Opin. Struct. Biol. 16, 362–367 (2006)

MURPAR: A Fast Heuristic for Inferring Parsimonious Phylogenetic Networks from Multiple Gene Trees

Hyun Jung Park and Luay Nakhleh

Dept. of Computer Science, Rice University, Houston, TX 77005
{hp6,nakhleh}@cs.rice.edu

Abstract. Phylogenetic networks provide a graphical representation of evolutionary histories that involve non-treelike evolutionary events, such as horizontal gene transfer (HGT). One approach for inferring phylogenetic networks is based on reconciling gene trees, assuming all incongruence among the gene trees is due to HGT. Several mathematical results and algorithms, both exact and heuristic, have been introduced to construct and analyze phylogenetic networks. Here, we address the computational problem of inferring phylogenetic networks with minimum reticulations from a collection of gene trees. As this problem is known to be NP-hard even for a pair of gene trees, the problem at hand is very hard. In this paper, we present an efficient heuristic, MURPAR, for inferring a phylogenetic network from a collection of gene trees by using pairwise reconciliations of trees in the collection. Given the development of efficient and accurate methods for pairwise gene tree reconciliations, MURPAR inherits this efficiency and accuracy. Further, the method includes a formulation for combining pairwise reconciliations that is naturally amenable to an efficient *integer linear programming* (ILP) solution. We show that MURPAR produces more accurate results than other methods and is at least as fast, when run on synthetic and biological data. We believe that our method is especially important for rapidly obtaining estimates of genome-scale evolutionary histories that can be further refined by more detailed and compute-intensive methods.

1 Introduction

One aspect of phylogenomic studies entails the reconstruction of evolutionary trees for different genomic regions and combining those trees to obtain an evolutionary history of the genomes under study. When events such as horizontal gene transfer (HGT) occur, the evolutionary history of the genomes is best represented by a *phylogenetic network*, which accounts simultaneously for vertical and horizontal inheritance of genetic material; e.g., see [2,11,15]. Several algorithms, both exact and heuristic, have been proposed for combining a pair of gene trees into a phylogenetic network (which is an NP-hard problem) [6,9,16,20,3,14,8,5,25,26,13]. Given that phylogenomic analyses involve a large number of gene trees, it is imperative to develop algorithms and tools that infer such networks from multiple gene trees.

In this paper, we address the computational problem of inferring a phylogenetic network from a set of gene trees. A few methods were introduced very recently to address this problem. Huson and Rupp [12] proposed a method for summarizing a collection of gene trees using *cluster networks*, which differ from the phylogenetic network model we address here. Beiko and Ragan [4] discussed aggregating inferred HGT events from pairwise tree comparisons, and discussed three strategies for this task; yet, they did not implement the strategies, nor did they study their performance. Iersel *et al.* [24] developed the CASS method, which is an efficient algorithm for inferring a minimal phylogenetic network that contains all the clusters of taxa displayed by the input gene trees, but not necessarily the input gene trees themselves. Further, Wu [27] recently introduced the PIRN algorithm for obtaining lower and upper bounds on the amount of reticulations necessary for reconciling a set of input gene trees. Finally, we introduced two methods for estimating the amount of reticulation, as well as inferring a phylogenetic network, from a collection of gene trees [17]. Both methods are based on obtaining estimates for the set of trees from pairwise distance calculations. In particular, the approach of combining pairwise reconciliations to obtain a solution for the entire set of trees, though *ad hoc*, showed good performance in simulation studies. Here, we discuss the suitability of combining pairwise reconciliations (reconciliations obtained from comparing pairs of trees in the set of all gene trees) to obtain a phylogenetic network that reconciles a set of gene trees. Further, we define the problem formally and provide an integer linear programming (ILP) solution for it. It is important to mention that this makes an improvement on the previous algorithm not only in terms of time, but also accuracy. Finally, we study the performance of the method, and compare it to other methods, on synthetic data sets and one biological data set. The results show that our method is fast in practice, and that it produces accurate estimates of the phylogenetic network. Our method makes two main assumptions about the input gene trees: (1) the trees are accurate (that is, we ignore incongruence among the trees due to error in the gene tree inference), and (2) reticulation is the only biological cause of gene tree incongruence. Of course, these two assumptions may be violated in practice; developing methods that relax these assumptions is the holy grail in this area, and is well beyond the scope of a single paper. Further, it is important to note that related methods also make these assumptions. We believe our method is best used to obtain a quick analysis assuming the two conditions hold, and then following up with careful inspection of the reticulations inferred by the method. Therefore, these strict assumptions notwithstanding, we believe the method amounts to an important contribution.

2 Phylogenetic Networks and Trees

In this section, we discuss the plausibility, as well as challenges, of obtaining a phylogenetic network that reconciles a set of gene trees by combining networks obtained from analyses of pairs of trees in the set. First, we begin by formally defining networks and the relationship between networks and trees.

Definition 1. *For a set of taxa $\mathscr{X}$, a phylogenetic $\mathscr{X}$-network, or $\mathscr{X}$-network, N is an ordered pair (G, f), where $G = (V, E)$ is a directed, acyclic graph (DAG) with $V = \{r\} \cup V_L \cup V_T \cup V_N$, where*

1. *$indeg(r) = 0$ (r is the root of N);*
   $\forall v \in V_L, indeg(v) = 1$ and $outdeg(v) = 0$ (V_L are the leaves of N);
2. *$\forall v \in V_T, indeg(v) = 1$ and $outdeg(v) \geq 2$ (V_T are the tree nodes of N);*
3. *$\forall v \in V_N, indeg(v) = 2$ and $outdeg(v) \geq 1$ (V_N are the reticulation nodes of N); and,*
4. *$E \subseteq V \times V$ are the network's edges. (we distinguish between reticulation edges, edges whose heads are reticulation nodes, and tree edges, edges whose heads are tree nodes or leaves).*

The mapping $f : V_L \to \mathscr{X}$ is a bijection function from V_L to the set of taxa $\mathscr{X}$.

A phylogenetic $\mathscr{X}$-tree is an $\mathscr{X}$-network in which $V_N = \emptyset$. While a network N represents the evolution of a set of genomes, these genomes can be partitioned into (non-recombining) regions $R_1, R_2, \ldots, R_k$, each of which has a treelike evolutionary history T_i. In other words, the set $\mathscr{T} = \{T_1, \ldots, T_k\}$ is a subset of the set of all trees *displayed* by the network N. More formally, $\mathscr{T} \subseteq \mathscr{T}(N)$, where $\mathscr{T}(N)$ is the set of *all* trees obtained as follows from N: (1) for each node of in-degree 2 remove one of the two incoming edges and (2) for each node u of in-degree and out-degree 1, remove u along with its incident edges, and add a new edge to connect u's parent to u's child (this step is repeated until no such nodes u remain).

3 Pairwise and Set-Wise Reconciliation of Trees

A main problem of interest is the following [15].

Problem 1. (SET-WISE HGT INFERENCE)

> **Input:** A collection of gene trees $\mathscr{G} = \{GT_1, \ldots, GT_k\}$, each modeling the evolutionary history of a genomic region of a set $\mathscr{X}$ of taxa.
> **Output:** A phylogenetic network N with the smallest number of reticulation nodes (a minimal network) such that $\mathscr{G} \subseteq \mathscr{T}(N)$.

When the input consists of exactly two trees, we refer to this as the PAIRWISE HGT INFERENCE problem.

We will now make two assumptions that we will use throughout this paper. First, a solution to the PAIRWISE HGT INFERENCE problem is obtained by solving for $SPR(T_1, T_2)$. In other words, we will take a smallest set Ξ of Subtree Prune and Regraft (SPR) moves that transform T_1 to T_2, and obtain N by adding Ξ to T_1. Second, in the PAIRWISE HGT INFERENCE problem, we will assume that the first tree is a species tree ST. In this problem, we will assume that a species tree is given, so that the distance from each tree in $\mathscr{G}$ to the species tree is computed. We discuss below a potential solution to the problem when a species tree is not given as part of the input.

A natural and simple candidate for solving the SET-WISE HGT INFERENCE problem by utilizing solutions for the pairwise problem is to take

$$SPR(ST, \mathscr{G}) = \cup_{gt \in \mathscr{G}} SPR(ST, gt) \qquad (1)$$

as an estimate of the solution of the *Set-wise HGT Inference* problem. Obviously, $|SPR(ST, \mathscr{G})|$ is larger than or equal to the number of reticulation nodes in a solution to the problem. An advantage of this approach is that fast exact algorithms and heuristics exist for obtaining $SPR(ST, gt)$, as described above, and taking the union of pairwise reconciliations is very simple computationally. Indeed, we developed a method, referred to here as M2, based on this idea, and showed that it yields good performance [17]. Nonetheless, several issues need to be resolved.

First, it is possible that the optimal solutions for $SPR(ST, gt)$ do not lead to the optimal solution for $SPR(ST, \mathscr{G})$, and cause overestimation in a "global" estimate of reticulations. For example, consider the HGT scenarios shown in Fig. 1. If we take the union of the four pairwise solutions computed by $SPR(ST, GT_i)$,

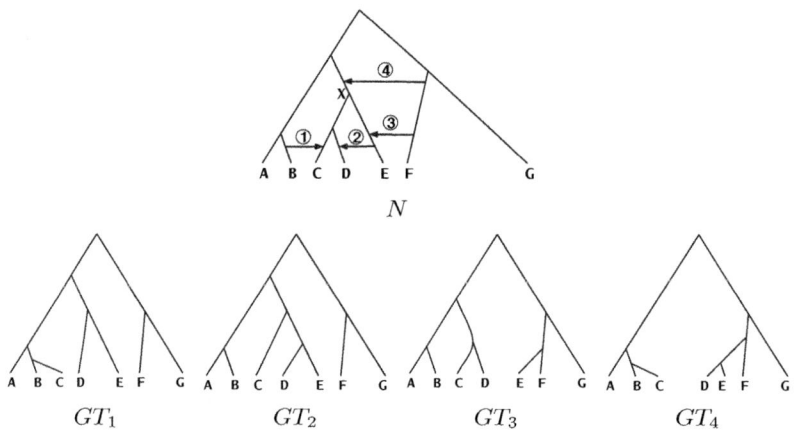

Fig. 1. A phylogenetic network with four independent HGT scenarios. The species tree ST in this case is the network N without the four HGT edges. The gene whose tree is GT_i ($1 \leq i \leq 3$) underwent HGT event (i), and the gene whose tree is GT_4 underwent HGT events (1) and (4). The combined effect of HGTs (1) and (4) on the gene tree topology is the same as the combined effect of HGTs (1), (2) and (3).

for $1 \leq i \leq 4$, we obtain a set of four HGT edges: $\Xi_1 = \{[B \rightarrow C], [E \rightarrow D], [F \rightarrow E], [F \rightarrow X]\}$. However, a smallest solution for the SET-WISE HGT INFERENCE problem given the species tree ST and the four gene trees contains three HGT edges, which is the set $\Xi_2 = \{[B \rightarrow C], [E \rightarrow D], [F \rightarrow E]\}$. In this case, the HGT edge $[F \rightarrow X]$ is not needed, since its effect can be simulated by the two HGT edges $[E \rightarrow D]$ and $[F \rightarrow E]$, once the HGT edge $[B \rightarrow C]$ is applied. However, notice that in this case, the solution that truly reflects what happened is Ξ_1, since the HGT event denoted by $[F \to X]$ did occur, even though its effect on the topologies of gene trees can be simulated by the other three HGT edges. In other words, while the union of pairwise solutions may not provide a smallest global solution, it may provide a solution that is closer to the true HGT scenarios that took place at the genomic level. Further, under these scenarios, where a smallest solution is a proper subset of the union of pairwise solutions, post-processing via gradual elimination of members of the union can yield a smallest solution. However, it is not guaranteed that the smallest solution is a subset of the union of pairwise reconciliations.

Second, $SPR(ST, gt)$ may not be unique; in fact, the number of possible solutions to the Pairwise HGT Inference problem can be exponential in the size of a solution [21]. To account for this issue, we need to consider all solutions, or a large number of them when obtaining all is computationally infeasible, for each pair of trees. Without accounting for multiple solutions, methods such as M2 [17] that agglomerate pairwise solutions obtain biased estimates.

A third issue that requires special attention is the following: while solutions to the PAIRWISE HGT INFERENCE problem may be acyclic (that is, the inferred HGT events, when added to the species tree, do not result in cycles), taking the union of solutions cannot guarantee acyclicity. When this occurs, the solution is not a phylogenetic network as given by Definition 1.

4 MURPAR

As mentioned above, our former method M2 [17] uses the approach given by Eq. (1), but it does not address the last two issues raised above. Here, we propose our heuristic MURPAR (MUlti-tree Reconciliation using PAirwise Reconciliations) for addressing these issues. Let ST be a species tree and $\mathscr{G} = \{GT_1, \ldots, GT_k\}$ be a collection of gene trees. Also let $S_i = \{s_{i,1}, s_{i,2}, \ldots, s_{i,m_i}\}$ be the set of all optimal pairwise solutions on the pair $\langle ST, GT_i \rangle$. Then, M2 counts the frequency with which each potential reticulation edge appears throughout $s_{i,j}, 1 \leq i \leq k, 1 \leq j \leq m_i$, and calculates the set of reticulation edges such that 1) it covers at least one solution for all trees, and 2) it maximizes its frequency value, in the assumption that the frequency reflects how often a reticulation edge is used in each solution and maximizing the frequency would result in a smallest set of reticulation edges.

However, with multiple solutions (e.g., obtained by using RIATA-HGT [16,20]), a reticulation edge occurring multiple times in $s_{i,j}, \forall 1 \leq j \leq m_i$, contributes more to the frequency than those occurring once. As a result, a solution would be biased towards the edges occurring multiple times in solutions. If we define $S = \bigcup_{1 \leq i \leq k, 1 \leq j \leq m_i} s_{i,j}$, then MURPAR seeks the smallest set $S' \subseteq S$ that satisfies the property $[\forall 1 \leq i \leq k, \exists s_{i,j} \in S_i \text{ s.t. } s_{i,j} \subseteq S']$. MURPAR solves this problem using integer linear programming (ILP). We define binary variables as follows:

(A) B_s, $\forall s \in S$. B_s will take value 1 if SPR move s is selected as an element of S', and 0 otherwise.
(B) P_{ij}, $\forall 1 \leq i \leq k, \forall 1 \leq j \leq m_i$. P_{ij} will take value 1 if all SPR moves in the optimal pairwise solution s_{ij} are selected, and 0 otherwise.

Finally, the ILP program is:

minimize $\sum B_s$
subject to
- $P_{ij} = \left[\wedge_{y \in s_{ij}} B_y\right], \forall 1 \leq i \leq k, \forall 1 \leq j \leq m_i$
- $[\vee_{1 \leq j \leq m_i} P_{ij}] = 1, \forall 1 \leq i \leq k$

Here, $\wedge$ and $\vee$ represent logical 'and' and logical 'or', respectively. All these constraints can be turned into linear constraints by introducing auxiliary variables as follows:

- $a = (b_1 \wedge b_2 \wedge \cdots \wedge b_p)$, where all variables are binary, can be turned into the linear inequalities $-1 \leq 2b_1 + 2b_2 + \cdots + 2b_p - 2pa \leq 2p - 1$.
- $(b_1 \vee b_2 \vee \cdots \vee b_p) = 1$, where all variables are binary, can be turned into the linear inequality $b_1 + b_2 + \cdots + b_p \geq 1$.

When a species tree ST is not given, we repeat MURPAR with $GT_i, 1 \leq i \leq k$ as the species tree and choose the smallest S'_i as S'.

Addressing cyclicity. As discussed above, the solution thus far may result in cyclic graphs, which are not phylogenetic networks. To address this issue, MUR-PAR post-processes the results to avoid the solutions with cycles using a straightforward cycle detection algorithm. If a minimal solution is found to have a cycle, MURPAR skips it and inspects the next minimal solution (minimal in terms of the number of reticulation nodes). While all solution candidates found by MUR-PAR may have cycles, and thus MURPAR returns no solution, we found through extensive simulations that this was never the case. Similarly, it was shown in [1] and confirmed in [23] that cycles may not be a major concern for reticulation detection algorithms when run on real data sets or synthetic data sets that are generated under realistic models.

5 Experimental Evaluation

5.1 Data

Simulations were conducted on 30- and 50-taxon phylogenies. For 30-taxon data sets, 10 random trees were generated using PHYL-O-GEN tool [18] as "species tree" under birth-death model, and 5 horizontal gene transfer events were simulated between pairs of branches on the species trees using Galtier's tool [7]. The simulation of horizontal gene transfer were conducted individually 10 times on each species tree, so totally 100 networks are generated from the simulation. Since Galtier's tool does not provide the details of simulated transfer events, we modified the tool to have it report the simulated transfers that it added. From the set of 32 gene trees contained in each network, 4, 8, 12, 16, 24, and 32 gene trees were randomly sampled and used as input to the methods.

For 50-taxon data sets, the same procedure as above was applied, except that the number of horizontal gene transfer events simulated was 10, and the sampling was made over 1024 gene trees.

For biological data, we used the Poacaea data set, which was originally sequenced by the Grass Phylogeny Group, and was used to test both CASS [24] and PIRN [27]. Binary trees were constructed for six loci: *ITS*, *ndhF*, *phyB*, *rbcL*, *rpoC* and *waxy* [19]. Since the gene trees had different sets of leaves, we selected the gene trees for *ndhF*, *phyB*, *rbcL*, *rpoC2* and *ITS*, and restricted them to 14 leaves that they have in common.

5.2 Methods and Accuracy Measure

The PAIRWISE HGT INFERENCE problem was solved using the RIATA-HGT method [16,20] as implemented in the PhyloNet software package [22]. Other pairwise inference tools including SPRIT [10] were tested as well, but results were almost identical; therefore, all results reported here are based on pairwise solutions obtained by RIATA-HGT. We used the GLPK ILP Solver to solve the ILP formulation of MURPAR. For comparison, we also ran PIRN [27] and Method M2 of [17].

We do not use CASS [24] for comparison. As indicated by the authors in [24], while CASS computes a minimal network N from an input set $\mathscr{C}$ of clusters of a set of gene trees $\mathscr{T}$, it is not guaranteed that $\mathscr{T} \subseteq \mathscr{T}(N)$. More formally, if $\mathscr{C}$ is the set of all clusters of taxa displayed by the trees in $\mathscr{T}$, the network N computed by CASS is the minimal network that displays all clusters in $\mathscr{C}$. It is important to note that if a network N displays all clusters of a set of trees, N does not necessarily display all the trees themselves. It is easy to see that if N is minimal for $\mathscr{C}$ and N' is minimal for $\mathscr{T}$ ($\mathscr{C}$ is the set of all clusters of trees in $\mathscr{T}$), then the number of reticulation nodes in N is smaller than or equal to that in N', because the problem with $\mathscr{C}$ is less restrictive than that with $\mathscr{T}$. An illustration of this issue is given in Fig. 2.

When a method is run on a collection $\mathscr{T} = \{T_1, \ldots, T_k\} \subseteq \mathscr{T}(N)$ displayed by network N, we record the number of reticulations that the method estimates. The accuracy of the methods is considered better as the estimated number becomes smaller. Besides the accuracy, we also assess the run time of the methods.

5.3 Results on Simulated Data

Running Times. Due to space limitations, we focus on the 30-taxon data sets; we observed very similar trends on 50-taxon data sets. We first assess the running times of the methods; It is important for a method to maintain reasonable computation speed in order to be of general application for large-scale data analysis. Through the experimental validation, we record the run times of the 100 cases for the methods by sample size in Fig. 3. In interpreting the figure, we focus on two properties regarding the computation speed, the overall run time and the frequency of outliers in time.

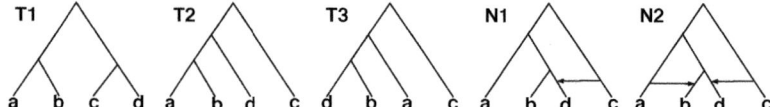

Fig. 2. Illustration of the difference between the formulation used by MUR-PAR, M2 [17] and PIRN [27] on the one hand, and the one used by CASS [24] on the other hand. For the input set of gene trees $\mathscr{T} = \{T1, T2, T3\}$, CASS computes a network with a single reticulation node ($N1$), since this network displays all clusters of the gene trees. However, MURPAR, M2, and PIRN compute minimal networks with two reticulation nodes, such as $N2$, since 2 is the minimum number of reticulation nodes required in a network that displays all the three gene trees.

In terms of the overall run time, it is clear that MURPAR runs fast across varying sample sizes. Compared with M2, the gain in speed of MURPAR is mainly attributed to the use of the ILP solver, since they share the underlying computation structure. The figure also shows that the gap between MURPAR and PIRN closes as the number of trees increases. However, the outliers in the case of PIRN are still much slower than those of MURPAR and M2. This point requires further elaboration. Holding the input size (in terms of the number of gene trees) fixed, the variance in the speed of MURPAR is very similar, whereas that of PIRN is very large. This somehow indicates dependence of the PIRN on the structure of the problem, and lack of dependence of MURPAR on such a structure. It may be that the smaller the number of gene trees, the fewer the constraints, and hence the larger the space that PIRN explores. In the case of MURPAR, the pairwise solutions constrain the search space significantly, giving the method gains in speed.

Accuracy. The numbers of reticulation nodes estimated by each of the methods are shown in Fig. 4. It is worth mentioning that even though the true networks were produced by adding 5 HGT edges, the true number of reticulations may be smaller, depending on the size of the input, since, for example, when only 4 gene trees are sampled, some HGT events may not be observable. Even though network N has $m > 1$ reticulation nodes, this does not necessarily mean that the collection $\mathscr{T}$ of trees, with $|\mathscr{T}| \geq 2$ will have all trees to allow for detecting m. For example, consider the collection $\mathscr{T}$ that has only a pair of trees whose SPR distance is 1. In this case, the number of detectable HGTs is 1, and not m. However, the number of detectable HGTs is expected to increase as more trees are given, and the results in the figure satisfy the expectation.

As more trees are sampled as the input, a naive method for the reticulation estimation would estimate more reticulations simply because the size of the data increases. M2 follows the expectation, in that the estimation gets larger with more trees. However, MURPAR overcomes this problem, even though it is based on the same idea. Rather, both the estimations of PIRN and MURPAR hardly increase with the input size. Between them, it is clear that as more trees are given,

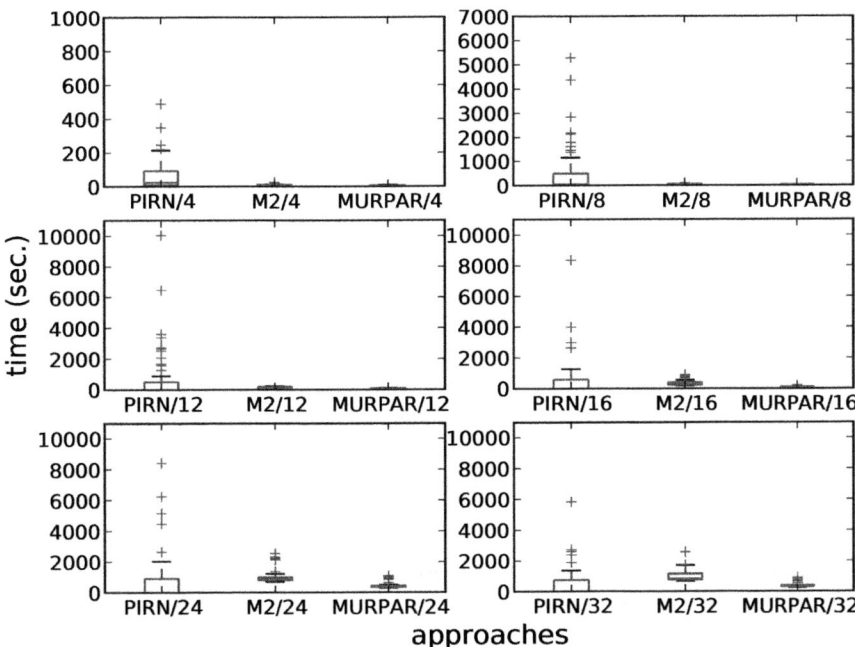

Fig. 3. Running times (in sec.) of the three methods PIRN, M2, and MUR-PAR. The numbers after the '/' are the numbers of gene trees in the input.

the estimate becomes more accurate in MURPAR than in PIRN, particularly for inputs of sizes ≥ 12. Even though the maximum difference between PIRN and MURPAR is one reticulation event on average, this difference gets larger for larger data sets.

5.4 Results on Biological Data

In order to evaluate how well the methods perform for biological data, we ran them on the five gene trees of the *Poaceae* data set (see Section 5.1 for details of this data set). Table 1 reports the estimated number of reticulations and the amount of time taken by the methods. Notice that we also ran PIRN in coarse mode (CoarsePIRN) on them (which is a faster, yet less accurate, version of PIRN).

PIRN identifies the lowest estimate of the number of reticulations, but it took the longest time to obtain the estimate. On the other hand, M2 and MUR-PAR obtained estimates that are higher by just one reticulation event, two and three orders of magnitude faster, respectively. In other words, MURPAR and M2 produce very accurate results within very short amounts of time. Between MURPAR and M2, the difference is negligible on this size of data. However, it is easy to see that the difference will grow with the increase of the input data, and

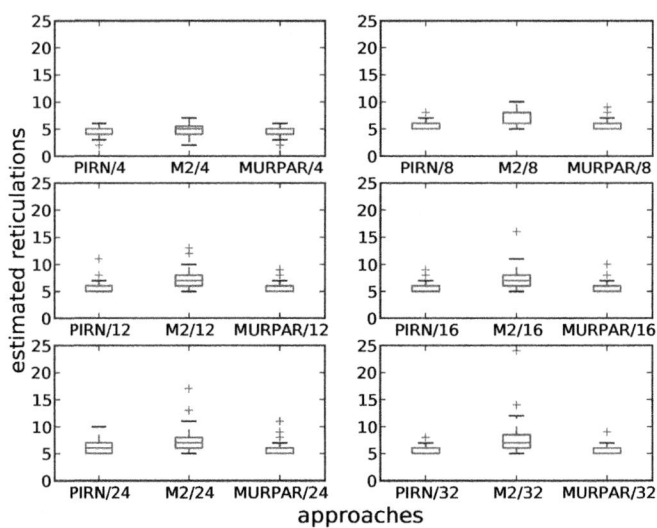

Fig. 4. Numbers of reticulations estimated by each of the three methods PIRN, M2, and MURPAR. The numbers after the '/' are the numbers of gene trees in the input.

Table 1. The number of estimated reticulation events and the run time (in seconds) of the methods on the five gene trees for *ndhF*, *phyB*, *rbcL*, *rpoC2*, and *ITS* in the Poaceae dataset

Approaches	#Reticulations	Time (sec.)
PIRN	13	2143
M2	14	16
MURPAR	14	8
CoarsePIRN	16	58

MURPAR will be preferred in large-scale data analysis. For PIRN, we also ran it in coarse mode. Notice that while PIRN in coarse mode is much faster than PIRN, and is comparable to M2 in terms of speed, the estimates it produces are higher than the other three methods. Considering the run times they take to work on the small input data (5 trees on 14 species), it is clear that PIRN would run the slowest in large-scale data analyses.

6 Conclusions

In this paper, we introduced MURPAR, a method for inferring a phylogenetic network from a collection of gene trees, under the assumption that all incongruence in the gene trees is due to reticulate evolutionary events. While MURPAR is not guaranteed to compute a minimal network, it produces an upper bound on the minimum number of reticulations required to reconcile all gene trees in the input. Performance analysis on both synthetic and biological data sets shows that the MURPAR method is both accurate and fast. Further, MURPAR's run time does not vary much within the same sample size, and has fewer outliers than other methods.

The idea of employing pairwise reconciliations in reconciling a set of gene trees has added advantages in that pairwise reconciliations can be computed in parallel or in a distributed fashion, thus speeding up the overall computation, and improvements to pairwise reconciliation methods will automatically translate into improvement of the MURPAR method. Direct interpretability of the results from the direct relationship of the solutions between SET-WISE HGT INFERENCE and PAIRWISE HGT INFERENCE is another advantage of MURPAR that it is easy to identify the dynamics between any gene trees in the tree set from the estimates of SET-WISE HGT INFERENCE.

Acknowledgement. This work was supported in part by NSF grant DBI-1062463, grant R01LM009494 from the National Library of Medicine, and an Alfred P. Sloan Research Fellowship to L.N. Further, the work was supported in part by the Shared University Grid at Rice funded by NSF under Grant EIA-0216467, and a partnership between Rice University, Sun Microsystems, and Sigma Solutions, Inc.

References

1. Addario-Berry, L., Hallett, M., Lagergren, J.: Towards identifying lateral gene transfer events. In: Proc. Eighth Pacific Symp. Biocomputing (PSB 2003), pp. 279–290 (2003)
2. Semple, C., Baroni, M., Steel, M.: A framework for representing reticulate evolution. Annals of Combinatorics 8, 391–408 (2004)
3. Beiko, R.G., Hamilton, N.: Phylogenetic identification of lateral genetic transfer events. BMC Evolutionary Biology 6, 15+ (2006)
4. Beiko, R.G., Ragan, M.A.: Untangling hybrid phylogenetic signals: Horizontal gene transfer and artifacts of phylogenetic reconstruction. Methods Mol. Biol. 532, 241–256 (2009)
5. Bordewich, M., Linz, S., John, K.S., Semple, C.: A reduction algorithm for computing the hybridization number of two trees. Evolutionary Bioinformatics 3, 86–98 (2007)
6. Bordewich, M., Semple, C.: On the computational complexity of the rooted subtree prune and regraft distance. Annals of Combinatorics 8, 409–423 (2004)
7. Galtier, N.: A model of horizontal gene transfer and the bacterial phylogeny problem. Systematic Biology 56(4), 633–642 (2007)
8. Goloboff, P.A.: Calculating SPR distances between trees. Cladistics 24, 591–597 (2007)
9. Hallett, M.T., Lagergren, J.: Efficient algorithms for lateral gene transfer problems. In: Proc. 5th Ann. Int'l Conf. Comput. Mol. Biol. (RECOMB 2001), pp. 149–156. ACM Press, New York (2001)

10. Hill, T., Nordstrom, K., Thollesson, M., Safstrom, T., Vernersson, A., Fredriksson, R., Schioth, H.: Sprit: Identifying horizontal gene transfer in rooted phylogenetic trees. BMC Evolutionary Biology 10(1), 42+ (2010)
11. Huson, D.H., Bryant, D.: Application of phylogenetic networks in evolutionary studies. Molecular Biology and Evolution 23(2), 254–267 (2006)
12. Huson, D.H., Rupp, R.: Summarizing Multiple Gene Trees Using Cluster Networks. In: Crandall, K.A., Lagergren, J. (eds.) WABI 2008. LNCS (LNBI), vol. 5251, pp. 296–305. Springer, Heidelberg (2008)
13. Linz, S., Semple, C.: A cluster reduction for computing the subtree distance between phylogenies. Annals of Combinatorics 15, 465–484 (2011)
14. MacLeod, D., Charlebois, R.L., Doolittle, F., Bapteste, E.: Deduction of probable events of lateral gene transfer through comparison of phylogenetic trees by recursive consolidation and rearrangement. BMC Evolutionary Biology 5 (2005)
15. Nakhleh, L.: Evolutionary phylogenetic networks: models and issues. In: Heath, L., Ramakrishnan, N. (eds.) The Problem Solving Handbook for Computational Biology and Bioinformatics, pp. 125–158. Springer, New York (2010)
16. Nakhleh, L., Ruths, D., Wang, L.-S.: RIATA-HGT: A Fast and Accurate Heuristic for Reconstructing Horizontal Gene Transfer. In: Wang, L. (ed.) COCOON 2005. LNCS, vol. 3595, pp. 84–93. Springer, Heidelberg (2005)
17. Park, H.J., Jin, G., Nakhleh, L.: Algorithmic strategies for estimating the amount of reticulation from a collection of gene trees. In: Proceedings of the 9th Annual International Conference on Computational Systems Biology, pp. 114–123 (2010)
18. Rambaut, A.: Phylogen: Phylogenetic tree simulator package (2002), http://evolve.zoo.ox.ac.uk/software/PhyloGen/main.html
19. Schmidt, H., Martin, W.: Phylogenetic Trees from Large Datasets Inaugural–Dissertation zur. PhD thesis, Heinrich-Heine-Universitt, Dsseldorf (2003)
20. Than, C., Nakhleh, L.: SPR-based tree reconciliation: Non-binary trees and multiple solutions. In: Proceedings of the Sixth Asia Pacific Bioinformatics Conference, pp. 251–260 (2008)
21. Than, C., Ruths, D., Innan, H., Nakhleh, L.: Confounding factors in HGT detection: Statistical error, coalescent effects, and multiple solutions. Journal of Computational Biology 14(4), 517–535 (2007)
22. Than, C., Ruths, D., Nakhleh, L.: PhyloNet: a software package for analyzing and reconstructing reticulate evolutionary relationships. BMC Bioinformatics 9, 322 (2008)
23. Tofigh, A., Hallett, M., Lagergren, J.: Simultaneous identification of duplications and lateral gene transfers. IEEE/ACM Transactions on Computational Biology and Bioinformatics, 1–19 (January 2011)
24. van Iersel, L., Kelk, S., Rupp, R., Huson, D.H.: Phylogenetic networks do not need to be complex: using fewer reticulations to represent conflicting clusters. Bioinformatics [ISMB] 26(12), i124–i131 (2010)
25. Wu, Y., Wang, J.: Fast Computation of the Exact Hybridization Number of Two Phylogenetic Trees. In: Borodovsky, M., Gogarten, J.P., Przytycka, T.M., Rajasekaran, S. (eds.) ISBRA 2010. LNCS, vol. 6053, pp. 203–214. Springer, Heidelberg (2010)
26. Whidden, C., Beiko, R.G., Zeh, N.: Fast FPT Algorithms for Computing Rooted Agreement Forests: Theory and Experiments. In: Festa, P. (ed.) SEA 2010. LNCS, vol. 6049, pp. 141–153. Springer, Heidelberg (2010)
27. Wu, Y.: Close lower and upper bounds for the minimum reticulate network of multiple phylogenetic trees. Bioinformatics [ISMB] 26(12), 140–148 (2010)

Large Scale Ranking and Repositioning of Drugs with Respect to DrugBank Therapeutic Categories

Matteo Re and Giorgio Valentini

DSI, Dipartimento di Scienze dell' Informazione,
Università degli Studi di Milano,
Via Comelico 39, 20135 Milano, Italia
{re,valentini}@dsi.unimi.it

Abstract. The ranking and prediction of novel therapeutic categories for existing drugs (drug repositioning) is a challenging computational problem involving the analysis of complex chemical and biological networks. In this context we propose a novel semi-supervised learning problem: ranking drugs in integrated bio-chemical networks according to specific DrugBank therapeutic categories. To deal with this challenging problem, we designed a general framework based on bipartite network projections by which homogeneous pharmacological networks can be combined and integrated from heterogeneous and complementary sources of chemical, biomolecular and clinical information. Moreover, we propose a novel method based on kernelized score functions for fast and effective drug ranking in the integrated pharmacological space. Results with 51 therapeutic DrugBank categories involving about 1300 FDA approved drugs show the effectiveness of the proposed approach.

1 Introduction

Drug development is a costly and failure-prone process [1]. In recent years a novel pharmacological research paradigm known as drug repositioning is emerging because of its ability to reduce development costs and to shorten paths to approval [2], which typically takes 10-15 years and upwards $1 billion [3], while revenues due to repurposed drugs can exceeds billions of dollars [4].

Drug repositioning, i.e. the prediction of novel therapeutic indications for existing drugs, is a challenging problem in modern computational biology. Computational approaches for drug repositioning focused mainly on small-scale applications, such as the analysis of specific classes of drugs or drugs for specific diseases [5, 6, 7, 8]. Large-scale applications, involving a relatively large number of drugs and diseases, count only a few examples [9, 10, 11, 12].

Different computational tasks related to the the drug repositioning problem have been proposed, ranging from clustering drugs either considering their pharmacophore descriptors [5] or Connectivity Map-based networks [10], to prediction of drug-target interactions [13, 14], or drug-disease associations [15, 11].

In this context, we propose a novel prediction task, i.e. the large-scale ranking of drugs with respect to DrugBank therapeutic categories[16]. We chose Drug-Bank categories since their associations to drugs are manually curated using medical literature such as PubMed, e-Therapeutics (www.e-therapeutics.ca) and STAT!Ref (AHFS) (online.statref.com), and because "at present, there is not a comprehensive and systematic representation of known drugs indications that would enable a fine-scale delineation of types of drug-disease relationships" [17]. For each considered DrugBank therapeutic category we provide a ranking of drugs, since this can allow the choice of top ranked "false positive" drugs as natural candidates for drug repositioning, while a pure classification approach cannot provide such preferential candidates.

To this end, we propose a novel and very fast semi-supervised network method based on kernelized score functions for ranking drugs according to their likelihood to belong to a given therapeutic category. Moreover, we propose a general framework based on bipartite networks projections for the construction of homogeneous pharmacological spaces. The nature of these network-structured projected spaces allows the application of prediction algorithms to homogeneous pharmacological spaces and improves the integration of different sources of chemical, biomolecular and clinical sources of information.

We evaluated the proposed approach by integrating three pharmacological similarity spaces accounting, respectively, for chemical similarity, drug-targets interaction similarity and drug-chemicals similarity, in order to rank a curated set of U.S. Food and Drug Administration (FDA) approved drugs according to the DrugBank therapeutic categories.

2 Methods

We propose $\psi NetPro$, Pharmacological Spaces Integration based on Networks Projections, a general approach to construct and integrate different pharmacological similarity spaces capturing different pharmacological characteristics of drugs, and a novel method for ranking drugs in the integrated pharmacological networks to discover new therapeutic indications for known drugs. In Section 2.1 we introduce the bipartite network projection method to construct homogeneous pharmacological spaces from inhomogeneous spaces represented though bipartite networks. In Section 2.2 we show how to construct and integrate different pharmacological spaces using different sources of chemical, biomolecular and pharmacological data, and finally in Section 2.3 we present our novel approach to rank drugs in pharmacological networks through kernel-based score functions.

2.1 Bipartite Networks Projection and Integration

Bipartite (or two-mode) networks are graphs composed by two types of vertices in which edges are established only between vertices belonging to different sets (Fig. 1 a). Bipartite networks can be transformed into one-mode networks (composed by a single type of nodes) by selecting one of the sets of nodes and

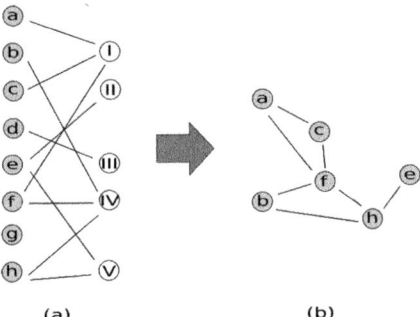

Fig. 1. Bipartite network projection: the two-mode network (a) is projected onto a one-mode network (b). Singleton nodes (i.e. d and g) are removed from the projected network.

linking two nodes from that set if the intersection of their neighborhoods in the two-mode network is not empty (Fig. 1 b).

More precisely, given a bipartite graph $G = <V, E>$, with two distinct sets of nodes $V_a, V_b \subset V$, $V_a \cup V_b = V$, $V_a \cap V_b = \emptyset$ and edges $(u,v) \in E \Rightarrow u \in V_a \wedge v \in V_b$, we may induce a projected graph $G_p = <V_p, E_p>$, with $V_p \subseteq V_a$, such that:

$$(u', u'') \in E_p \iff \exists v \in V_b \text{ s.t. } (u', v) \in E \wedge (u'', v) \in E \quad (1)$$

This operation is commonly referred to as "binary mode projection" and is suitable for the induction of a similarity space between vertices $v \in V_a$ (Fig. 1). The binary mode projection produces one-mode networks containing binary edges, but more complex projection schemes can assign edge weights according to the degree of nodes and the edge weights in the bipartite two-mode network. In our experiments we adopted the binary projection technique, since the bipartite drug-target data downloaded from the DrugBank database are unweighted, and for homogeneity we applied a binary projection also to the other considered data (see Section 2.2 for more details).

The bipartite network projection scheme may induce different pharmacological spaces depending on the nature of the bipartite network (e.g. drug-protein or drug-chemicals interaction bipartite networks), but the projected networks correspond to homogeneous pharmacological spaces representing different notions of induced pharmacological similarity between drugs. These spaces may be integrated using appropriate network integration methods and proper normalization techniques. For instance, we adopted the normalized graph Laplacian L [18] to make comparable the pharmacological networks $G = <V, E>$ represented through the corresponding symmetric adjacency matrices W:

$$L = D^{-\frac{1}{2}}(D - W)D^{-\frac{1}{2}} = I - D^{-\frac{1}{2}}WD^{-\frac{1}{2}} \quad (2)$$

where D is a diagonal matrix with elements $d_{ii} = \sum_j w_{ij}$, I is the identity matrix and w_{ij} are the elements of the matrix W.

In our setting we integrated multiple networks with a simple technique that assures a high coverage of the drugs included in the integrated pharmacological network, without penalizing drugs for which a specific source of data is unavailable. More precisely, given a set of n pharmacological networks $G^d = <V^d, E^d>, 1 \leq d \leq n$, constructed through appropriate bipartite graph projections, the integrated pharmacological network $\bar{G} = <\bar{V}, \bar{E}>$, with $\bar{V} = \bigcup_d V^d$ and $\bar{E} \subseteq \bigcup_d E^d$, can be derived by averaging the normalized edge weights only when data for the corresponding pair of drugs is actually available. In other words, if w_{ij}^d represents the weight of the edge $(v_i, v_j) \in E^d$, the weight $\bar{w}_{ij}$ of the edge $(v_i, v_j) \in \bar{E}$ is computed as follows:

$$\bar{w}_{ij} = \frac{1}{|D(i,j)|} \sum_{d \in D(i,j)} w_{ij}^d, \qquad D(i,j) = \{d | v_i \in V^d \wedge v_j \in V^d\} \qquad (3)$$

It is worth noting that other network integration methods may lead to better results (e.g. weighted integrated networks that take into account the information content of each source of data), but we applied this simple approach only to show the feasibility and effectiveness of the proposed overall approach.

2.2 Construction of Pharmacological Networks

We constructed three pharmacological similarity networks reflecting the pairwise chemical structure similarity between drugs ($\Phi_{chemsim}$), the similarity between drugs derived from common protein targets ($\Phi_{drugtarget}$) and the pairwise similarity from chemical-chemical interactions ($\Phi_{chemint}$) between the considered drugs and other chemicals involved in their pharmacological activity.

Chemical and pharmacological data bases. Data for the computation of $\Phi_{chemsim}$ and $\Phi_{drugtarget}$ have been obtained from DrugBank [16], while data for $\Phi_{chemint}$ have been extracted from the STITCH database [19]. DrugBank is a unique bioinformatics/chemoinformatics resource that combines detailed drug (i.e. chemical) data with comprehensive drug target (i.e. protein) information. In the current release DrugBank contains detailed information about 6707 drug entries including 1436 FDA-approved small molecule drugs. In order to construct a highly reliable drugs set we selected from DrugBank the largest set of FDA approved drugs targeting at least one FDA approved target. This led to the definition of a collection composed by 1253 drugs.

STITCH integrates data distributed over many databases. For instance, the chemical-chemical interaction networks stored in STITCH includes information about the impact of genetic variation on drug response and from the Comparative Toxicogenomics Database (which contains more than 8500 direct chemical-disease relationships), thus ensuring the existence of drug-drug relationships induced by common genetics and/or toxicogenomics disease-association profiles [20, 21].

Constructing pharmacological spaces from different sources of data. For $\Phi_{chemsim}$ we directly computed the structural chemical similarities between each pair of drugs, while for the other pharmacological spaces we applied the projection techniques described in Section 2.1.

The simplest similarity space, $\Phi_{chemsim}$, is based on chemical structure similarities and was obtained by computing the Tanimoto similarity scores between each pair of drugs in the reference set [22]. The scores were obtained by comparing the simplified molecular input line entry specification (SMILES) annotations contained in DrugBank entries [23]. The obtained adjacency matrix was then converted to a binary matrix by thresholding the similarity scores according to the procedure reported in [13].

The second considered similarity space, $\Phi_{drugtarget}$, was obtained by creating a bipartite network between the drugs and all the FDA approved targets, according to the information stored in DrugBank. Once constructed, this network has been projected onto a one mode network and processed according to the procedures described in Section 2.1.

The third pharmacological similarity space ($\Phi_{chemint}$) has been constructed by processing the chemical-chemical interactions stored in the STITCH 2.0 database [24]. This dataset is expected to be informative because these interactions are obtained by considering many sources of information (i.e. metabolic pathways, binding experiments, phenotypic effects and drug-target relationships). The adjacency matrix was converted to a binary matrix by thresholding the interaction scores to 0.7 in order to ensure a high confidence in the selected STITCH chemical interactions. The thresholding led to a final coverage of 50% of the drugs in our reference set.

Progressive integration of pharmacological networks. We progressively integrated the computed pharmacological networks in order to add different and complementary sources of information and to maintain a high-coverage of drugs for large-scale drug repositioning. To this end we considered at first the $\Phi_{chemsim}$ space alone (that is the space with the highest drug coverage), then we progressively integrated the other two pharmacological spaces characterized by a lower coverage, that is respectively $\Phi_{drugtarget}$ and $\Phi_{chemint}$. These progressively enriched pharmacological networks have been represented through the corresponding adjacency matrices W_1, W_2 and W_3, where the numeric index indicates the number of different integrated pharmacological networks. Despite the three networks having the same number of nodes/drugs (1253), our "progressive integration" strategy yields to a significant increment in the number of the edges, that grow from 13010, to 43827 and 96711 respectively in W_1, W_2 and W_3, thus resulting in a high-coverage and a large-scale setting of the drug repositioning problem.

2.3 Ranking Methods for Drug Therapeutic Category Prediction

By using the adjacency matrices W corresponding to the graphs $G = <V, E>$ obtained by bipartite network projection and integration (Sect. 2.1), we dispose of networks in the pharmacological space well-suited for ranking the drugs $v \in V$ according to their likelihood to belong to a specific therapeutic category C. To this aim we can exploit the pharmacological similarities between pairs of drugs $v_i, v_j \in V$, represented by the weights $w_{ij} > 0$ of the edges $(i,j) \in E$, the overall topology of the integrated pharmacological spaces, and a subset of drugs $V_C \subset V$ belonging to a priori known therapeutic category C.

In our experiments we compared results obtained with drug ranking algorithms based on random walks on graphs with our novel proposed method that can be interpreted as a kernelized extension of the classical random walks.

Random Walks. Random walk (RW) algorithms [25] rank drugs by exploring and exploiting the topology of the pharmacological network: random walks across the network are performed starting from a subset $V_C \subset V$ of drugs belonging to a specific therapeutic category C by using a transition probability matrix $\boldsymbol{Q} = \boldsymbol{D}^{-1}\boldsymbol{W}$, where $\boldsymbol{W}$ is the adjacency matrix, and $\boldsymbol{D}$ is a diagonal matrix with diagonal elements $d_{ii} = \sum_j w_{ij}$. The elements q_{ij} of $\boldsymbol{Q}$ represent the probability of a random step from v_i to v_j. If $\boldsymbol{p}_t$ represents the probability vector of finding a "random walker" at step t in the nodes $v \in V$, then the probability at step $t+1$ is:

$$\boldsymbol{p}_{t+1} = \boldsymbol{Q}^T \boldsymbol{p}_t \qquad (4)$$

The initial probability of belonging to set of drugs corresponding to a given therapeutic category can be set to $p_o = 1/|V_C|$ for the drugs $v \in V_C$ and to $p_o = 0$ for the drugs $v \in V \setminus V_C$, and the update (4) is iterated until convergence. We could observe that the random walker could progressively "forget" the a priori information available for the therapeutic category C, by iteratively walking across the overall network. To avoid this problem, we could try to apply the random walk with restart (RWR) algorithm: at each step the random walker can move to one of its neighbours or can restart from its initial condition with probability θ:

$$\boldsymbol{p}_{t+1} = (1-\theta)\boldsymbol{Q}^T\boldsymbol{p}_t + \theta \boldsymbol{p}_o \qquad (5)$$

With both RW and RWR methods at the steady state we can rank the vector $\boldsymbol{p}$ to prioritize drugs according to their likelihood to belong to the therapeutic category under study.

Score Functions Based on Kernelized Random Walks. In this section we propose a novel similarity-based method that on the one hand embeds in a kernel function the random walk strategy and on the other hand uses this kernel within a properly defined kernelized similarity score functions to rank drugs according to the topology of the pharmacological network.

More precisely, we can define a distance measure $D(v, V_C)$ between a drug $v \in V$ and the set of the drugs $x \in V_C$ in a reproducing kernel Hilbert space $\mathcal{H}$, according to a suitable mapping $\phi : V \to \mathcal{H}$. For instance, we can consider the minimum euclidean distance in the Hilbert space $\mathcal{H}$ between a drug $v \in V$ and the set of drugs V_C belonging to a specific therapeutic category:

$$D_{NN}(v, V_C) = \min_{x \in V_C} \left\| \phi(v) - \phi(x) \right\|^2 \tag{6}$$

By recalling that $< \phi(\cdot), \phi(\cdot) > = K(\cdot, \cdot)$, where $K : V \times V \to \mathbb{R}$ is a kernel function associated to the mapping ϕ, we can choose in principle any valid kernel, but in this context it is meaningful to use a *random walk kernel* [18] constructed from the adjacency matrices $\boldsymbol{W}_1$, $\boldsymbol{W}_2$ and $\boldsymbol{W}_3$, since it provides a similarity measure that takes into account direct and indirect relationships between drugs in the pharmacological space. The Gram matrix $\boldsymbol{K}$ associated to the random walk kernel function $K(\cdot, \cdot)$ is obtained from the adjacency matrix $\boldsymbol{W}$ of the pharmacological network:

$$\boldsymbol{K} = (a-1)\boldsymbol{I} + \boldsymbol{D}^{-\frac{1}{2}} \boldsymbol{W} \boldsymbol{D}^{-\frac{1}{2}} \tag{7}$$

where $\boldsymbol{I}$ is the identity matrix, $\boldsymbol{D}$ is a diagonal matrix with elements $d_{ii} = \sum_j w_{ij}$ and a is a value larger than 1.

By developing the square (6) we can derive the following similarity measure:

$$Sim_{NN}(v, V_C) = - \min_{x \in V_C} \left[K(v,v) - 2K(v,x) + K(x,x) \right] \tag{8}$$

By assuming an equal auto-similarity $K(x, x)$ for all $x \in V$, we can simplify (8), thus achieving the *nearest neighbours score* S_{NN}:

$$S_{NN}(v, V_C) = - \min_{x \in V_C} -2K(v,x) = 2 \max_{x \in V_C} K(v,x) \tag{9}$$

It is easy to see that a different notion of distance based on the first k nearest-neighbours leads to the definition of the *k-nearest neighbours score* S_{kNN}:

$$S_{kNN}(v, V_C) = 2 \sum_{x \in I_k(v)} K(v,x) \tag{10}$$

where $I_k(v) = \{ x \in V_C | x$ is ranked among the first k in V_C according to $K(v, x)\}$. In a similar way we can also derive the *average score* similarity measure S_{AV} based on the average distance D_{AV} with respect to to the set of drugs V_C belonging to the C therapeutic category:

$$S_{AV}(v, V_C) = \frac{2}{|V_C|} \sum_{x \in V_C} K(v,x) \tag{11}$$

It is worth noting that the S_{AV} score resembles the one proposed by Borgwardt and others in the context of gene function prediction from synthetic lethality networks: from this standpoint our approach can be viewed as an extension of the algorithm proposed in [26].

3 Experiments

3.1 Experimental Setup

We propose a novel learning problem in the context of drug ranking and repositioning: the prediction of the therapeutic category of drugs according to the annotations provided by DrugBank 3.0.

Table 1. Average AUC results across therapeutic classes of the compared ranking methods using different pharmacological networks W_1, W_2 and W_3

	RW	RWR	S_{AV}	S_{NN}	S_{kNN}
W_1	0.6846	0.8037	0.8262	0.8074	0.8277
W_2	0.5780	0.9171	0.9232	0.9066	0.9230
W_3	0.5334	0.9258	0.9312	0.9129	0.9299

In order to obtain the therapeutic category labels we parsed the DrugBank entries belonging to our reference set (1253 FDA approved drugs, see Section 2.2) by extracting all the drug category annotations excluding the chemical categories (categories reflecting the chemical nature of the considered compounds). We finally removed from our therapeutic categories set all the classes associated to less than 15 drugs obtaining 51 therapeutic classes, in order to exclude classes with too few positive examples to assure reliable predictions. We evaluated the proposed ranking method by using a 5-folds cross validation scheme repeated 10 times. As the output of the proposed methods is a continuous score for each drug-therapeutic category pair, we computed the Area Under the ROC curve (AUC), and the precision at fixed recall levels averaged across all the considered therapeutic classes.

3.2 Results

Table 1 shows the average AUC across therapeutic classes. We can observe that both RWR and kernelized score function methods achieve good results (for several classes the AUC is 1 or very close to 1 when the most informative network W_3 is used – data not shown), while the classical RW substantially fails in these ranking tasks, since it explores too remote relationships between drugs, thus introducing noise in prediction results. More interestingly, independently of the considered methods (apart from RW), the average AUC increases as new pharmacological spaces are added: most of the increment is achieved when we integrate 2 pharmacological spaces (W_2), but note that the apparently small increment obtained, e.g. by S_{kNN}, when we pass from 2 to 3 integrated pharmacological spaces is actually statistically significant according to the Wilcoxon ranks sum test (p-$value$< 0.01). These results are also confirmed by the precision at different recall levels outcomes (Fig. 2): we can observe an increment in performance whenever we move from W_1 to W_2 and W_3, no matter the method we apply. For lack of space we reported only RWR and S_{kNN} results, but with the other methods (except RW that performs poorly also with this metric) we can observe similar trends.

Comparing AUC results between the different methods, according to the Wilcoxon rank sum test, there is no statistically significant difference between S_{AV} and S_{kNN}, while both S_{AV} and S_{kNN} achieve significantly better results than RWR and S_{NN} (at 0.005 significance level), independently of the considered pharmacological space. Considering precision at fixed recall levels, S_{NN} performs significantly worse than the other methods. S_{AV} and S_{kNN} achieve always better or equal results than RWR with both W_1 and W_2 pharmacological

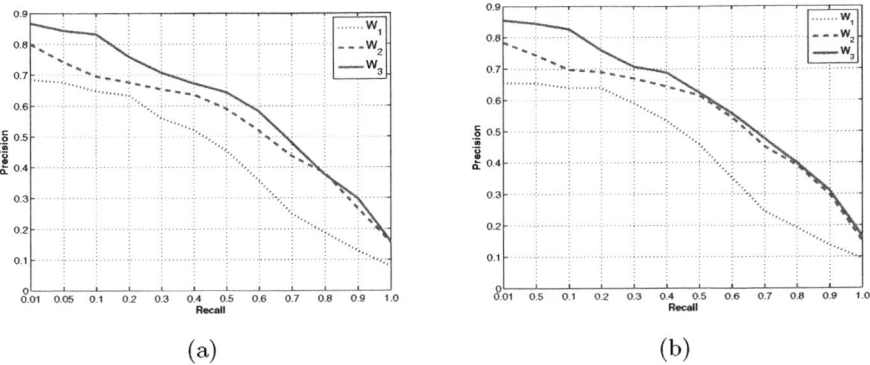

Fig. 2. Precisions at fixed recall levels, with $\boldsymbol{W}_1$, $\boldsymbol{W}_2$ and $\boldsymbol{W}_3$ pharmacological networks. (a) Random Walks with Restart (RWR); (b) S_{kNN}. Results are averaged across the 51 therapeutic DrugBank classes.

Table 2. Computational time requirements of the compared methods with the $\boldsymbol{W}_3$ network, using an Intel i7-860 2.80 GHz processor

	RW	RWR	S_{AV}	S_{NN}	S_{kNN}
time (sec.)	13840	645	5	5	12

networks, while with the most informative $\boldsymbol{W}_3$ network no significant difference between methods can be registered at any recall level (0.05 significance level, Wilcoxon ranks sum test).

Table 2 reports the empirical computational complexity of the different methods for the completion of the entire experimental scheme (5-folds CV repeated 10 times for each of the 51 therapeutic categories). Results show that kernelized score methods are significantly faster than RW and RWR methods.

Cross-validated average results across classes show that our proposed method is able to recover therapeutic classes of drugs. A thorough analysis of the results relative to each therapeutic category is out of the scope of this investigation, but in order to show the potential of the proposed method we report the analysis of the top ranked false positives predicted in three drug categories. All the ranking results show an AUC increment due to the progressive networks integration, and we chose among them three of the classes with the largest AUC improvement. "Antidyskinetics" drugs are used in the treatment of motor disorders. In this ranking task we obtained 0.730, 0.887 and 0.923 average AUC using the W_1, W_2 and W_3 networks respectively. The first top ranked negative (L-Tryptophan, DrugBank id: DB00150) was reported to be effective in preventing levodopa-induced motor complications in the treatment of patients affected by Parkinson disease [27], and hence could be associated to the "Antidyskinetics" category. In the ranking task associated with the "Anti HIV Agents" category we achieved respectively 0.753, 0.900 and 0.943 AUC results using our progressively integrated networks. The first top ranked negative was Darunavir (DB01264) and, according to the associated DrugBank entry, it is indicated in the treatment of HIV, but not annotated as "Anti HIV Agents", probably since just annotated as "HIV Protease Inhibitors". The top ranked false positive in the task associated with the "GABA Modulators" (AUC 0.941, 0.972 and 0.995) is Adinazolam (DB00546). This drug, and the four top ranked false positives in this task are benzodiazepines, a class of substances known to modulate the effect of GABA [28, 29].

4 Conclusions

Results show that in the context of the drug repositioning problem the construction and integration of informative pharmacological spaces is at least relevant as the design and the choice of proper label ranking algorithms. Indeed the best precision at a given recall results are obtained with the integrated and most-informative pharmacological network W_3, independently of the method used (Fig. 2). With the simplest and least-informative pharmacological space W_1, based on direct chemical similarities between drugs, S_{AV} and S_{kNN} significantly outperform the other methods, and this is true also with the W_2 network. This means that the process of integration of multiple pharmacological spaces by projection of drug-target and drug-chemicals bipartite networks plays a crucial role to improve the information content of the original simple direct chemical similarity space between drugs. Interestingly enough, important increment in performances are also obtained in the ranking of drugs belonging to difficult-to-predict therapeutic classes such as the "Antiparkinson agents" (W_1 AUC : 0.7486, W_2 AUC : 0.8930, W_3 AUC : 0.9316, results obtained using S_{kNN} with $k = 19$). Results averaged across classes show that our proposed approach is able to correctly rank known drugs with respect to their known therapeutic categories. Moreover a preliminary analysis of the top-ranked false positives shows that our proposed methods can discover potential drug candidates for novel therapeutic indications.

We would like also to emphasize that kernelized score ranking methods could be applied to significantly larger drug networks, due to their low computational complexity and scalability (Table 2). Indeed in our experiments we considered about a thousand of FDA-approved drugs, but the same approach could be applied to thousands of investigational compounds, thus finding initial therapeutic indications for unknown drugs. Moreover, we could apply the same network projection and integration approach to enrich the pharmacological space with new information coming from annotated side-effects (as the one stored in public databases such as SIDER [30]), or from manually curated pathways databases such as Reactome [31], or from large collections of gene expression signatures as the ones included in the Connectivity Map public repository [9], or also from data obtained through Next Generation Sequencing techniques, one of the most promising biotechnologies for drug discovery and development [32].

Even if using simple binary projections we obtained high performances in term of AUC, to better exploit the fine-grained information stored in the aforementioned databases, in the future work we plan to experiment with real-valued network projections, to take into account the weights eventually associated to the edges of the bipartite network.

Acknowledgments. We thank the reviewers for their comments and suggestions. The authors gratefully acknowledge partial support by the PASCAL2 Network of Excellence under EC grant no. 216886. This publication only reflects the authors' views.

References

[1] DiMasi, J., et al.: New drug development in the United States from 1963 to 1999. Clinical Pharmacology and Therapeutics 69(5), 186–196 (2001)
[2] Ashburn, T., et al.: Drug repositioning: identifying and developing new uses for existing drugs. Nature Reviews 3(8), 28–55 (2004)
[3] DiMasi, J., et al.: The price of innovation: new estimates of drug development costs. Journal of Health Economics 22(2), 151–185 (2003)
[4] Anand, G.: How drug's rebirth as treatment for cancer fueled price rises: oncedemonized thalidomide boosts celgene's sales; patients see costs soar. The Wall Street Journal 15(A1) (2004)
[5] Noeske, T., Sasse, B., Strak, H., et al.: Predictiong compound selectivity by selforganizing maps: cross-activities of metabotropic glutamate receptor antagonists. Chem. Med. Chem. 1, 1066–1068 (2006)
[6] Wei, G., Twomey, D., Lamb, J., et al.: Gene expression-based chemical genomics identifies rapamycin as a modulator of MCL1 and glucocorticoid resistance. Cancer Cell 10, 331–342 (2006)
[7] Kotelnikova, E., Yuryev, A., Mazo, I., Daraselia, N.: Computational approaches for drug repositioning and combination therapy design. Journal of Bioinformatics and Computational Biology 8, 593–606 (2010)
[8] Li, J., Zhu, X., Chen, J.: Building disease-specific drug-protein connectivity maps from molecular interaction networks and pubmed abstracts. PLoS Computational Biology 5, e1000450 (2009)
[9] Lamb, J., et al.: The Connectivity Map: Using gene-expression signatures to connect small molecules, genes, and disease. Science 313(5795), 1929–1935 (2006)
[10] Iorio, F., Bosotti, R., Scacheri, E., Mithbaokar, P., Ferriero, R., Murino, L., Tagliaferri, R., Brunetti-Pierri, N., Isacchi, A., di Bernardo, D.: Discovery of drug mode of action and drug repositioning from transcriptional responses. PNAS 107(33), 14621–14626 (2010)
[11] Gottlieb, A., Stein, G., Ruppin, E., Sharan, R.: PREDICT, a method for inferring novel drug indications with application to personalized medicine. Molecular Systems Biology 7, 496 (2011)
[12] Sirota, M., et al.: Discovery and preclinical validation of drug indications using compendia of public gene expression data. Sci. Transl. Med. 96(3), 96–97 (2011)
[13] Keiser, M., Setola, V., Irwin, J., et al.: Predicting new molecular targets for known drugs. Nature 462, 175–181 (2009)
[14] Yamanishi, Y., Kotera, M., Kaneisha, M., Goto, S.: Drug-target interaction prediction from chemical, genomic and pharmacological data in an integrated framework. Bioinformatics (ISMB) 26, 246–254 (2010)
[15] Chiang, A., Butte, A.: Systematic evaluation of drug-disease relationships to identify leads for novel drug uses. Clin. Pharmacol. Ther. 86, 507–510 (2009)

[16] Knox, C., Law, V., Jewison, T., Liu, P., Ly, S., Frolkis, A., Pon, A., Banco, K., Mak, C., Neveu, V., Djoumbou, Y., Eisner, R., Guo, A., Wishart, D.: DrugBank 3.0: a comprehensive resource for 'omics' research on drugs. Nucleic Acids Res. 39, D1035–D1041 (2011)

[17] Dudley, J., Desphonde, T., Butte, A.: Exploiting drug-disease relationships for computational drug repositioning. Briefings in Bioinformatics 12(4), 303–311 (2011)

[18] Smola, A.J., Kondor, R.: Kernels and Regularization on Graphs. In: Schölkopf, B., Warmuth, M.K. (eds.) COLT/Kernel 2003. LNCS (LNAI), vol. 2777, pp. 144–158. Springer, Heidelberg (2003)

[19] Kuhn, M., von Mering, C., Campillos, M., Jensen, L., P. B.: STITCH: interaction networks of chemicals and proteins. Nucleic Acids Res. 36, D684–D688 (2008)

[20] Gong, L., et al.: PharmGKB: an integrated resource of pharmacogenomic data and knowledge. Curr. Protoc. Bioinformatics 14(17) (2008)

[21] Davis, A., et al.: The Comparative Toxicogenomics Database: update 2011. Nucleic Acids Res. 39, D1067–D1072 (2011)

[22] Nikolova, N., Jaworska, J.: Approaches to measure chemical similarity - a review. QSAR Comb. Sci. 22(9-10), 1006–1026 (2003)

[23] Weininger, D.: Smiles, a chemical language and information system. Journal of Chemical Information and Modeling 28(31) (1988)

[24] Kuhn, M., Szklarczyk, D., Franceschini, A., Campillos, M., von Mering, C., Jensen, L., Beyer, A., Bork, P.: STITCH 2: an interaction network database for small molecules and proteins. Nucleic Acids Res. 38, D552–D556 (2010)

[25] Lovasz, L.: Random Walks on Graphs: a Survey. Combinatorics, Paul Erdos is Eighty 2, 1–46 (1993)

[26] Lippert, G., Ghahramani, Z., Borgwardt, K.: Gene function prediction from synthetic lethality networks via ranking on demand. Bioinformatics 26(7), 912–918 (2010)

[27] Sandyk, R., Fisher, H.: L-tryptophan supplementation in parkinson's disease. Int. J. Neurosci. 45((3-4), 215–219 (1989)

[28] MacDonald, R., Jeffery, L.: Benzodiazepines specifically modulate GABA-mediated postsynaptic inhibition in cultured mammalian neurones. Nature 271, 563–564 (1976)

[29] Hanson, S., Czajkowski, C.: Structural mechanisms underlying benzodiazepine modulation of the $GABA_A$ receptor. The Journal of Neuroscience 28(13), 3490–3499 (2008)

[30] Kuhn, M., Campillos, M., Letunic, I., Jensen, L., P., B.: A side effect resource to capture phenotypic effects of drugs. Mol. Syst. Biol. 6(343) (2010)

[31] Croft, D., O'Kelly, G., Wu, G., Haw, R., Gillespie, M., Matthews, L., Caudy, M., Garapati, P., et al.: Reactome: A database of reactions, pathways and biological processes. Nucleic Acids Res. 39, D691–D697 (2010)

[32] Woollard, P., Mehta, N., Vamathevan, J., Van Horn, S., Bonde, B., Dow, D.: The application of next-generation sequencing technologies to drug discovery and development. Drug Discovery Today 16(11-12), 512–519 (2011)

Score Based Aggregation of microRNA Target Orderings

Debarka Sengupta[1,*], Ujjwal Maulik[2], and Sanghamitra Bandyopadhyay[1]

[1] Machine Intelligence Unit, Indian Statistical Institute, Kolkata, India
debarka_r@isical.ac.in
[2] Dept. of Comp. Sc. and Eng., Jadavpur University, Kolkata, India

Abstract. Rank aggregation refers to the task of combining different orderings of an identical set of objects to obtain a consensus ranked list. Other than meta-search in web mining, in last few years, this technique has successfully been employed to address problems arising from bioinformatics domain. Consensus ranking of disease related genes, miRNA targets are ample examples in this context. It can be argued that scores are more informative than mere ranks. Existing score based aggregation techniques are evolutionary in nature and consume significant amount of time. We, for the first time propose a Markov chain for score based aggregation ranked lists. The proposed method is found out-performing the existing methods in terms of time consumption (by far) and performance when used in context of microRNA (miRNA) target ranking. The supplementary materials are uploaded at: $http://www.isical.ac.in/ \sim bioinfo\_miu/rankfuse.rar$

1 Background

Rank fusion has a profound history of advancements with its root back to the latter half of the eighteenth century, when Borda proposed *election by order of merit* followed by Condorcet's proposal of pairwise majority voting known as *Condorcet's criterion* [Grazia, 1953]. Borda suggests the simplest possible consensus ranking where the order is determined by taking simple average of ranks from different lists whereas, *Condorcet's criterion* permits A to be ranked higher than B in the consensus list if the majority of the lists vote for A's precedence over B. The method for which the underlying principle is to obtain the consensus, is called the *positional rank aggregation method* and the one for which the underlying principle follows the rule of majority, is called the *Condorcet method*.

Rank appears naturally in every discipline of science. For example disease candidate genes can ranked using different statistical methods [Chen, 2011]. Combination of multiple rankings is therefore a reasonable task in many circumstances. Besides meta search of web [Dwork *et al.*, 2001], in the past few years rank fusion has also been successfully used in meta-analysis of several bioinformatics data, generated by today's high-throughput techniques. For example

---

* Corresponding author.

rank fusion is used for combining the results from different microarray studies in [DeConde et al., 2006, Pihur et al., 2008]. Ordered lists of putative target genes suggested by different microRNA target prediction algorithms are considered for fusion by [Lin and Ding, 2008]. In [Pihur et al., 2007] rank fusion is used for ranking some clustering methods evaluated separately by some validation measures over biological data. In [Sengupta et al., 2010], a measure is proposed using the notion of rank aggregation, which evaluates an ordering w.r.t., the rest. This measure has been shown to be useful in judging the performance of the microRNA target prediction algorithms. Very recently a Markov chain has been proposed for rank aggregation while allowing user to set certain confidence scores to each of the input rankings [Sengupta et al., 2012]. A new, improved objective function has recently been proposed for rank aggregation with a promise of improved consensus ordering of miRNA-targets [Sengupta et al., 2011].

In course of the advancements many heuristics have come into picture for aggregating the ranked lists by combinatorial approximation. Differen variants Markov chains have successfully been employed for rank aggregation scenarios (though in context of meta search) [Dwork et al., 2001]. Composite ranking method based on $p-values$ has been proposed by [Zhou et al., 2006]. *Cross-entropy Monte Carlo* (CE) method for integrating top k lists is proposed in [Lin and Ding, 2008]. Very recently an R package *RankAggreg*, has been compiled by [Pihur et al., 2009] which employs Genetic algorithm (GA) and CE for rank aggregation. Though there are a number of heuristics used for rank fusion in context biological rankings, it's only the work of [Pihur et al., 2007] where scores are considered during fusion. Of note, score based implementations of the evolutionary techniques such as GA and CE are overly slow even for a set of lists of moderate size.

In the present article we put forth a Markov chain for score based rank aggregation. The proposed approach is compared with the evolutionary techniques in context of consensus miRNA-target ranking. It is found that the proposed method performs at least as good as the rest in terms of ranking validated miRNA targets upwards in the list.

2 Foundation

2.1 The General Goal of Rank Aggregation and Logic Behind Consideration of Scores

Sometimes the central opinion of all is more credible than any of the individuals. By rank aggregation we try to land at a consensus ranking when multiple rankings are given for a fixed set of elements. Technically this is achieved by minimizing the disagreement of the candidate consensus ranking with the input rankings. Scores are often more informative than ranking. In practice, difference in ranks does not depict difference in scores. Disagreement of two lists are popularly measured by Kendall's tau or Spearman's footrule distances. In

[Pihur et al., 2007], score based variants of these functions are proposed. In the present work similar sort of formulation is used by Markov chain to land at the consensus ranking.

2.2 Rank Aggregation

Let L be the set of input rankings i.e., $L = \{L_1, L_2, ..., L_n\}$ and S be the union set of all ranked elements i.e., $S = \bigcup_{x=1}^{n} L_x$. Each of the input lists consists of k ranked elements. Let τ be any k sized ordered subset of S. Then the objective is to find τ^* such that its cumulative distance with all given lists is minimized. Formally it can be written as follows:

$$\tau^* = \arg\min \phi(\tau), \tau \subset S = (1, 2, ..., m)$$
$$= \arg\min \sum_{L_x \in L} d(\tau, L_x), \tau \subset S, \quad (1)$$

where d is distance measure like *Spearman's footrule* or the tailored *Kendall's tau distance* (discussed in the next subsection) between two *top-k* lists. τ^* is referred to as the *top-k* list of the aggregated candidate targets set, and $y^* = \phi(\tau^*)$ is the minimum value of the objective function ϕ. The above definition is valid for *top-k* lists and full lists as well. However, in this article we use full lists for aggregation.

2.3 Distance Functions with Their Score Based Analogs

In rank fusion a suitable distance is minimized between τ and the input lists. Choosing appropriate distance attributes greatly to the overall performance of the rank fusion. Pros and cons of the popular distances, in context of rank fusion can be found in [Dwork et al., 2001, Pihur et al., 2009, Fagin et al., 2003].

Spearman's Footrule Distance. The *Spearman's footrule distance* between L_A and L_B is given as follows:

$$S(L_A, L_B) = \sum_{e \in L_A \bigcup L_B} |r^{L_A}(e) - r^{L_B}(e)|, \quad (2)$$

where r is a function that, for a particular list, takes a specific element e as the argument and tracks its position in the concerned list. In case of *top-k* lists the following addition is necessary: for each $e \in L_A \bigcup L_B$ $r^{L_A}(e) = r(e)I(e \in L_A) + (k+1)I(e \notin L_A)$, where $r(e)$ is its original rank, and I is an indicator function and that takes the value of 1 if the argument in parentheses is true, and 0 otherwise. Similarly r^{L_B} can also be computed.

The weighted or score based analog of the Spearman's distance is formulated as follows:

$$S'(L_A, L_B) = \sum_{e \in L_A \cup L_B} |M(r^{L_A}(e)) - M(r^{L_B}(e))| \times$$
$$|r^{L_A}(e) - r^{L_B}(e)|$$

where for example $M(r^{L_A}(e))$ returns the score of e given in the list L_A.

Kendall's Tau Distance. Much popular *Kendall's tau distance* counts the pairwise disagreements between two lists. Below is an intuitive, tailored version of the distance applicable to both full and partial lists [Pihur et al., 2009].

$$K(L_A, L_B) = \sum_{\forall i,j \in L_A \cup L_B} K_{ij} \qquad (3)$$

where,
$$K_{ij} = 1 \ if \ [r^{L_A}(i) - r^{L_A}(j)][r^{L_B}(i) - r^{L_B}(j)] > 0$$
$$= 0 \ if \ [r^{L_A}(i) - r^{L_A}(j)][r^{L_B}(i) - r^{L_B}(j)] < 0$$
$$= p \ if \ [r^{L_A}(i) - r^{L_A}(j)][r^{L_B}(i) - r^{L_B}(j)] = 0$$

The weighted or score based analog of the Kendall's distance is formulated as follows:

$$K'(L_A, L_B) = \sum_{\forall i,j \in L_A \cup L_B} K_{ij} \times |M(r^{L_A}(i)) - M(r^{L_B}(j))|$$

The third case appears only when none of i or j exists in any of the lists i.e., $r^{L_x}(i) = r^{L_x}(j) = k + 1$. The value of p may lie between 0 and 1 i.e., $0 < p < 1$. The following three choices are common in practice 0, 0.5 and 1. Further detailed on these measures can be found in [Pihur et al., 2009, Dwork et al., 2001]. The scaled Kendall's tau distance can retrieved by dividing the measure by C_2^k where k is the size of each list.

Aggregation obtained by optimizing the Kendall's distance is called the Kemeny Optimal Aggregation (KOA) and the final list corresponds to the true geometric median of the input lists. KOA is preferred over the rest as it has a reasonable maximum likelihood interpretation [Young, 1988]. Moreover it is neutral, consistent and obeys Condorcet property [Young and Levenglick, 1978]. Apart from these we need to discuss another property called Extended Condorcet Criterion (ECC) which has a great role in *spam fighting*. Spam is basically the event of having a web page undeservedly highly ranked as a result of some illegitimate manipulation. Spams arise in web search engines. Similar problem arises in bioinformatics as well when some of bio-molecules are unduly ranked high or low because of some experimental error or human error. ECC makes sure that x must be ranked above y if there exists a partition (C, C') of S such that for any $i \in C$ and $j \in C'$ the majority prefer i to j, then i

must be ranked above j [Truchon, 1998]. Locally Kemeny Optimal Aggregation satisfies the ECC [Dwork et al., 2001]. By applying the local Kemenization procedure, we can obtain a ranking that is maximally consistent with the Borda ordering but in which the Condorcet winners are at the top of the list. A ranking R is Locally kemeny Optimal if no swap of adjacent elements is possible that produces a ranking R' such that $\sum_{x=1}^{n} K(R, L_x) \geq \sum_{x=1}^{n} K(R', L_x)$. Local Kemenization can be obtained in polynomial time as follows: at the mth iteration insert the mth element x in the bottom of the list, and bubble it up until there is an element y such that the majority places y over x (see [Online material , 2008] for for worked our example). It is generally suggested to use the Local Kemenization over any initial aggregation using any rank fusion method [Dwork et al., 2001]. By doing this exercise it is made sure that that the spams are pushed to the bottom of the final list. Quite importantly we observed that a simple Local Kemenization with its simple majoritarian treatment is inappropriate in case of a score based rank fusion. Therefore we modified the kemenization accordingly. In spite of having so many desirable properties KOA has limitations in terms of its computational intractability. It is shown by [Dwork et al., 2001] that the KOA is NP-Hard even when $K = 4$. A few attempts like [Dwork et al., 2001, Lin and Ding, 2008, DeConde et al., 2006] etc. have been proved useful for KOA. All these approaches have their own merits and demerits. For example, Markov chains are many times quicker than its evolutionary evolutionary rivals like GA and CE. However GA or CE may search more extensively over infinitely long run. However in the next section we describe a Markov chain for score based rank fusion followed by weighted local Kemenization. We use the proposed technique to combine full lists of miRNA-targets and compare with the performance of the existing evolutionary techniques.

3 Formulation of the Proposed Algorithm

A (homogeneous) Markov chain for a system is specified by a set of states $T = \{1, 2, ..., |T|\}$ and a $|T| \times |T|$ non negative stochastic matrix (so that sum of each row is 1) M. The system begins in some initial state in T and moves from one state to another state during each step. Such transitions are guided by M, the transition matrix. At any step, if the system is in state i, it shifts to state j with probability M_{ij}. If the current state is given as a probability distribution, the probability distribution of the next state is given by the product of the vector representing the current state distribution and M. The start state of the system is usually chosen according to some distribution x (usually, the uniform distribution) on T. After m steps, the state of the system is distributed according to xM^m. Under some conditions (excluded), irrespective of the start distribution x, within a few steps the system eventually reaches an unique stationary distribution where the state distribution does not change. It can be shown that the stationary distribution is given by the principal left eigenvector y of M, i.e., $yM = \lambda y$. A power-iteration algorithm can quickly obtain a reasonable approximation of y. The entries in y define a natural ordering on T.

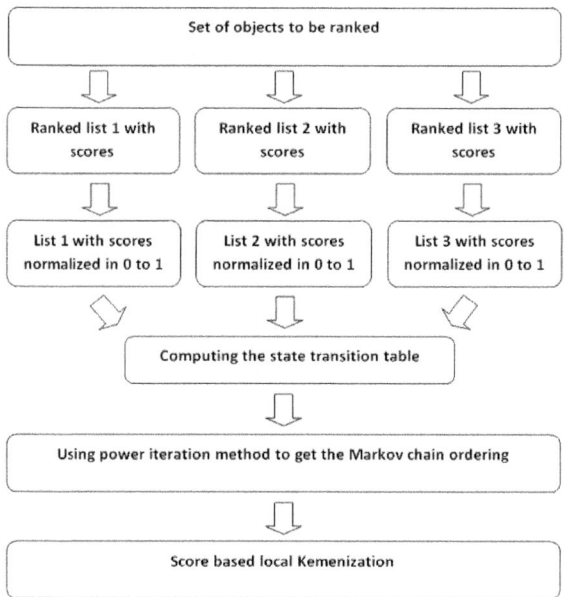

Fig. 1. Flowchart of the proposed technique

Followings are the steps to prepare the transition table which is further used for computing the principal left eigenvector:

- **Step 1:** First of all take the union of the given $top - k$ lists: $S = \bigcup_{x=1}^{n} L_x$.
- **Step 2:** Scores in each list are normalized within the interval $[0, 1]$

$$M^*(r^{L_x}(i)) = \frac{M^({r^{L_x}(i)}) - min(M(r^{L_x}(j)))}{max(M(r^{L_x}(j))) - min(M(r^{L_x}(j)))}$$
$$where, i, j = 1, 2, 3, ..., k$$

- **Step 3:** For each pair of items i and j in S, let the preference for j over i, $m^*_{ij} = 1$ if

$$\sum_{\forall L_x \in L} I(r^{L_x}(j) < r^{L_x}(i)) \times |M^*(r^{L_x}(i)) - M^*(r^{L_x}(j))| >$$
$$\sum_{\forall L_x \in L} I(r^{L_x}(i) < r^{L_x}(j)) \times |M^*(r^{L_x}(i)) - M^*(r^{L_x}(j))|$$

where I is an indicator function that returns 1 if the statement in the respective parentheses are true and 0 otherwise. We need to recall that for any object $r^{L_x}(e) = k + 1$ if e does not belong to L_x

- **Step 4:** Define the transition matrix $M = \{m_{ij}\}$ as follows: $m_{ii} = m^*_{ij}/|S|$ when $i \neq j$ and $m_{ij} = 1 - \sum_{i \neq j} m_{ij}$ otherwise.
- **Step 5:** Make the transition matrix M ergodic by multiplying each element by $1 - \varepsilon$ and then adding $\varepsilon/|S|$ to each element, where ε is a small, positive number. It is recommended to set $\varepsilon = 0.15$.

Once the state transition matrix is created, as mentioned earlier a simple power iteration method can find out the principal left eigenvector of the matrix which suggests a natural ordering of the elements based on descending eigenvalues. Computation of the state transition matrix for a toy example is available with the supplementary material.

3.1 Extension of the Local Kemenization Technique for Allowing Scores

We have already discussed the local Kemenization algorithm in brief in the *Foundation* section. It can be immediately realized that a simple majority voting principle does not count the deviations in scores. Hence we replace that by weighted majority voting where weights. Any ordered pair (j, i) is majority voted if

$$\sum_{\forall L_x \in L} I(r^{L_x}(j) < r^{L_x}(i)) \times |M^*(r^{L_x}(i)) - M^*(r^{L_x}(j))| >$$
$$\sum_{\forall L_x \in L} I(r^{L_x}(i) < r^{L_x}(j)) \times |M^*(r^{L_x}(i)) - M^*(r^{L_x}(j))|$$

where I is an indicator function that returns 1 if the statement in the respective parentheses are true and 0 otherwise. This is similar to the consideration during the state transition table creation. A flowchart of the proposed rank aggregation technique is furnished in Figure 1.

4 Results

4.1 An Artificial Simulation

For evaluating the performance of the proposed methods, we executed a simulation study. In this we created an arbitrary ranked list of size 10 with scores scaled within $[0, 1]$. Then we allowed each element's score to randomly fluctuate within the bound:

$$M(r^{L_x}(i)) - 0.1 < M^{new}(r^{L_x}(i)) < M(r^{L_x}(i)) + 0.1$$

Then it is necessary that we normalize them again within the interval $[0, 1]$ because the scores 0 or 1 may, after fluctuation exceed the $[0, 1]$ interval. By this process a ranked lists is prepared from the parent list. Then this newly created list is treated in similar way to generate the next list and so on. By this process we make sure that positional deviations of the elements are not always restricted

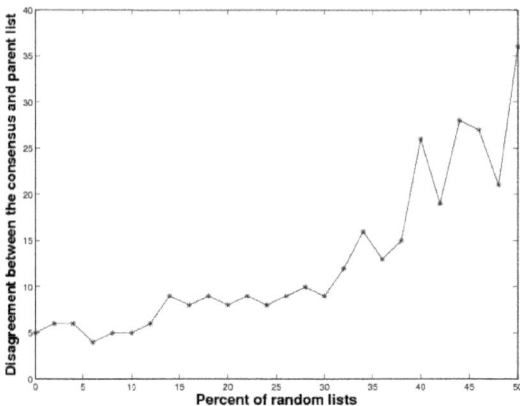

Fig. 2. The plot shows the varying Kendall's tau distance (simple) with increasing fraction of random impurities

within small bounds. These lists when aggregated, are expected to mimic the parent list closely. Now we replaced a fraction of the lists with random ones and aggregated using our algorithm. The percentage of replaced lists varied from 0% to 50% with an interval of 2. So in total 26 runs were executed. The GA or CE based algorithms could not be considered for comparison as they fall slower with the increasing number of lists. The performance of our method in terms of retaining the agreement with the parent list, is shown in Figure 2. It is found that the Kendall's tau distance between the consensus and the parent list never exceeded 10 (maximum could be 45) up to 30% of allowed impurities.

4.2 Application on miRNA Target Prediction

The microRNA guided RNA silencing pathway is a recently discovered process that regulates gene expression by acting upon messenger RNA (mRNA) [Bartel, 2004]. It has earned great attention of the scientific community in recent years since it is found to be strongly related to cell proliferation, apoptosis and

Table 1. Validated examples within the top 20 of the consensus lists produced by different algorithms

Rank Aggregation Method	miRNA			
	has-mir-1	has-mir-16	has-mir-124	has-mir-373
Proposed	11	8	12	10
GA+Kendall's tau distance	11	7	11	10
CE+Kendall's tau distance	10	8	10	9
GA+Spearman's distance	6	7	11	8
CE+Spearman's distance	9	8	9	10

Table 2. Top-20 of the consensus lists along with the validated targets within (Only for hsa-mir-16, the number validated targets within top 15 of the consensus list is reported)

		Top-20 of the consensus list	Validated among top-20 of the consensus lists
hsa-mir-1	GA+Kendall	DDX5, BDNF, BCL2, LOC285636, NETO2, ANKIB1, OTX2, PREX1, MEX3C, RNUXA, ARHGEF18, HSPD1, BACH2, MET, RAB11FIP2, SFRS10, ADPGK, TNKS2, HNRNPU, LIN7C	DDX5, BDNF, BCL2, NETO2, ANKIB1, PREX1, ARHGEF18, HSPD1, RAB11FIP2, LIN7C
	CE+Kendall	NETO2, DDX5, OTX2, LOC285636, BDNF, MEX3C, BCL2, ARHGEF18, LASP1, PREX1, ANKIB1, MET, HSPD1, KCNJ2, BACH2, RNUXA, ADAR, SMEK2, UBE4A, FOXP1	NETO2, DDX5, BDNF, BCL2, ARHGEF18, LASP1, PREX1, ANKIB1, HSPD1, KCNJ2, ADAR
	Proposed	OTX2, NETO2, DDX5, LOC285636, PREX1, BDNF, MEX3C, RNUXA, KCNE1, C11orf61, MAP4K3, HDAC4, UBE4A, HIAT1, LASP1, ANKIB1, BCL2, ARHGEF18, BACH2, KCNJ2	NETO2, DDX5, PREX1, BDNF, KCNE1, HDAC4, LASP1, ANKIB1, BCL2, ARHGEF18, KCNJ2
hsa-mir-16	GA+Kendall	TPPP3, WBP11, EIF2C4, CDC37L1, ZCCHC2, VEGFA, CARD8, CCNE1, RNF111, CFL2, DLL1, CD28, KIF21A, C10orf104, CDC14B	TPPP3, VEGFA, CARD8, CCNE1, CFL2, C10orf104, CDC14B
	CE+Kendall	VEGFA, TPPP3, WBP11, CFL2, AKAP11, EIF2C4, ZCCHC2, HMGA1, CCNE1, C10orf104, CDC37L1, CDC14B, RAB10, CDC37L1, YWHAQ, NEBL	VEGFA, TPPP3, CFL2, HMGA1, CCNE1, C10orf104, RAB10, CDC14B
	Proposed	VEGFA, TPPP3, CDC14B, YWHAQ, WBP11, CFL2, BCL2, HMGA1, C10orf104, NEBL, DLL1, PDCD4, BTG2, KIF21A, ZCCHC2	VEGFA, TPPP3, CDC14B, CFL2, BCL2, HMGA1, C10orf104, PDCD4
hsa-mir-124	GA+Kendall	ANTXR2, CDCA7L, ATF7IP, CGN, SERP1, MYO10, RHOG, FA2H, CHSY1, PRRX1, TSPAN15, SEMA6D, ZFAND3, TTL, ARRDC1, LRRC57, RAB34, ARHGAP29, PLSCR3, RDH10	CDCA7L, SERP1, MYO10, RHOG, FA2H, CHSY1, TSPAN15, ARRDC1, ARHGAP29, PLSCR3, RDH10
	CE+Kendall	ANTXR2, CDCA7L, CGN, MYO10, FA2H, ZFAND3, SEMA6D, ARHGAP29, PLSCR3, SERP1, RDH10, ATF7IP, RHOG, RAB34, CDCA7L, PRRX1, EDEM1, NID1, WTAP, SERTAD4, CHSY1	MYO10, FA2H, ARHGAP29, PLSCR3, SERP1, RDH10, RHOG, CDCA7L, NID1, CHSY1
	Proposed	ANTXR2, ATF7IP, FA2H, NID1, LEMD3, ARRDC1, NFATC1, MYO10, CDCA7L, CGN, RHOG, SEMA6D, SERP1, PLSCR3, PRRX1, SERTAD4, RAB34, CDK6, TSPAN15, RDH10	FA2H, NID1, ARRDC1, NFATC1, MYO10, CDCA7L, RHOG, SERP1, PLSCR3, CDK6, TSPAN15, RDH10
hsa-mir-373	GA+Kendall	SYDE1, STK4, AOF1, PLAG1, FYCO1, TUSC2, ZBTB41, GLTP, OLFM3, C10orf104, CNOT6, TBC1D2, MKRN1, ADAM9, FAM13C1, FGD5, C5orf41, NCOA7, SLC25A23, YTHDF3	STK4, AOF1, FYCO1, TUSC2, GLTP, CNOT6, TBC1D2, MKRN1, ADAM9, SLC25A23
	CE+Kendall	TUSC2, PLAG1, STK4, AOF1, FYCO1, GLTP, ADAM9, SYDE1, CNOT6, FGD5, ZBTB41, NCOA7, LHX6, OLFM3, C10orf104, RPIA, YTHDF3, AOF1, CRTC2, VPS26A	TUSC2, STK4, FYCO1, GLTP, ADAM9, CNOT6, RPIA, AOF1, VPS26A
	Proposed	PLAG1, ZBTB41, SYDE1, STK4, FAM13C1, ADAM9, TBC1D2, MKRN1, TUSC2, GLTP, C10orf104, FGD5, TNFAIP1, FYCO1, OLFM3, AOF1, CNOT6, C5orf41, NCOA7, YTHDF3	STK4, ADAM9, TBC1D2, MKRN1, TUSC2, GLTP, TNFAIP1, FYCO1, AOF1, CNOT6

tumorigenesis [Chan and Krichevsky, 2005, Esquela-Kerscher and Slack, 2006]. Although a large number of human microRNAs have been reported, many of their mRNA targets are yet unexplored. TargetScan [Krek et al., 2005], TargetMiner [Bandyopadhyay and Mitra, 2009], miRanda [John et al., 2004], PITA [Kertesz et al., 2007], RNAHybrid [Rehmsmeier, 2004] etc are some popular target prediction algorithms.

It is common knowledge that different target prediction programs provide different predictions, and the volume of overlap among different lists of predicted targets is often very poor [Sethupathy et al., 2006]. It is anticipated that integration of the target lists obtained from different target prediction algorithms would fetch the putative targets upwards in the consensus list [Lin and Ding, 2008].

In this context, to evaluate the credibility of our proposed measure we identified 50 arbitrary mRNAs targets of hsa-mir-1, hsa-mir-16, hsa-mir-124 and hsa-mir-373 which are commonly predicted by three algorithms namely: miRANDA, PITA and RNAHybrid. Among them 25 are experimentally validated and rest 25 are not (information pertaining to experimental validations are collected from miRecords [Xiao et al., 2009]. One exception is hsa-mir-16 for which we had to use 20 validated targets and rest of the targets just predicted). The ranked lists corresponding to the algorithms were found while sorting targets based on the interaction scores produced by the respective algorithms (see supplementary material for the ranked lists). To this end we employed our proposed technique followed by the GA and CE based fusion algorithms. The counts of experimentally validated examples within the top 20 mRNA targets of the consensus lists are shown in Table 1. Of note, the average time consumed by WMC was $\sim 1.9 \times 10^{-1}$ seconds.

It is apparent from the results that our algorithm performed at least as good as the GA or CE algorithms with respect to the experimentally validated targets. So, the proposed method can be treated as an alternative to the evolutionary computing based techniques which consume a lot of time and sometimes the user needs to wait for infinitely long time. In Table 2 the experimentally validated target genes (within top 20 of the consensus list) for the respective miRNAs are shown along with the top 20 of the consensus list.

5 Conclusion

In the present article we proposed a novel rank fusion technique that uses both rank and score of an element for optimizing the consensus ranking. This has huge importance while aggregating biological rankings as biological studies are more concerned of scores the subjects obtain in respective experimental set up. The score based modification to the local Kemenization algorithm stood relevant with respect to the score consideration. The score based Markov chain is not only preferred for its polynomial time complexity ($O(nk)$ in practice where n refers to the number of elements in each list and k the number of lists) [Dwork et al., 2001] but also for its ability of ranking many of the the important elements at the top of the consensus list. Through an artificial simulation we found that the proposed method performs reasonably well in noisy environment, which indicates the spam reduction quality of the algorithm. Our future work will be definitely to test the proposed technique for partial list aggregation and to find what percentage of overlap among the given lists is necessary to land at meaningful consensus list.

Acknowledgement. D.S. and S.B. gratefully acknowledge the financial support from the grant no. DST/SJF/ET-02/2006-07 under the Swarnajayanti Fellowship scheme of the Department of Science and Technology, Government of India. The authors also acknowledge the technical assistance of Mr. Indranil Aich, who was a project linked personnel at the Machine Intelligence Unit of Indian Statistical Institute when this work was done.

References

[Dwork et al., 2001] Dwork, C., Kumar, R., Naor, M., Sivakumar, D.: Rank Aggregation Methods for the Web. In: Proc. 10th International World Wide Web Conference, pp. 613–620 (2001)

[Fagin et al., 2003] Fagin, R., Kumar, R., Sivakumar, D.: Comparing top k lists. SIAM Journal of Discrete Mathematics 7, 134–160 (2003)

[Pihur et al., 2009] Pihur, V., Datta, S., Datta, S.: RankAggreg, an R package for weighted rank aggregation. BMC Bioinformatics (2009)

[Lin and Ding, 2008] Lin, S., Ding, J.: Integration of Ranked Lists via Cross Entropy Monte Carlo with Applications to mRNA and miRNA Studies. Biometrics (2008), doi:10.1111/j.1541-0420.2008.01044.x

[Pihur et al., 2007] Pihur, V., Datta, S., Datta, S.: Weighted rank aggregation of cluster validation measures: a Monte Carlo cross-entropy approach. Bioinformatics 23(13), 1607–1615 (2007)

[Pihur et al., 2008] Pihur, V., Datta, S., Datta, S.: Finding cancer genes through meta-analysis of microarray experiments: Rank aggregation via the cross entropy algorithm. Genomics 92, 400–403 (2008)

[DeConde et al., 2006] DeConde, R., Hawley, S., Falcon, S., Clegg, N., Knudsen, B., Etzioni, R.: Combining results of microarrayexperiments: a rank aggregation approach. Proc. Stat. Appl. Genet. Mol. Biol. 5, Article15 (2006)

[Krek et al., 2005] Krek, A., Grun, D., Poy, M.N., Wolf, R., Rosenberg, L., Epstein, E.J., MacMenamin, P., da Piedade, I., Gunsalus, K.C., Stoffel, M., Rajewsky, N.: Combinatorial microRNA target predictions. Nature Genetics 37, 495–500 (2005)

[John et al., 2004] John, B., Enright, A.J., Aravin, A., Tuschl, T., Sander, C., Marks, D.S.: Human microRNA targets. PLOS Biology 2, 1862–1879 (2004)

[Zhou et al., 2006] Zhou, J., Lin, S., Melfi, V., Verducci, J.: Composite MicroRNA target predictions and comparisons of several prediction algorithms. MBI Technical Report, No. 51 (2006)

[Even et al., 1998] Even, G., Naor, J., Schieber, B., Sudan, M.: Approximating minimum feedback sets and multicuts in directed graphs. Algorithmica 20(2), 151–174 (1998)

[Goldenberg, 1989] Goldenberg, D.E.: Genetic Algorithms in Search, Optimization and Machine Learning. Addison Wesley, Reading (1989)

[Chan and Krichevsky, 2005] Chan, J.A., Krichevsky, A.M., et al.: MicroRNA-21 is an antiapoptotic factor in human glioblastoma cells. Cancer Res. 65, 6029–6033 (2005)

[Esquela-Kerscher and Slack, 2006] Esquela-Kerscher, A., Slack, F.J.: Oncomirs - microRNAs with a role in cancer. Nat. Rev. Cancer 6, 259–269 (2006)
[Bartel, 2004] Bartel, D.P.: MicroRNAs: genomics, biogenesis, mechanism, and function. cell 116(2), 281–297 (2004)
[Kertesz et al., 2007] Kertesz, M., Iovino, N., et al.: The role of site accessibility in microRNA target recognition. Nat. Gen. (2007), doi:10.1038/ng2135
[Sethupathy et al., 2006] Sethupathy, P., Megraw, M., Hatzigeorgiou, A.G.: A guide through present computational approaches for the identification of mammalian microRNA targets. Nature Methods (2006), doi:10.1038/NMETH954
[Truchon, 1998] Truchon, M.: An extension of the Condorcet criterion and Kemeny orders. cahier 98-15 du Centre de Recherche en Economie et Finance Appliquees (1998)
[Young, 1988] Young, H.P.: Condorcet's theory of Voting. Amer. Political Sci. Review 82, 1231–1244 (1988)
[Young and Levenglick, 1978] Young, H.P., Levenglick, A.: A consistent extension of Condorcet's election principle. SIAM J. on Applied Math. 35(2), 285–300 (1978)
[Bandyopadhyay and Mitra, 2009] Bandyopadhyay, S., Mitra, R.: Targetminer: MicroRNA Target Prediction with Systematic Identification Of Tissue Specific Negative Examples. Bioinformatics (2009), doi:10.1093/bioinformatics/btp503
[Xiao et al., 2009] Xiao, F., Zuo, Z., Cai, G., Kang, S., Gao, X., Li, T.: miRecords: an integrated resource for microRNA-target interactions. Nucleic Acids Res. 37, D105–D110 (2009)
[Sengupta et al., 2010] Sengupta, D., et al.: A novel measure for evaluating an ordered list: application in microRNA target prediction. In: ISB 2010, India (2010), ISBN:978-1-60558-722-6, http://doi.acm.org/10.1145/1722024.1722067
[Grazia, 1953] Grazia, A.D.: Mathematical Derivation of an Election System. Isis 44 (1953)
[Rehmsmeier, 2004] Rehmsmeier, M., Steffen, P., Höchsmann, M., Giegerich, R.: Fast and effective prediction of microRNA/target duplexes. RNA 10, 1507–1517 (2004)
[Sengupta et al., 2012] Sengupta, D., Maulik, U., Bandyoapdhyay, S.: Weighted Markov Chain Based Aggregation of Bio-molecule Orderings. IEEE-TCBB (2012), doi:10.1109/TCBB.2012.28
[Sengupta and Bandyopadhyay, 2011] Sengupta, D., Bandyoapdhyay, S.: Participation of microRNAs in human interactome: extraction of microRNA microRNA regulations. Molecular Biosystems (2011), doi:10.1039/C0MB00347F
[Sengupta et al., 2011] Sengupta, D., Maulik, U., Bandyoapdhyay, S.: Entropy steered Kendall's tau measure for a fair Rank Aggregation. NCEATCS (2011), doi:10.1109/NCETACS.2011.5751397
[Chen, 2011] Chen, J., Aronow, B.J., Jegga, A.G.: Disease candidate gene identification and prioritization using protein interaction networks. BMC Bioinformatics (2009), doi:10.1039/C0MB00347F, doi:10.1186/1471-2105-10-73
[Online material, 2008] Rank Aggregation Methods for the Web, www.cs.uc.edu/~annexste/Courses/cs728-2008/Lecture11.ppt

Modeling Complex Diseases Using Discriminative Network Fragments

(Invited Keynote Talk)

Ambuj K. Singh

Department of Computer Science
Program in Biomolecular Science & Engineering
University of California at Santa Barbara
Santa Barbara, CA 93106
ambuj@cs.ucsb.edu

Abstract. A number of complex diseases are network-based where a set of pathways needs to be perturbed. Understanding the logic of such perturbations from high-throughput datasets as next-generation sequencing or gene expression can help in elucidating the nature of diseases and their multiple states.

I will discuss a recent approach that mines multiple annotated networks to find the small network fragments that drive the state of the entire network. In this approach, a gene/protein interaction network is used as the underlying structure and high throughput data is used to annotate the nodes of the network. The global state of a network is annotated as normal or diseased.

Mining discriminative subgraphs from large networks is a powerful mechanism for identifying network components that are influential in determining the global network state. It is different from learning classification models in a number of ways. The first difference is in semantics. A traditional classifier operates on unstructured data where each feature represents an axis in a high-dimensional space. In our problem, each feature (or annotated node in a graph) is constrained within a structure and the network event being modeled evolves through the global structure. A traditional classifier only analyzes the statistical significance of a feature and ignores the underlying structure. Second, in our problem, the goal is to mine discriminative subgraphs, each of which aids in predicting the global network state. The proposed technique operates at a level of abstraction of discriminative subgraphs instead of individual nodes.

To achieve the desirable properties highlighted above, we design a technique for mining network-constrained decision trees that learn network-encoded logic functions to predict the global network state. To tackle the exponential subgraph search space, we formulate the idea of an Edit Map, on which we perform Metropolis-Hastings sampling to drastically reduce the computation cost. We have performed extensive experiments to evaluate the efficiency and effectiveness of our method. Our results show that the proposed algorithm achieves an accurate approximation of the optimal answer set. Furthermore, the method outperforms the current state-of-the-art classifiers developed for gene/protein interaction networks.

Novel Multi-sample Scheme for Inferring Phylogenetic Markers from Whole Genome Tumor Profiles

Ayshwarya Subramanian[1], Stanley Shackney[2], and Russell Schwartz[1,3]

[1] Department of Biological Sciences, Carnegie Mellon University,
Pittsburgh, PA 15213
[2] Oncotherapeutics, Pittsburgh PA 15243
[3] Lane Center for Computational Biology, Carnegie Mellon University,
Pittsburgh, PA 15213

Abstract. Computational cancer phylogenetics seeks to enumerate the temporal sequence of aberrations in tumor evolution, thereby delineating the evolution of possible tumor progression pathways, molecular subtypes and mechanisms of action. We previously developed a pipeline for constructing phylogenies describing evolution between major recurring cell types computationally inferred from whole-genome tumor profiles. The accuracy and detail of the phylogenies, however, depends on the identification of accurate, high-resolution molecular markers of progression, i.e., reproducible regions of aberration that robustly differentiate different subtypes and stages of progression. Here we present a novel hidden Markov model (HMM) scheme for the problem of inferring such phylogenetically significant markers through joint segmentation and calling of multi-sample tumor data. Our method classifies sets of genome-wide DNA copy number measurements into a partitioning of samples into normal (diploid) or amplified at each probe. It differs from other similar HMM methods in its design specifically for the needs of tumor phylogenetics, by seeking to identify robust markers of progression conserved across a set of copy number profiles. We show an analysis of our method in comparison to other methods on both synthetic and real tumor data, which confirms its effectiveness for tumor phylogeny inference and suggests avenues for future advances.

Keywords: Bioinformatics, cancer, phylogenetics, multi-sample, array comparative genomic hybridization (aCGH).

1 Introduction

Analysis of cancer genomes using high-throughput genomic methods has revealed a high degree of variability at the genetic level [18, 8] in otherwise histopathologically indistinguishable tumors. In the process, such analyses have identified specific molecular markers and pathways associated with the onset and progression of cancers in specific tissue types, classification of molecular subtypes and patient sub-populations [25, 26]. This information can inform the design of targeted therapeutics and diagnostic strategies [2]. Analysis of tumor progression has nonetheless been hindered by high heterogeneity in tumor progression pathways, even in tumors impacting similar regulatory pathways and biological processes [30, 12]. Heterogeneity can occur between patients or within a single patient, where sub-populations of cells may correspond to different states or even pathways of tumor progression. Computational cancer phylogenetics provides a strategy for making sense of the complexity of tumor evolution by identifying recurring pathways of tumor evolution both within and across patients through the use of phylogenetic inference algorithms. Such methods, however, require some mechanism for identifying discrete states of progression and estimating evolutionary distances among them. In the case of character-based phylogeny approaches, this process involves identifying robust markers of progression whose presence or absence can be used to track tumor evolution. In the present work, we focus specifically on the problem of marker inference from array comparative genomic hybridization (aCGH) data providing genome-scale DNA copy number measurements. For these data, the problem corresponds to finding discrete genomic regions of DNA gain or loss that can serve as markers of tumor progression.

Existent methods for aCGH analysis include algorithms for smoothing, segmentation and combined segmentation and classification of both single- [20, 10, 5, 31] and multi-sample data [21, 24, 32–34, 4, 16]. Such methods can be highly effective at identifying discrete copy number variations in such data, but are poorly suited to the problem of phylogenetic inference because they do not constrain solutions to common markers across tumor data. They thus provide no straightforward way to infer a set of robust markers with defined boundaries across patients and progression states for use in phylogenetic inference. A similar objective was considered by Picard et al. for the method, CGHSeg [19] , addressing the problem of joint segmentation and calling of multiple samples, primarily as a way of improving accuracy of assignment using similarities between data. This method, though, was also not designed for the purpose of phylogenetic inference, and is inefficient for the data characteristics needed for these purposes, especially the combination of large numbers of markers with defined boundaries across a modest number of discrete samples characteristic of whole-genome datasets.

In previous work, we developed an approach to the problem of tumor phylogenetics based on the use of mixture models to infer discrete states of progression recurrent across tumor samples [23, 29]. We subsequently used this mixture modeling approach as the basis for a pipeline for tumor phylogeny inference [27]. For this pipeline, we developed a multi-sample segmentation method based on a simple statistical test applied to fixed-length windows of probes heuristically merged to identify amplicons from a set of inferred mixture components. The unmixing procedure in its present formulation can only reliably infer amplifications and, hence, we focus only on copy number amplifications in this work. An additional statistical test would then call presence or absence of each amplicon in each component, converting the components into discrete character arrays suitable for character-based phylogenetic inference. Validation on a set of components derived from real breast tumor data [15] showed the marker selection method to be reasonably effective at finding known breast cancer amplicons suitable for use as phylogenetic markers. The segmentation step, however, showed a poor ability to resolve fine-scale structure within amplicons, limiting the number of phylogenetic markers and the ability of the method to discriminate between subtle changes in nearby markers. In addition, separating segmentation from calling left no way to guarantee that amplicons detected in the segmentation stage would in fact be called differently in different components and thus become useful markers for phylogenetics.

The present work is aimed at developing an improved marker detection method designed to maintain the advantages of our prior work in using multi-sample segmentation from mixture components to identify a robust set of common markers usable across samples, while adapting ideas from prior single-sample methods to improve fine-scale resolution of amplicon structure. The method uses a novel HMM scheme to do joint segmentation and calling of markers simultaneously from a set of mixture components. It is thus similar in character to the method of Picard et al. [19] although with fewer assumptions about shared features of amplicons across samples. In addition, it shares some similarity to the FLLat method [16], which also performs multi-sample aCGH analysis, although with a distinct computational model based on a fused Lasso optimization. Our new approach allows joint segmentation and thus detection of phylogenetically useful markers across mixture components. In contrast to our prior work, the use of the HMM scheme also allows the method to detect changes in assortments of amplicons across components within regions of amplification. We analyze the method on both simulated and real data and compare it to related methods heuristically adapted to the problem of phylogenetic calling. The results reveal the method to give superior performance at phylogenetic reconstruction for high but biologically reasonable levels of experimental noise.

2 Approach

Our model is based on a generalization of the use of HMMs to multi-sample data for the purpose of finding a common marker set across a set of samples. It accomplishes this task by treating states of the HMM as tuples of amplification states across samples, with each copy number probe assigned one state. Any contiguous region of common state in which at least one component is called amplified can then serve as a single marker for phylogenetic inference.

We assume that we are presented with a whole genome tumor data set $\mathbf{X}$ consisting of real-valued measured copy number ratios x_{ij} for a set of observed components (cell types or stages of progression) $i = 1, 2, ...m$ and genomic coordinates $j = 1, 2, ...n$. The HMM then defines the joint probability distribution of the sequence x_{ij} in the observed matrix $\mathbf{X}$ in terms of a latent matrix $\mathbf{H}$ of discrete amplification states across components. We assume two possible copy number states for each x_{ij}: 0 (normal/diploid) and 1 (aberrated/aneuploid/polyploid). Each component is then classified by a row $\mathbf{H}_i$ of $\mathbf{H}$, a binary vector of amplification states across the n copy number probes. Each state of the HMM corresponds to a column $\mathbf{H}_j$ of $\mathbf{H}$, a binary vector of amplification states of all m components at a single probe position. The resulting time complexity is $\mathcal{O}(2^{2m}n)$. While the number of states is exponential in m, it is manageable for our purposes because the number of mixture components is small, typically 4-10, regardless of the number of samples analyzed. A Markov chain produced by the model would correspond to a string of state assignments across the probes in genomic order.

We simplify the Markov model by assuming that all transition probabilities between consecutive probes take on one of four values:

- p_{na}, the probability of going from the normal state to any one of the $2^m - 1$ aberrant states, which we define also to be the probability of going from any one aberrant state to any other
- p_{aa}, the probability of staying in the same aberrant state
- p_{nn}, the probability of staying in the normal state
- p_{an}, the probability of going from an aberrant state to normal

We set $p_{aa} = \frac{w-1}{w}$ to enforce an average amplicon width w, where we assume in the present work that $w = 10$. We set p_{na} by the formula

$$p_{na} = \left(\frac{p}{n*m}\right)\left(\frac{1}{2^m - 1}\right)$$

where p is a penalty set to 0.1 in the present work, effectively penalizing the model for assigning large numbers of amplicons by creating a prior expectation of 0.1 amplicons occurring by chance across the entire data set. The other two transition probabilities are then fixed by p_{aa} and p_{na}:

$$p_{nn} = 1 - (2^m - 1) * p_{na} \text{ and } p_{an} = 1 - (2^m - 1) * p_{aa}$$

We define emission probabilities O by assuming x_{ij} comes from either a normal diploid distribution or an aberrant aneuploid distribution:

$$P(O_d|H) = \phi(x; \mu_d, \sigma) \text{ and } P(O_a|H) = \phi(x; \mu_a, \sigma)$$

where we assume here that diploid data has a mean $\mu_d = 1$, corresponding to a mean ratio of one between observed data and a diploid control, while aneuploid data is modeled as having a mean ratio $\mu_a = 4$ relative to a diploid control. The initial state probability π for all aberrated states is assumed to be $q = (p/(2^m - 1)/n)$ leaving an initial probability of the normal state of $1 - (2^m - 1)q$. The resulting model was implemented in MATLAB.

3 Experimental Methods

3.1 Synthetic Data

To assess accuracy on data of known ground truth, we simulated a series of aCGH data sets across a range of assumed experimental noise levels. We assumed a log-normal noise model $Y_{ij} = M_{ij} + \mathcal{N}(0, \sigma)$ for each sample i ($i = 1, 2, \ldots, m$) and aCGH probe position j ($j = 1, 2, \ldots, n$). Here, Y_{ij} is the simulated copy number ratio in the log domain, M_{ij} is the amplification model and $\mathcal{N}(0, \sigma)$ is Gaussian noise. We modeled the distribution of copy numbers in tumor data by an exponential distribution $M_{ij} = 1 + \mathbb{1}(j \subset S_i)\text{Exp}(\lambda)$ where $\mathbb{1}$ is the indicator function for the presence of site j in an amplicon S_i. We estimated the exponential rate λ from the real component data in Sec. 3.2 using the mean of observed probe values above 5, to minimize contamination by non-amplified probes. We then simulated a series of components to model tumor evolution over a complete binary tree of depth three. Beginning from an all-diploid root, we simulated amplicons of fixed width $w = 20$ in a hypothetical data set of 1161 probes (to match the proportion of amplifications in the real data) in 6 components, adding one new amplicon per non-root node to those present in the node's parent to model acquisition of successive amplicons over succeeding generations of progression. Amplicons were placed uniformly at random within the genome, rejecting and rerunning any placement that resulted in two amplicons within w probes of one another. We then generated observed signal values for amplified and non-amplified sites by the log-normal noise model described above. This process was repeated for 200 replicates each at noise levels $\sigma = 0$ to 1.8 in increments of 0.1.

For each replicate, we ran the HMM algorithm as described in Sec. 2. For comparison, we tested the same data on two alternatives: the single-sample method Circular Binary Segmentation (CBS) [17] using the MATLAB function *cghcbs* and the multisample *multiseg* function in the R package CGHSeg [19]. The CBS output was called at a threshold of log(1.5) into amplified or normal. CGHSeg returns called values for each sample. Downstream analysis was performed to extract and merge probes called amplified in at least one sample to yield recurrent markers with common boundaries, each of which serve as a character for the phylogeny inference. Phylogenetic trees were inferred by adding an all-diploid root to the set of character states and then running unweighted maximum parsimony inference using PAUP [28].

Quality of the methods was measured on three tasks. First, accuracy of amplicon detection across samples was quantified by the sensitivity, defined as fraction of genuinely amplified markers assigned to an amplicon, and specificity, defined as the fraction of markers assigned to an amplicon that were in fact amplified. Second, accuracy of marker assignment to amplicons was measured, quantified by the fraction of amplicons correctly called as amplified or non-amplified for all components. Finally, accuracy of phylogeny inference was assessed, quantified by the Branch Score Distance [11] using the *treedist* function of PHYLIP [6], a measure of agreement between the true and inferred phylogenies.

3.2 Experiments: Real Data

We further demonstrated our methods on real data derived from a publicly available (NCBI GEO GSE16672) primary ductal breast carcinoma aCGH dataset [15]. This data was chosen because the cell sorting and sectioning methods underlying the tumor data extraction were developed specifically to aid phylogenetic analysis, making them well suited to our purposes, and because the data contains multiple samples per tumor, making them especially useful for studies of tumor heterogeneity and mixture analysis. The raw data comprises 87 tumor sectors obtained from 14 ductal breast cancer tumors run on a high-density ROMA platform with 83,055 probes. We confined our analysis to the twenty-two autosomal chromosomes, reducing the dataset to 78,874 probes. We converted the raw aCGH data from log to linear domain, denoised it with a total variation denoising and then subjected it to an unmixing analysis to infer 6 components, or putative tumor progression states, as described in [29]. We then ran our method as described in Sec. 2 using PAUP for maximum parsimony tree building as with the simulated data.

4 Results and Discussion

4.1 Synthetic Data

The results of our method and the two comparative methods on simulated data are summarized in Figure 1. Fig. 1(a,b) shows accuracy at the level of amplicon assignment. Fig. 1(a) shows that our method has a lower sensitivity than either of the comparative methods, a result that is expected because our method is tuned to recognize only markers robustly supported across the data sets. Fig. 1(b) shows that all three methods have a high specificity for amplicon calling, with no false positive calls until relatively high levels of noise. At high noise levels, CGHseg is most prone to false positive calls, CBS least prone, and our own method intermediate between the two. Our method thus appears likely to produce a smaller marker set but one less prone to spurious markers than CGHseg. CBS yields higher accuracy at the marker level than either multi-sample method. Manual examination of the the results suggested two prominent kinds of errors in marker assignment: a failure to identify low-copy-number amplicons as amplified, leading to false negative errors, and a tendency to collapse pairs of amplicons close to one another on the genome into a single amplicon. Figure 2 illustrates these two kinds of errors on two sample simulated data sets. Our method may be prone to both error types in part because it treats amplification as a binary state rather than a gradation of amplification levels, a tradeoff intended to reduce the state space of the HMM.

Fig. 1(c) shows accuracy of calling amplification states within detected amplicons. All three methods closely track the sensitivity plot of Fig. 1(a) up to a noise level of about 1.0, suggesting that each is highly accurate in calling states given the amplicons at low to moderate noise levels. In this range, our method suffers primarily from the fact that it fails to identify some amplicons and thus cannot call them correctly. At higher noise levels, all three methods begin to exhibit noticeably lower accuracy at marker calling than at amplicon assignment, suggesting that both failure to identify amplicons and miscalled states within amplicons contribute appreciably to the error rate. Our method falls off in accuracy most precipitously in this range, perhaps because its HMM-based calling relies on individual probe values rather than entire amplicons, leading to more efficient but less accurate estimates of amplification levels.

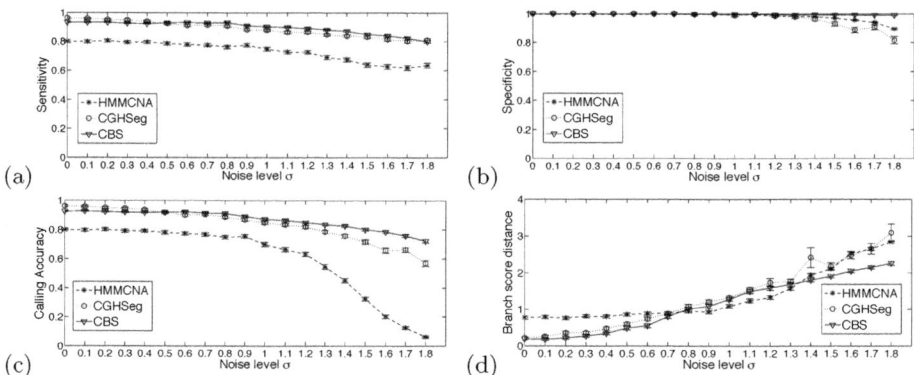

Fig. 1. Accuracy of our method (HMMCNA), CBS, and CGHseg on simulated data. (a, b) Accuracy in amplicon assignment, classified by the sensitivity (a) and specificity (b) of correctly assigning markers. (c) Calling accuracy, measured by the fraction of amplified markers assigned the correct amplification state. (d) Tree-building accuracy, quantified by the branch-score distance between the true and observed tree. All measures are reported as functions of the log-normal noise level σ, averaged over 200 independent runs per noise level.

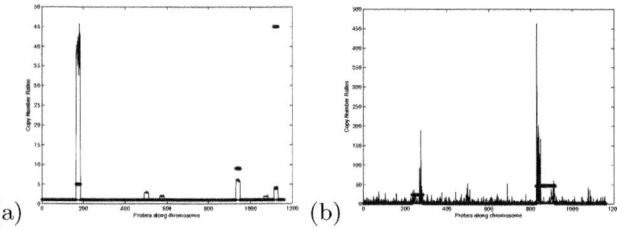

Fig. 2. Illustration of common assignment errors. Thin lines show simulated observed copy numbers and thick bars show assigned states, with those above one on the y-axis corresponding to states amplified in at least one component. (a) Failure to detect low-copy-number amplicons at $\sigma = 0.1$. (b) Merging of nearby amplicons at $\sigma = 1.6$.

Fig. 1(d) shows the accuracy at inferring phylogenetic trees, which is the specific goal of our method. The picture shows a more complicated profile than the prior plots. At low noise levels ($\sigma < 0.7$), CBS and CGHseg both perform similarly to one another and better than our method. At moderate noise levels (σ about 0.7–1.4) our method produces the best accuracy of the three by the BSD measure. At very high noise levels, both our method and CGHseg rapidly degrade in performance, while CBS remains relatively tolerant of high noise. It thus appears that our method is best suited for performance at moderate noise levels while CBS outperforms both multi-sample methods at either low or very high noise levels. An obvious question is where actual aCGH data sets fall along this noise range, but that question does not have a definitive answer because the noise level is highly specific to the platform and samples, as well as any preprocessing applied to them. In addition, the true noise level is difficult to measure empirically because noise will be conflated with true copy number variations when one analyzes the variance of a data set. $\sigma = 0.7$ corresponds to a signal-to-noise ratio of approximately 0.1 by the method of Nowak et al. [16], at the high end of the noise levels they examined but reasonable for aCGH data. We can thus conclude that our method does provide an advantage over the existing methods in accurate phylogeny reconstruction in the presence of high but plausible noise levels.

4.2 Real Data

We next applied our method to mixture components derived from the real breast cancer data set of Navin et al. [15] both for further validation and to illustrate its value in predicting progression on real tumor samples. The HMM method found 81 marker amplicons, triple the 27 detected by our prior method [27]. We have no absolute ground truth from which to judge the results on the real data, but can evaluate their plausibility by comparison to genes with known association with breast or other cancers. The analysis is complicated by the fact that some inferred amplicons are quite large and include many genes, which might be presumed to be predominantly passenger genes irrelevant to the progression process. We used the UCSC genome Table browser NCBI build 35 (corresponding to the aCGH array platform build) to find 9091 genes housed within the genome regions of the 83 amplicons. We next proceeded to identify among the 9091 genes those that appear in the Catalogue Of Somatic Mutations In Cancer (COSMIC) Database v. 57 [3]. We found 5382, or 59.5%, of the genes to be associated with some type of cancer in COSMIC. Of these, 613 are specifically annotated as breast cancer related out of 1735 breast cancer associated genes in COSMIC. To test whether these numbers suggest an enrichment for breast cancer-associated genes in our amplicons, we performed a chi-square test of significance of enrichment of our gene set for breast cancer markers relative to the full 23307 unique Refseq-curated human genes in NCBI build 35. The results were weakly significant enrichment (p-value 0.029, chi-square value 4.78). Anecdotally, this new set of amplicons carries several important markers not identified by our earlier method, notable among them being ESR1, BRAF, KRAS, FGFR, JUN and JAK2.

Figure 3 provides a visual comparison of results of our method to those of CBS and CGHSeg, using chromosome 17. Our method and CGHSeg produce similar results, although with some fine-scale amplicons picked out by CGHSeg that are not detected by our method. CBS produced considerably more breakpoints than either other method. Over the entire genome, CBS produced 1425 distinct marker segments, a much higher number than our own method. We cannot definitely say to what degree these extra breakpoints reflect better sensitivity to true variations versus spurious breaks due to experimental noise. CGHSeg has substantially higher computational cost and could not complete analysis of the full genome in two weeks of processing and we therefore do not provide a full comparison to that method. It should be noted, though, that neither of these

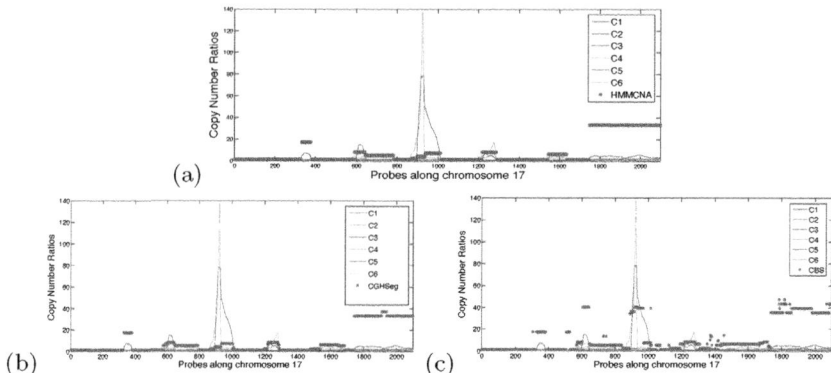

Fig. 3. Segmentation of chromosome 17 using mixture components of Navin et al. (a) Our method, HMMCNA. (b) CGHSeg. (c) CBS.

methods are designed to work with mixture components of the sort for which our method was developed, which might be expected to conform poorly to their error models. We also note that our method is much faster than the alternatives. Using chromosomes 8 and 17 as representatives, our method required 8.28s and 2.93s, CGHSeg 102.6h and 5.66h, and CBS 1.91h and 1.02h.

Next, we analyzed the phylogenetic tree obtained from the markers, summarized in Figure 4. Nodes correspond to putative stages of progression and edges to amplicons gained during discrete steps of progression. To choose a label for each amplicon, we attempted to select one most plausible driver gene for that amplicon by first favoring breast cancer-specific genes identified from COSMIC, using genes cited by Navin et al. [15] in their own analysis of their data to break ties, and next using other COSMIC genes not specifically connected to breast cancer. These heuristics were able to label all 83 amplicons. We also annotated any known cancer amplifications listed in the Cancer Gene Census [7].

The resulting tree is shown in Figure 4. The tree exhibits homoplasy (recurrent mutation) at amplicons 4, 43, 46, 68, 77 and 81 but no reversion of markers, a result we believe to improve upon that of our prior method [27], which exhibited both homoplasy and reversions. While the homoplasy might reflect genuine convergence of distinct progression pathways, it could also be explained by false positive calling errors or errors in phylogeny inference due to the maximum parsimony assumption. Analyzing the tree in more detail reveals several features of note. The progression pathway from the root to C2 occurs with the gain of breast cancer gene KRAS. While this pathway thus appears to be the most obviously associated with RAS abnormalities, it should be kept in mind that that these are most often gene mutations, and HRAS is more commonly deranged than KRAS in breast cancer. We also see a gain of ESR1 in the pathway from root to C4, which identifies this pathway to be most closely associated with estrogen receptor expression. It should be kept in mind, though, that ESR1 amplification occurs in only a small percentage of breast tumors [9]. The transition from Steiner node

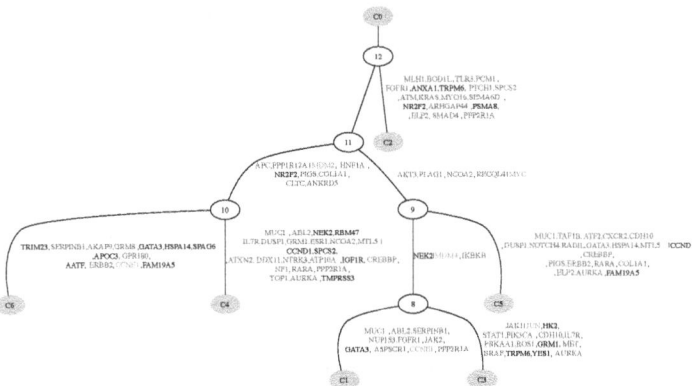

Fig. 4. Maximum parsimony tree inferred from real breast cancer data. Edges are labeled with putative driver genes, with COSMIC breast cancer genes in red and the gene census amplifications in green. Green nodes are observed components and white to inferred ancestral states, also known as Steiner nodes.

10 to component C6 occurs with the co-amplification of ERBB2 and CCNE. It has been reported recently that co-occurrence of ERBB2 and CCNE1 leads to Herceptin therapy resistance in HER2 overexpressed breast cancer [13, 22]. This may suggest a Herceptin resistant arm of breast tumorigenesis. Progression pathways to C1, C3, and C5 share common amplifications of breast tumor genes AKT3 and MYC gained in Steiner node 9. The progression pathway to C5 from Steiner node 9 with the gain of ERBB2 and CCND1 suggests a second distinct arm of ERBB2 gain. The co-amplification of CCND1, MYC, and HER2 in later stages of breast cancer has been noted in the literature [1, 14]. The transition to C3 also occurs with the gain of JAK1 and STAT1. The progression pathway to C1 occurs with the gain of JAK2. Thus, the tree is consistent with the hypothesis that amplification of JAK/STAT pathway family members is a late occurrence in breast cancer evolution.

5 Conclusion

We have developed a novel method for joint segmentation and calling of multi-sample genome-scale DNA copy number data, designed specifically for use in tumor phylogenetics. The method uses a novel multi-sample HMM approach to identify consistent markers across a set of samples, typically mixture components inferred from raw tumor data, for use as markers for phylogenetic inference. Comparison with a state-of-the-art multi-sample scheme and a leading single-sample scheme shows that our method can trade off weaker performance by more conventional measures of CGH calling accuracy for superior performance at the specific task of tumor phylogenetics in high but realistic noise domains. Further, the method substantially improves on our own prior work for the problem of phylogenetic inference from inferred mixture components, tripling the number of markers available for phylogeny inference and detecting several important progression markers not previously found from these data. In future work, we hope to improve on the current approach through a more realistic model of amplification distributions including handling of genomic deletions, algorithmic improvements to avoid combinatorial increase in state size with components, and improvements in the upstream unmixing and downstream phylogenetic inference steps. The method may also have value for other applications of copy number data in phylogenetics and related problems.

Acknowledgments. A.S., S.S., and R.S. were supported in this work by U.S. National Institutes of Health awards 1R01CA140214 and 1R01AI076318.

References

1. Al-Kuraya, K., Schraml, P., Torhorst, J., et al.: Prognostic relevance of gene amplifications and coamplifications in breast cancer. Cancer Research 64(23), 8534–8540 (2004)
2. Ashworth, A., de Bono, J.S.: Translating cancer research into targeted therapeutics. Nature (2010)
3. Bamford, S., Dawson, E., Forbes, S., et al.: The COSMIC (catalogue of somatic mutations in cancer) database and website. Br. J. Cancer (2004)
4. Beroukhim, R., Getz, G., Nghiemphu, L., et al.: Assessing the significance of chromosomal aberrations in cancer: Methodology and application to glioma. Proceedings of the National Academy of Sciences 104(50), 20007–20012 (2007)
5. Eilers, P.H.C., de Menezes, R.: Quantile smoothing of array CGH data. Bioinformatics 21(7), 1146–1153 (2005)
6. Felsenstein, J.: PHYLIP - Phylogeny Inference Package (Version 3.2). Cladistics 5, 164–166 (1989)
7. Futreal, P.A., Coin, L., Marshall, M., et al.: A census of human cancer genes. Nat. Rev. Cancer 4(3), 177–183 (2004)
8. Golub, T.R., Slonim, D.K., Tamayo, P., et al.: Molecular classification of cancer: class discovery and class prediction by gene expression monitoring. Science 286, 531–537 (1999)
9. Horlings, H.M., Bergamaschi, A., Nordgard, S.H., et al.: ESR1 gene amplification in breast cancer: a common phenomenon? Nat. Genet. (2008)
10. Hsu, L., Self, S.G., Grove, D., et al.: Denoising array-based comparative genomic hybridization data using wavelets. Biostatistics 6(2), 211–226 (2005)
11. Kuhner, M.K., Felsenstein, J.: A simulation comparison of phylogeny algorithms under equal and unequal evolutionary rates. Molecular Biology and Evolution 11(3), 459–468 (1994)
12. Miller, L.D., Smeds, J., George, J., et al.: An expression signature for p53 status in human breast cancer predicts mutation status, transcriptional effects, and patient survival. Proceedings of the National Academy of Sciences of the United States of America 102(38), 13550–13555 (2005)

13. Mittendorf, E.A., Liu, Y., Tucker, S.L., et al.: A novel interaction between HER2/neu and cyclin E in breast cancer. Oncogene 29, 3896–3907 (2010)
14. Moelans, C.B., de Weger, R.A., Monsuur, H.N., et al.: Molecular profiling of invasive breast cancer by multiplex ligation-dependent probe amplification-based copy number analysis of tumor suppressor and oncogenes. Mod. Pathol. (2010)
15. Navin, N., Krasnitz, A., Rodgers, L., et al.: Inferring tumor progression from genomic heterogeneity. Genome Research 20, 68–80 (2010)
16. Nowak, G., Hastie, T., Pollack, J.R., Tibshirani, R.: A fused lasso latent feature model for analyzing multi-sample aCGH data. Biostatistics 12(4), 776–791 (2011)
17. Olshen, A.B., Venkatraman, E.S., Lucito, R., Wigler, M.: Circular binary segmentation for the analysis of array based DNA copy number data. Biostatistics 5(4), 557–572 (2004)
18. Perou, C.M., Sorlie, T., Eisen, M.B., et al.: Molecular portraits of human breast tumors. Nature 406, 747–752 (2000)
19. Picard, F., Lebarbier, E., Hoebeke, M., Rigaill, G., Thiam, B., Robin, S.: Joint segmentation, calling, and normalization of multiple CGH profiles. Biostatistics 12(3), 413–428 (2011)
20. Picard, F., Robin, S., Lavielle, M., et al.: A statistical approach for array CGH data analysis. BMC Bioinformatics 6 (2005)
21. Pique-Regi, R., Ortega, A., Asgharzadeh, S.: Joint estimation of copy number variation and reference intensities on multiple DNA arrays using GADA. Bioinformatics 25(10), 1223–1230 (2009)
22. Scaltriti, M., Eichhorn, P.J., Cortes, J., et al.: Cyclin E amplification/overexpression is a mechanism of trastuzumab resistance in HER2+ breast cancer patients. Proceedings of the National Academy of Sciences (2011)
23. Schwartz, R., Shackney, S.: Applying unmixing to gene expression data for tumor phylogeny inference. BMC Bioinformatics 11, 42 (2010)
24. Shah, S.P., Cheung, K.J., Johnson, N.A., et al.: Model-based clustering of array cgh data. Bioinformatics 25(12), i30–i38 (2009)
25. Sorlie, T., Perrou, C.M., Tibshirani, R., et al.: Gene expression profiles of breast carcinomas distinguish tumor subclasses with clinical implications. Proc. Natl. Acad. Sci. USA 98, 10869–10864 (2001)
26. Sotiriou, C., Neo, S.Y., McShane, L.M., et al.: Breast cancer classification and prognosis based on gene expression profiles from a population-based study. Proc. Natl. Acad. Sci. USA 100, 10393–10398 (2003)
27. Subramanian, A., Shackney, S., Schwartz, R.: Inference of tumor phylogenies from genomic assays on heterogeneous samples. In: Proc. ACM-BCB 2011 (2011)
28. Swafford, D.: PAUP*. Phylogenetic Analysis Using Parsimony (*and other methods). Version 4 (2002)
29. Tolliver, D., Tsourakakis, C., Subramanian, A., et al.: Robust unmixing of tumor states in array comparative genomic hybridization data. Bioinformatics 26(12), i106–i114 (2010)
30. van't Veer, L.J., Dai, H., van de Vivjer, M., et al.: Gene expression profiling predicts clinical outcome of breast cancer. Nature 415, 530–536 (2002)
31. Wang, K., Li, M., Hadley, D., et al.: Penncnv: An integrated hidden markov model designed for high-resolution copy number variation detection in whole-genome snp genotyping data. Genome Research 17(11), 1665–1674 (2007)

32. Wiel, V.D., Mark, A., Brosens, R., et al.: Smoothing waves in array CGH tumor profiles. Bioinformatics 25(9), 1099–1104 (2009)
33. Wu, L.Y., Chipman, H.A., Bull, S.B., Briollais, L., Wang, K.: A Bayesian segmentation approach to ascertain copy number variations at the population level. Bioinformatics 25(13), 1669–1679 (2009)
34. Zhang, N.R., Senbabaoglu, Y., Li, J.Z.: Joint estimation of DNA copy number from multiple platforms. Bioinformatics 26(2), 153–160 (2010)

Algorithms for Knowledge-Enhanced Supertrees

André Wehe[1], J. Gordon Burleigh[2], and Oliver Eulenstein[1]

[1] Department of Computer Science, Iowa State University, Ames, IA 50011, USA
[2] Department of Biology, University of Florida, Gainesville, FL 32609, USA

Abstract. Supertree algorithms combine smaller phylogenetic trees into a single, comprehensive phylogeny, or supertree. Most supertree problems are NP-hard, and often heuristics identify supertrees with anomalous or unwanted relationships. We introduce knowledge-enhanced supertree problems, which seek an optimal supertree for a collection of input trees that can only be assembled from a set of given, possibly incompatible, phylogenetic relationships. For these problems we introduce efficient algorithms that, in a special setting, also provide exact solutions for the original supertree problems. We describe our algorithms and verify their performance based on the Robinson Foulds (RF) supertree problem. We demonstrate that our algorithms (i) can significantly improve upon estimates of existing RF-heuristics, and (ii) can compute exact RF supertrees with up to 17 taxa.

Keywords: Evolutionary biology, supertree, Robinson-Foulds.

1 Introduction

There is growing interest in using phylogenetic trees to investigate evolutionary and ecological processes, including the maintenance of biodiversity [11,9] and the effects of global change [10,24,22]. These analyses often require extremely large trees that include species from many disparate clades. Most traditional phylogenetic analyses build trees for either relatively few closely related taxa or exemplar taxa from distantly related clades. Thus, there is a need for methods that can synthesize the existing phylogenetic data. Supertree problems, which take input trees with partially overlapping taxon sets and seek a tree (the supertree) containing all of the taxa in the input trees, provide one such approach (e.g. [17,3,4]).

Most standard supertree problems are NP-complete [2,15,26]. Like other NP-complete problems in general, exact solutions can be obtained by exhaustive search. However, run-times typically become prohibitive already for very small instances. Heuristic searches, often based on local search algorithms, are necessary to estimate solutions for larger data sets. While heuristic solutions appear to work well in many cases, one persistent criticism of supertree methods is that the resulting supertrees often contain relationships that appear to be anomalous, or at least unwanted. In some cases, this may result from the failure of a supertree heuristic to find an optimal solution [1,8]. In other cases, these relationships may be a symptom of undesirable properties of the supertree problem.

One approach to avoid problematic relationships and to improve the efficiency of heuristic searches is to provide a set of possible phylogenetic relationships from which the supertree has to be assembled. These phylogenetic relationships might be provided by existing supertrees or other phylogenetic hypotheses, and can be incompatible. Focusing the supertree search on these candidate relationships would not only prevent the inclusion of undesired relationships in the supertrees, which are excluded from the set of candidates, but it also can tremendously reduce the search space for supertree solutions.

Here we introduce knowledge-enhanced supertree problems that construct supertrees from credible phylogenetic knowledge. Given a collection of input trees and a set of possibly contradictory phylogenetic relationships, a knowledge-enhanced supertree problem seeks an optimal supertree only in the tree space described by the phylogenetic relationships. We provide efficient algorithms for these problems that, in a special setting, can compute exact solutions for the original supertree problems. We focus on the RF supertree problem to describe our algorithms and to verify their performance in practice. We demonstrate that our knowledge-enhanced algorithms improve upon RF supertrees estimated by the heuristic of Bansal et al. [1]. Further, we show that our algorithms can compute, for the first time, exact RF supertrees for up to 17 taxa within 9 hours.

Related Work. Knowledge-enhanced searches have been an effective approach to guide local search heuristics for supertree, and other phylogenetic, problems. In these approaches, unresolved trees or clades are passed as constraints to the heuristic's local search method (e.g. RAxML [19], PAUP [20], and DupTree [23]), which forces the solution supertree to agree with the constraints. Although such constraints can be effective, they may be inefficient. The constraint search must consider all possible topologies consistent with the constraints, rather than a specific set of possible incompatible topologies. Furthermore, these constrained approaches are still heuristics and therefore risking suboptimal solutions.

In this work, we introduce knowledge enhanced supertree problems and provide such a definition based on the Robinson-Foulds (RF) supertree problem. The RF supertree problem seeks a binary supertree that minimizes the sum of the RF distances between every input tree and the supertree. This problem is NP-hard [2], and therefore, it has been addressed by local search heuristics [1,8]. These heuristics allow for time efficient estimation of RF supertrees for data sets with hundreds of taxa, but experiments suggest that they may become stuck at locally optimal solutions [1,8]. The RF supertree problem can also be solved exactly with exhaustive search, but this becomes prohibitively time consuming for already small instances (e.g. approximately 8 taxa). Exact algorithms that are substantially faster than exhaustive search have been developed for several other supertree problems [25,21,7]. However, no such algorithms have been described for the RF supertree problem.

Our Contribution. We define knowledge-enhanced RF supertree problems, and provide efficient algorithms to solve them. *Knowledge-enhanced RF supertree problems* seek an optimal RF supertree for a given collection of input trees that can only be assembled from a given set of, possibly contradictory, phylogenetic relationships. These relationships can be described either as clusters or sibling-clusters. *Sibling-clusters* are two disjoint clusters whose parent in a rooted phylogenetic tree is the union of these clusters. Such phylogenetic relationships are available in abundance, for example, as previously inferred phylogenetic estimates, including estimates from supertree heuristics. We describe efficient algorithms that solve the knowledge-enhanced RF supertree problems for a collection of n taxa and m input nodes for either c clusters or t siblings clusters in time $O\left(c^2 nm\right)$ and $O\left(tnm\right)$ respectively. We verify the performance of these algorithms for large empirical collections of trees. Our algorithms significantly improve on estimates of these collections that were computed by the RF heuristic from Bansal et al [1].

We also describe an exact algorithm for the RF supertree problem that is a special version of the knowledge-enhanced algorithms. Our exact algorithm improves on the best known (naïve) solution for the RF supertree problem by a factor of $\sqrt{n^n}/3^n$, where n is the number of taxa in the supertree. This substantial improvement allows, for the first time, to compute exact RF supertrees with up to 17 taxa.

Finally, we show how our new algorithmic approaches for the RF supertree problem lead to efficient algorithms for the knowledge-enhanced versions of the following supertree problem: the duplication and loss problem, the deep coalescence problem, and the MRP problem. The exact algorithms for the original supertree problems are a special setting of the corresponding knowledge-enhanced algorithms.

2 Basic Notation and Preliminaries

In the interest of brevity some proofs have been omitted, but are included in a technical report that is available on request from the authors. Let T be a rooted tree. We denote the vertex set, edge set, and leaf set of T by $V(T)$, $E(T)$, and $\mathsf{Le}(T)$ respectively. The root of T is denoted by $\mathsf{Rt}(T)$. Given a vertex $v \in V(T)$, we denote the children of v by $\mathsf{Ch}_T(v)$, and the parent of v by $\mathsf{Pa}_T(v)$. Two vertices in T are called *siblings* (of each other) if they have the same parent. We write (u, v) to denote the edge $\{u, v\} \in E(T)$ where $u = \mathsf{Pa}_T(v)$. The set of *internal vertices* of T, denoted $I(T)$, is defined to be the set $V(T) \setminus \mathsf{Le}(T)$. We call T *full binary* if each vertex $v \in I(T)$ has exactly two children.

Given a vertex-set $U \subseteq V(T)$, we denote by $T(U)$ the unique subtree of T that spans U with the minimum number of vertices. Furthermore, the *restriction* of T to U, denoted by $T_{|U}$, is the tree that is obtained from $T(U)$ by suppressing all non-root vertices of degree two. The *subtree* of T *rooted at* $v \in V(T)$, denoted by T_v, is defined to be $T_{|U}$, for $U := \{u \in \mathsf{Le}(T) \mid v$ is on the shortest path between $\mathsf{Rt}(T)$ and $u\}$.

Given a vertex $v \in V(T)$ we define the *cluster of* v to be $\mathcal{C}_T(v) := \mathsf{Le}(T_v)$, and the *cluster presentation* of T is defined as $\mathcal{C}(T) := \{\mathcal{C}_T(v) \mid v \in V(T)\}$.

Let X be a label set. A tree T is called an X-tree if $\mathsf{Le}(T) = X$, and called a *partial X-tree* if $\mathsf{Le}(T) \subseteq X$. The set of all X-trees is denoted by $\mathcal{T}(X)$, and the set of all X-trees that are full binary is denoted by $\mathcal{B}(X)$.

The following conventions are used for convenience throughout the manuscript. We write the set-operation $\cup$ as $\dot\cup$ when the corresponding intersecting sets are disjoint. Unless noted otherwise, the term tree refers to a rooted tree that has no vertices of degree two other than its root.

The definition for the *Robinson-Foulds (RF) distance* [16] is expressed as the dissimilarity of a pair of trees on the same taxon set. For rooted trees this dissimilarity is based on clusters. For the RF supertree problem the input trees are partial X-trees, so trees may share only some of the taxa with the supertree. In this case the minus RF distance can be computed (e.g. [1,8]), which is the *RF distance* on a pair of trees restricted to their shared taxon set as follows:

Definition 1. *[minus RF distance]*
Let S be an X-tree, and T be a Y-tree where $Y \subseteq X$. We define $RF(T,S) := |\mathcal{C}(T) \triangle \mathcal{C}(S_{|Y})|$, where $\triangle$ denotes the symmetric difference. Given a set P of partial X-trees, we define the minus RF distance from P to S as $RF(P,S) := \Sigma_{T \in P} RF(T,S)$, and the minus RF distance of P in the scope of $\mathcal{B}(X)$ as $RF(P) := \min_{S \in \mathcal{B}(X)} RF(P,S)$.

Problem 1. [RF supertree (NP-hard [2])]
Instance: A set P of partial X-trees.
Find: The score $RF(P)$ and a full binary tree $S \in \mathcal{B}(X)$ such that $RF(P) = RF(P,S)$.

The Cluster Similarity Problem. Here we first introduce the *cluster-similarity* measure, which is a measure complementary to the RF distance. We then introduce the *cluster-similarity* problem (Problem 2). This *cluster-similarity* measure has been implicitly used in [1], and Problem 1 is equivalent to the Problem 2.

Definition 2. *[c-similarity]*
Let S be an X-tree, and T be a Y-tree where $Y \subseteq X$. The c(luster)-similarity from T to S is defined as $R(T,S) := |\mathcal{C}(T) \cap \mathcal{C}(S_{|\mathsf{Le}(T)})|$. Given a set P of partial X-trees, we define the c-similarity from P to S as $R(P,S) := \Sigma_{T \in P} R(T,S)$, and the c-similarity of P in the scope of $\mathcal{B}(X)$ as $R(P) := \max_{S \in \mathcal{B}(X)} R(P,S)$.

Corollary 1. *[c-similarity to RF conversion]*
For X-trees T and S, the conversion between the RF and c-similarity measures is $RF(T,S) = |\mathcal{C}(T)| + |\mathcal{C}(S)| - 2R(T,S)$.

Problem 2. [c-similarity]
Instance: A set P of partial X-trees.
Find: The similarity score $R(P)$, and a full binary tree $S \in \mathcal{B}(X)$ such that $R(P) = R(P,S)$.

Refinements for the Cluster Similarity Problem. We are introducing two new problems that are refining the candidate trees for a solution of the c-similarity problem. First, we introduce the Problem 3, which is considering only candidate trees that have a cluster-presentation which is a subset of a given refining cluster set. Problem 4 is further narrowing down the refinement to candidate trees that can only have sibling clusters of a given set of sibling clusters. To guarantee solvability of these refined problems, we require that at least one candidate tree can be represented by the given refining sets, and call such refining sets *complete*. For each refined problem, we first introduce definitions necessary to state the problem, and then define the problem.

Definition 3. *[cluster refined candidate trees]*
Let X be a label set, and K be a set of subsets of X. We define $\mathcal{BC}(X, K) := \{T \in \mathcal{B}(X) \mid \mathcal{C}(T) \subseteq K\}$, and call K cluster-complete when $\mathcal{BC}(X, K) \neq \emptyset$. Given a set P of partial X-trees, we define the c-similarity of P refined by K as $RC(P, K) := \max_{S \in \mathcal{BC}(X, K)} R(P, S)$.

Problem 3. [cluster refined c-similarity]
Instance: Set P of partial X-trees and a cluster-complete set K of subsets of X.
Find: The similarity $RC(P, K)$ and a full binary tree $S \in \mathcal{BC}(X, K)$ where $RC(P, K) = R(P, S)$.

Definition 4. *[sibling-refined candidate trees]*
The sibling presentation of a full binary tree T is defined as $\mathsf{Sb}(T) := \{(A, B) \mid \exists \text{ siblings } u, v \in V(T) : \mathsf{Pa}_T(u) = \mathsf{Pa}_T(v), A := \mathcal{C}_T(u), B := \mathcal{C}_T(v)\}$. Let X be a label set, and L be a set of bi-partitions of subsets of X. We define $\mathcal{BS}(X, L) := \{T \in \mathcal{B}(X) \mid \mathsf{Sb}(T) \subseteq L\}$, and call L sibling-complete when $\mathcal{BS}(X, L) \neq \emptyset$. Given a set P of partial X-trees we define the c-similarity of P refined by L as $RS(P, L) := \max_{T \in \mathcal{BS}(X, L)} R(P, T)$.

Problem 4. [sibling refined c-similarity]
Instance: A set P of partial X-trees and a sibling-complete set L of bi-partitions of subsets of X.
Find: The similarity $RS(P, L)$ and a full binary tree $S \in \mathcal{BS}(X, L)$ such that $RS(P, L) = R(P, S)$.

3 Structural Properties of the c-Similarity Problems

The c-similarity problem and the refined c-similarity problems share an optimal substructure. Similar substructures have been shown for the gene tree parsimony problems [21,25,12]. Here, we first phrase this optimal substructure for the c-similarity measure, and next we follow up with the recurrences for each of the c-similarity problems.

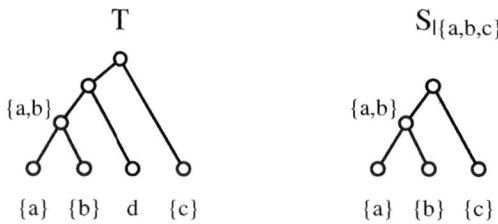

Fig. 1. This example shows the c-similarity for the Y-tree T and X-tree S restricted to the set $Z := \{a, b, c\}$. The blue nodes indicate 4 common clusters $\{a\}, \{b\}, \{c\}$, and $\{a, b\}$, hence the restricted c-similarity is $R_{|Z}(T, S) = 4$. These clusters imply an upper bound on the Robinson-Foulds distance for every X-tree S containing the subtree $S_{|Z}$.

Optimal Substructure of the c-Similarity Problem

Definition 5. *[restricted c-similarity]*
Let S be an X-tree, T be a Y-tree where $Y \subseteq X$, and $Z \subseteq X$. The c-similarity from T to S restricted by Z is defined as $R_{|Z}(T, S) := |\mathcal{C}(T) \cap \mathcal{C}(S_{|Y \cap Z})|$. Given a set P of partial X-trees we define (i) the c-similarity from P to S restricted by Z as $R_{|Z}(P, S) := \Sigma_{T \in P} R_{|Z}(T, S)$, and (ii) the c-similarity of P restricted by Z in the scope of $\mathcal{B}(X)$ as $R_{|Z}(P) := \max_{S \in \mathcal{B}(X)} R_{|Z}(P, S)$.

For any X-tree S with subtree $S_{|Z}$ the restricted c-similarity establishes the upper bound $RF(T, S) \leq |\mathcal{C}(T)| + |\mathcal{C}(S)| - 2R_{|Z}(T, S)$ for the Robinson-Foulds distance in a straightforward manner. Figure 1 shows the restricted c-similarity measure by example.

Proposition 1. *[optimal substructure]*
Let P be a set of partial X-trees, and S be a full binary X-tree such that $R_{|X}(P) = R_{|X}(P, S)$. Then $R_{|Z}(P) = R_{|Z}(P, S_v)$ where $Z = \mathsf{Le}(S_v)$ for any vertex $v \in V(S)$. For brevity we omit the proof.

Recurrences for the c-Similarity Problems

We describe the recurrences for the c-similarity problems introduced in Section 2. The recurrences in this section follow from the optimal substructure given in Proposition 1. Throughout this section let P be a set of partial X-trees, $Y \subseteq X$, and $Y \neq \emptyset$.

Definition 6. *[cluster induced c-similarity score]*
Let T be a Z-tree. We define $\Gamma(T, Y) := 1$, if $Y \cap Z \in \mathcal{C}(T)$, and $\Gamma(T, Y) := 0$, otherwise. Let P be a set of partial X-trees. We define $\Gamma(P, Y) := \Sigma_{T \in P} \Gamma(T, Y)$.

Proposition 2. *[c-similarity recurrence]*
$$R_{|Y}(P) := \begin{cases} \Gamma(P, Y), & \text{if } |Y| = 1, \\ \Gamma(P, Y) + \max_{(A,B) \in \Pi(Y)} \left(R_{|A}(P) + R_{|B}(P) \right), & \text{otherwise,} \end{cases}$$
where $\Pi(Y)$ is the set of all non-trivial bi-partitions for Y.

Definition 7. *[restricted cluster refined c-similarity]*
Let K be a cluster-complete set of subsets of X, and $Z \subseteq X$. The c-similarity of P refined by K, and restricted by Z is defined as $RC_{|Z}(P,K) := \max_{S \in \mathcal{BC}(X,K)} R_{|Z}(P,S)$.

Proposition 3. *[cluster refined recurrence]*
Let K be a cluster-complete set of subsets of X. Then, we have the following recurrence:
$$RC_{|Y}(P,K) := \begin{cases} \Gamma(P,Y), & \text{if } |Y| = 1, \\ \Gamma(P,Y) + \omega, & \text{otherwise,} \end{cases}$$
where $\omega := \max_{(A,B) \in \Pi(Y):\ \{A,B\} \subseteq K} \left(RC_{|A}(P,K) + RC_{|B}(P,K) \right)$.

Definition 8. *[restricted sibling refined c-similarity]*
Let L be a sibling-complete set of bi-partitions of subsets of X, and $Z \subseteq X$. The c-similarity of P refined by L, and restricted by Z is defined as $RS_{|Z}(P,L) := \max_{S \in \mathcal{BS}(X,L)} R_{|Z}(P,S)$.

Proposition 4. *[sibling refined recurrence]*
Let L be a sibling-complete set of bi-partitions of subsets of X. Then, we have the following recurrence:
$$RS_{|Y}(P,L) := \begin{cases} \Gamma(P,Y), & \text{if } |Y| = 1, \\ \Gamma(P,Y) + \omega, & \text{otherwise,} \end{cases}$$
where $\omega := \max_{(A,B) \in L:\ A \dot\cup B = Y} \left(RS_{|A}(P,L) + RS_{|B}(P,L) \right)$.

4 Dynamic Programming for RF and c-Similarity Supertree Problems

Following from Corollary 1, a solution for the c-similarity supertree problems also provides a solution for the equivalent RF supertree problems. Here we present a dynamic programming (DP) solution for the c-similarity problem and the refined c-similarity problems introduced in Section 2. First, we describe the algorithm for the recurrence of the sibling refined c-similarity problem, and then based on this algorithm, we follow up the solutions for the other c-similarity problems. Algorithm 1 computes the recurrence of the sibling refined c-similarity problem. For brevity we omit the proof of correctness for this standard DP algorithm.

Theorem 1. *[sibling refined c-similarity]*
Let P be a set of partial X-trees and L be a sibling-complete set of bi-partitions of subsets of X. The c-similarity problem, Problem 4, can be solved in $O(tnm)$ time, where $t = |L|$, $n = |X|$, and $m = |V(P)|$.

Proof. Let S be the sibling refined RF supertree that is constructed by following the recursive dynamic programming structure of the refined c-similarity recurrence. Then, S is the solution for Problem 4.

Complexity: The computational complexity for Algorithm 1 is dominated by computing the restricted c-similarity for non-trivial clusters. Part 1: Computing

Algorithm 1. This dynamic program (DP) solves the sibling refined c-similarity problem, Problem 4 by computing the DP table $\mathcal{R}$.

Input:
Set P of partial X-trees
Sibling-complete set L of bi-partitions of subsets of X
Output:
The table $\mathcal{R}$ of restricted c-similarity scores $R_{|Y}(P, L)$,
where Y is an element of the X-complete set $\{A \cup B | (A, B) \in L\}$
Algorithm:
Part 1. Initialize the DP table $\mathcal{R}$ for trivial clusters.
$\mathcal{R} \leftarrow \emptyset$ with the default value of $-\infty$ for unassigned rows
for $v \in X$ do
 $\quad \mathcal{R}[\{v\}] \leftarrow \Gamma(P, Y)$
end for
Part 2. Compute the DP table $\mathcal{R}$ for non-trivial clusters.
$(A_1, B_1), (A_2, B_2), \ldots, (A_t, B_t)$ is a sorted list of L
 $\quad$ such that $|A_i \dot\cup B_i| \leq |A_{i+1} \dot\cup B_{i+1}|$ for $1 \leq i < t$
for $i := 1$ to t do
 $\quad \mathcal{R}[A_i \dot\cup B_i] \leftarrow \max(\mathcal{R}[A_i \dot\cup B_i], \mathcal{R}[A_i] + \mathcal{R}[B_i] + \Gamma(P, A_i \dot\cup B_i))$
end for
return $\mathcal{R}$

---

the restricted c-similarity for trivial clusters takes $O(m)$ time, by counting the different leaf nodes in P. Part 2: Sorting Q takes $O(tn)$ time. The access time for an element in $\mathcal{R}$ is $O(n)$ and $\Gamma(P, A \dot\cup B)$ is computed in $O(nm)$ time, so computing the restricted c-similarity for a non-trivial cluster takes $O(nm)$ time in a loop of $O(t)$ iterations. Therefore, the computational complexity bound of Algorithm 1 is $O(tnm)$ time.

The dynamic programming algorithm for Theorem 1 can be adapted to also solve Problem 3 and Problem 2. Next, we construct different sibling-complete sets for each problem. Table 1 summarizes the different inputs.

Theorem 2. *[cluster refined c-similarity]*
Let P be a set of partial X-trees and K be a cluster-complete set of bi-partitions of subsets of X. The cluster refined c-similarity problem can be solved in $O(c^2 nm)$ time, where $c = |K|$, $n = |X|$, and $m = |V(P)|$.

Proof. Let S be a solution X-tree for Problem 3, i.e. a full binary tree $S \in \mathcal{BC}(X, K)$ such that $RC(P, K) = R(P, S)$. It follows that $\mathcal{C}(S) \subseteq K$, so for each pair of siblings $u, v \in V(S)$ there exists (A, B) such that $A, B \in K$ and $\mathcal{C}_S(u) = A$ and $\mathcal{C}_S(v) = B$. Construct $L := \{(A, B) | A, B \in K, A \cap B = \emptyset\}$ for the input of Algorithm 1, and it follows that $\text{Sb}(S) \subseteq L$.

Complexity: The size of L is $O(c^2)$. It follows, the computational complexity for Problem 3 is bounded by $O(c^2 nm)$ time.

Table 1. Inputs for RF supertree problems

supertree Problem	sibling-complete set	size
(a) exact	$\{\Pi(Y) \mid Y \in \mathcal{P}(X), Y \neq \emptyset\}$	$O\left(3^{|X|}\right)$
(b) cluster refined	$\{(A,B) \mid A,B \in K, A \cap B = \emptyset\}$	$O\left(|K|^2\right)$
(c) sibling refined	L	$|L|$

The input for Algorithm 1 for the exact and refined c-similarity supertree problems, (a) Problem 2, (b) Problem 3, and (c) Problem 4. X is the taxa set, $\Pi(Y)$ the set of all non-trivial bi-partitions for Y, $\mathcal{P}(X)$ be the power set of set X, K is a cluster-complete set of subsets of X, and L is a sibling-complete set of bi-partitions of subsets of X.

Theorem 3. *[c-similarity]*
Let P be a set of partial X-trees. The c-similarity problem can be solved in $O(3^n nm)$ time, where $n = |X|$ and $m = |V(P)|$.

Proof. Let S be a solution X-tree for Problem 2; i.e. a full binary tree $S \in \mathcal{B}(X)$ such that $R(P) = R(P,S)$. Let $\mathcal{P}(X)$ be the power set of set X. Construct $L := \{\Pi(Y) \mid Y \in \mathcal{P}(X), Y \neq \emptyset\}$ for the input of Algorithm 1. For all $T \in \mathcal{B}(X)$, $\mathsf{Sb}(T) \subseteq L$. It follows, that $\mathsf{Sb}(S) \subseteq L$.

Complexity: There are $\frac{1}{2}(3^n - 2^{n+1} + 1) = O(3^n)$ unique sibling clusters in L. It follows, the computational complexity for Problem 2 is bounded by $O(3^n nm)$ time.

Currently, the best-known (naive) exhaustive search algorithm for the RF supertree problem requires searching through all possible $(2n-3)!!$ rooted supertrees. Thus, the best naive solution for the RF supertree problem is solvable in $O((2n-3)!! nm)$ time. We obtain the speed up of $\Theta\left(\sqrt{n^n}/3^n\right)$ as follows:
$\frac{(2n-3)!! nm}{3^n nm} = \frac{(2n-3)!!}{3^n} \leq \frac{n^{n/2}-4}{3^n} = \Theta\left(\sqrt{n^n}/3^n\right)$ \\ for $n > 0$.

Different Supertree Objectives

Our DP algorithms for the efficient RF supertree search can also be directly applied to other supertree objectives, these include gene duplications and loss, deep coalescence, and Maximum Parsimony (MP) for the MRP problem. The requirement on the supertree objective is the objective specific definition of the restricted score, equivalent to the restricted c-similarity for RF. That implies: given a subtree $S_{|Z}$, the Z-restricted score has to invoke an upper bound on the score if the subtree $S_{|Z}$ is contained in the supertree. For gene duplication and loss, this restricted score is the number of duplications and losses induced by $S_{|Z}$, for deep coalescence the restricted score is the number of embedded lineages induced by $S_{|Z}$, and for MP the restricted score is the number of character state changes induced by $S_{|Z}$. The specific definitions of the restricted scores for these objectives are straightforward, and omitted for brevity.

Table 2. The improvement of the sibling refinement method on RF supertrees constructed by the SPR local search heuristic

Data Set	RF Method	rooted RF Distance	Improvement
Saxifragales	SPR local search	2308	48
(959 taxa; 51 trees)	Refinement method	2260	2.08 %
Marsupial	SPR local search	1510	16
(272 taxa; 158 trees)	Refinement method	1492	1.06 %
Gymnosperm	SPR local search	4343	231
(950 taxa; 78 trees)	Refinement method	4112	5.32 %

5 Experiments

We first examined if knowledge-enhanced searches can improve upon solutions from RF supertree heuristics using the SPR local search [1]. We used three large data sets, one published data set from marsupials [6] and two unpublished data sets from the gymnosperm and Saxifragales plant clades. The gymnosperm trees were made from RAxML analyses of the gene alignments found in [5], and the Saxifragales trees were made with RAxML analyses of gene alignments assembled from GenBank. The RAxML analyses all used the GTRCAT model, and the tree data sets will be available from the dryad data repository (http://datadryad.org). In all the experiments, we first ran the SPR-heuristic to obtain a set of RF supertrees. For each data set, we used 10 different starting trees, built by random stepwise addition, and for each starting tree, we repeated the RF-SPR using a ratchet heuristic (see [1]) 25 times for each search. This resulted in 250 RF supertrees for each data set. These analyses took between 6.5 hours (for the marsupial data set) and 10 days (for the gymnosperm data set).

We ran knowledge-enhanced searches using the 250 RF supertrees from the SPR analysis as the constraint set. The cluster refinement method finished only for the marsupial data set (in 5 days). However, the sibling refinement method runs took between 18 seconds (for the marsupial data set) and 1.5 minutes (for the gymnosperm data set). In all three data sets, the constrained RF search improved the RF score of the best RF supertrees from the SPR search. The improvements were between 1.06 and 5.32% (see Table 2). This experiment demonstrates that the solutions to the RF supertree problem from the SPR local search are suboptimal, and given the proper set of knowledge-enhanced supertrees, the sibling refinement method can rapidly improve upon RF supertrees made with SPR heuristics.

Next, we examined the performance of our exact RF supertree solution on a data set with 699 gene trees representing 12 plant taxa. The taxa include an outgroup moss, *Physcomitrella patens*, nine gymnosperm taxa including cycad Cycas rumphii, Gnetales taxa *Gnetum gnemon* and *Welwitschia mirabilis*, and the conifers *Cryptomeria japonica, Pseudotsuga menziesii, Picea glauca, Picea sitchensis, Pinus pinaster*, and *Pinus taeda*, and finally two angiosperm (*Arabidopsis thaliana* and *Oryza sativa*). To build the gene trees, we obtained amino acid alignments for genes from Phytome v.2, an online comparative genomics database for plants [13]. We sampled 699 genes that had exactly one sequence from at least four of our selected taxa. We built maximum likelihood gene trees from each alignment RAxML-VI-HPC v.2.2.3 [18] with the JTTCAT amino acid substitution model [14]. The gene trees will be available on the dryad data repository. It took 13.2 minutes to compute the exact RF solution for this data set on a Macintosh Powerbook with a 2 GHz Intel Core 2 Duo processor and 2 GB memory. The RF score was 528, and the supertree, which indicated the Gnepine phylogenetic hypothesis for seed plants, was consistent with many recent phylogenetic studies. While this represents the largest empirical data set with an exact solution for the RF supertree problem, we also were able to compute exact solutions for data sets with up to 17 taxa and 100 random topologies in 9 hours.

6 Conclusion

Knowledge-enhanced RF supertree problems can rapidly construct supertrees from sets of existing, possibly contradictory, phylogenetic relationships. This approach can be used to obtain an exact solution to a supertree problem, while ensuring that the solution only includes the given phylogenetic relationships.

Our experiments demonstrated that, not only do our exact algorithms far exceed the best-known (naïve) solution for solving the RF supertree problem, the knowledge-based supertree searches provide a fast and simple approach to improve upon suboptimal solutions for local search heuristics for the RF supertree problem. Although we have focused on algorithms for RF supertrees in this study, we demonstrated that our algorithms are also adaptable for other supertree problems like the gene duplication, duplication and loss, deep coalescence, and MRP problems.

Our results suggest several future directions for research. Although the knowledge-enhanced supertree analyses can construct extremely large supertrees in a reasonable amount of time, the quality of the resulting supertrees depends on the provided phylogenetic relationships. In our experiments, acquiring the set of phylogenetic relationships took far longer than the knowledge-enhanced RF supertree search. Thus, the key to the application of knowledge-enhanced searches is the ability to build rapidly many high quality relationships, and our future work will explore best methods to do this.

References

1. Bansal, M.S., Burleigh, J.G., Eulenstein, O., Fernández-Baca, D.: Robinson-foulds supertrees. Algor. Mol. Biol. 5, 18 (2010)
2. Barthelemy, J.P., McMorris, F.R.: The median procedure for n-trees. Journal of Classification 3, 329–334 (1986)
3. Bininda-Emonds, O.R.P.: The evolution of supertrees. Trends Ecol. Evol. 19, 315–322 (2004)
4. Bininda-Emonds, O.R.P., Gittleman, J.L., Steel, M.A.: The (super) tree of life: procedures, problems, and prospects. Annu. Rev. Ecol. Syst. 33, 265–289 (2002)
5. Burleigh, J.G., Barbazuk, W.B., Davis, J.M., Morse, A.M., Soltis, P.S.: Exploring diversification and genome size evolution in extant gymnosperms through phylogenetic synthesis. Journal of Botany, 292857 (2012)

6. Cardillo, M., Bininda-Emonds, O.R.P., Boakes, E., Purvis, A.: A species-level phylogenetic supertree of marsupials. Journal of Zoology 264, 11–31 (2004)
7. Chang, W., Burleigh, J.G., Fernandez-Baca, D.F., Eulenstein, O.: An ILP solution for the gene duplication problem. BMC Bioinformatics 12(suppl.1), S14 (2011)
8. Chaudhary, R., Burleigh, J.G., Fernández-Baca, D.: Fast Local Search for Unrooted Robinson-Foulds Supertrees. In: Chen, J., Wang, J., Zelikovsky, A. (eds.) ISBRA 2011. LNCS, vol. 6674, pp. 184–196. Springer, Heidelberg (2011)
9. Davies, T.J., Fritz, S.A., Grenyer, R., Orme, C.D.L., Bielby, J., Bininda-Emonds, O.R.P., Cardillo, M., Jones, K.E., Gittleman, J.L., Mace, G.M., Purvis, A.: Phylogenetic trees and the future of mammalian biodiversity. Proc. Natl. Acad. Sci. USA 105, 11556–11563 (2008)
10. Edwards, E.J., Still, C.J., Donoghue, M.J.: The relevance of phylogeny to studies of global change. Trends Ecol. Evol. 22, 243–249 (2007)
11. Forest, F., Grenyer, R., Rouget, M., Davies, T.J., et al.: Preserving the evolutionary potential of floras in biodiversity hotspots. Nature 445, 757–760 (2007)
12. Hallett, M.T., Lagergren, J.: New algorithms for the duplication-loss model. In: RECOMB, pp. 138–146 (2000)
13. Hartmann, S., Lu, D., Phillips, J., Vision, T.J.: Phytome: A platform for plant comparative genomics. Nucleic Acids Research 34, D724–D730 (2006)
14. Jones, D.T., Taylor, W.R., Thornton, J.M.: The rapid generation of mutation data matrices from protein sequences. Comp. Appl. Biosci. 8, 25–282 (1992)
15. Ma, B., Li, M., Zhang, L.: From gene trees to species trees. SIAM J. Comput. 30, 729–752 (2000)
16. Robinson, D.F., Foulds, L.R.: Comparison of phylogenetic trees. Mathematical Biosciences 53, 131–147 (1981)
17. Sanderson, M.J., Purvis, A., Henze, C.: Phylogenetic supertrees: Assembling the trees of life. Trends Ecol. Evol. 13, 105–109 (1998)
18. Stamatakis, A.: Raxml-vi-hpc: Maximum likelihood-based phylogenetic analyses with thousands of taxa and mixed models. Bioinformatics 22, 2688–2690 (2006)
19. Stamatakis, A., Hoover, P., Rougemont, J.: A fast bootstrapping algorithm for the raxml web-servers. In Systematic Biology 57(5), 758–771 (2008)
20. Swofford, D.L.: PAUP*: Phylogenetic analysis using parsimony (*and other methods), version 4.0b10 (2002)
21. Than, C., Nakhleh, L.: Species tree inference by minimizing deep coalescences. PLoS Comput. Biol. 5(9), e1000501 (2009)
22. Thuiller, W., Lavergne, S., Roquet, C., Boulangeat, I., Lafourcade, B., Araujo, M.B.: Consequences of climate change on the tree of life in europe. Nature 470, 531–534 (2011)
23. Wehe, A., Bansal, M.S., Burleigh, J.G., Eulenstein, O.: Duptree: A program for large-scale phylogenetic analyses using gene tree parsimony. Bioinformatics 24(13), 1540–1541 (2008)
24. Willis, C.G., Ruhfel, B., Primack, R.B., Miller-Rushing, A.J., Davis, C.C.: Phylogenetic patterns of species loss in thoreauś woods are driven by climate change. Proc. Nat. Acad Sci. USA 105, 17029–17033 (2009)
25. Yu, Y., Warnow, L., Nakhleh, T.: Algorithms for mdc-based multi-locus phylogeny inference: beyond rooted binary gene trees on single alleles. J. Comput Biol. 18(11), 1543–1559 (2011)
26. Zhang, L.: Inferring a species tree from gene trees under the deep coalescence cost. In: RECOMB, pp. 192–193 (2000)

Improvement of BLASTp on the FPGA-Based High-Performance Computer RIVYERA

Lars Wienbrandt, Daniel Siebert, and Manfred Schimmler

Department of Computer Science
Christian-Albrechts-University of Kiel
24098 Kiel, Germany
{lwi,dsi,masch}@informatik.uni-kiel.de

Abstract. NCBI BLASTp plays the major role of protein database searches already for years. However, with today's growth of sequence database sizes, it becomes more inefficient with standard PC architectures. One solution to address this problem was already presented in our previous implementation, published in [16], taking advantages of the massive parallelization provided by the FPGA-based high-performance computer RIVYERA [3].

The analysis of bottlenecks in our BLASTp pipeline showed the urgent need to speed up the two-hit finder component, as well as the postprocessing on the PC. After a complete redesign of the two-hit finder and the insertion of a new "gapped extension" filter, we achieve a speedup of up to 376, compared to one thread of a fully utilized 2x Intel Xeon E5520 PC system at 2.26GHz running original NCBI BLASTp v. 2.2.25+. This is about two times the performance of our previous implementation.

1 Introduction

The originally published BLAST algorithm [5] was developed to speed up biological sequence alignment, an essential method to perform searches in biological sequence databases. As an example, the functionality or effect of a newly sequenced protein may be predicted by comparison to already known sequences in the database.

Since BLAST is a heuristic algorithm, it shows a significant speedup compared to other non-heuristic alignment algorithms as the well-known Smith-Waterman [14] or Needleman-Wunsch [12]. The major tradeoff has to be made in terms of alignment quality. Smith-Waterman and Needleman-Wunsch produce an optimal alignment while BLAST approximates the optimum result. However, since todays exponential growth of database sizes as well as an increasing length of the query sequences, an optimal alignment becomes unfeasable. Even the original BLAST becomes too inefficient to conform with a biologists workflow. Thus, new solutions are required reducing runtime while maintaining the same level of alignment quality. With continuing the development and maintenance of BLAST by NCBI [1], several enhancements, such as the two-hit method and gapped BLAST [6], reduce computation time and improve result quality. However, the execution of the BLAST algorithm still suffers from long runtimes, especially for large query sets and databases, and the utilization of fast computing architectures becomes inevitable.

CUDA-BLASTp [10] is a novel solution using GPU-based acceleration. The authors achieve a speedup of up to 6 on an nVidia GeForce GTX 280 graphics card compared to single CPU-thread of an Intel Core i7-920. In contrast, we have already provided an efficient solution concerning this problem with special purpose hardware. We have implemented an FPGA-based hardware acceleration for BLASTp using a low-cost Xilinx Spartan3-5000 device, while taking advantages of a massive parallelization utilizing 128 of such FPGAs in the hardware architecture RIVYERA [16]. The implemented BLASTp pipeline is closely oriented to the one presented by Kasap et al. [9]. Sotiriades et al. [15] published a hardware implementation of BLAST as well, with the advantage to be freely configurable to use with BLASTn, BLASTp, BLASTx, tBLASTn and tBLASTx. However, it misses superior advantages by omitting the usage of gapped BLAST or the two-hit method. Mahram and Herbordt followed another approach [11]. They used a single Altera Stratix-III FPGA connected to 4.5GB DRAM to speed up the process of the ungapped and gapped extension. Recently, Guo et al. published an FPGA-based prototype implementation using a multi-hits detection strategy [7].

This idea finally led us to further analyze our design for major bottlenecks in the dataflow. One major bottleneck has been discovered in the two-hit method, since its runtime complexity is quadratic in the number of hits to be analyzed. Only parallelization of this component and a bounded search space has made the absolute runtime acceptable. A second bottleneck, regards the postprocessing step to generate the final alignments on the PC. The sheer amount of candidates from the ungapped extension steps of all concurrently executed BLASTp pipelines is simply too high to be processed efficiently on the PC. Thus, it becomes the slowest component in the entire pipeline.

This paper describes our new version of RIVYERA BLASTp. The ideas are based on the FPGA-based BLASTp pipeline presented in Mercury BLASTp [8] by Jacob et al. The applied modifications include the replacement of the quadratic two-hit method to a linear one, and the utilization of free FPGA resources to perform a modified Needleman-Wunsch alignment on a banded matrix. This last step acts as a gapped extension filter, reducing the number of alignment candidates to be processed by the host PC.

We integrated our design completely into the NCBI BLASTp software version 2.2.25+. Therefore, it can be executed transparent as any usual NCBI BLASTp query and delivers the accustomed result format. Hence, biologists do not require special skills to handle this solution.

We tested our new design with three different query sets on the NCBI *RefSeq* database [2]. In comparison to an actual high-performance PC equipped with two Intel Xeon E5520 at 2.26GHz and 48GB RAM, utilized with all 16 threads, we gained a total speedup of up to 23.5, i.e. a speedup of 376 against one thread. Therefore, one FPGA reaches the same performance as three CPU threads. This is a factor of about 2 compared to our previous implementation. Regarding

GPUs, we again achieve a speedup of more than 22 compared to CUDA-BLASTp (author related results on an nVidia GeForce GTX280 graphics card).

2 BLAST

The Basic Local Alignment Search Tool (BLAST) was originally published 1990 by Altschul et al. in [5]. Now, NCBI provides one of the most commonly used versions [1], significantly improved by the *two-hit method* and a gapped alignment strategy [6]. BLAST is generally used to perform sequence database searches to find sequence similarities, either for DNA (BLASTn) or protein (BLASTp) sequences. Several other variations like BLASTx, tBLASTn, tBLASTx or PSI-BLAST exist, but all with a similar core algorithm which is described in the following.

BLAST is organized in several steps. In the first step, the query sequence is being preprocessed to find k-words. k-words are short sequences of size k (k-mers) which are similar to k-mers of the query sequence. In the following, this is also referred to as *neighborhood*. The value k is fixed, but different for either BLASTn ($k = 11$) or BLASTp ($k = 3$). A k-mer is declared similar to a k-mer of the query sequence if the score of a direct comparison, calculated according to a scoring matrix (such as BLOSUM62), exceeds a predefined threshold value.

Afterwards, exact matches (*hits*) of the neighborhood in the database sequences are located. The two-hit method analyzes each pair of hits to hold the same distance to each other in the query sequence and the subject sequence. The pairs satisfying this restriction are referred to as *two-hits*.

To save runtime and memory the distance between every two tested hits is bounded to a certain parameter A. Additionally, the value k is applied as lower bound to omit overlapping hits ($k = 3$ for BLASTp). Therefore, the following equation shows the condition for a two-hit whereby s_0 and s_1 state the location of two hits in the subject and q_0 and q_1 their locations in the query respectively:

$$k \leq q_1 - q_0 = s_1 - s_0 < A \qquad (1)$$

Each two-hit is being further examined by an *ungapped extension* process, i.e. extending both hits in both directions, calculating a similarity score of the current part of the subject and query sequence. In fact, the NCBI implementation first closes the gap between the two hits from right to left, then extends from the first hit of the pair (in positional order) to the left and afterwards extends from the second hit to the right. The extension stops for each direction if the score declines a certain cut-off distance below the so far reached maximum. This method is referred to as *X-drop* mechanism. Figure 1 shows an example of the NCBI ungapped extension.

A *high-scoring pair (HSP)* refers to the two positions where the maximum scores were reached for each direction in the ungapped extension process. To introduce gapped alignment, HSPs are being analyzed by a slightly modified version of the Needleman-Wunsch algorithm [12]. First, the alignment is bound to the positional range of the ungapped extension, and second, in contrast to

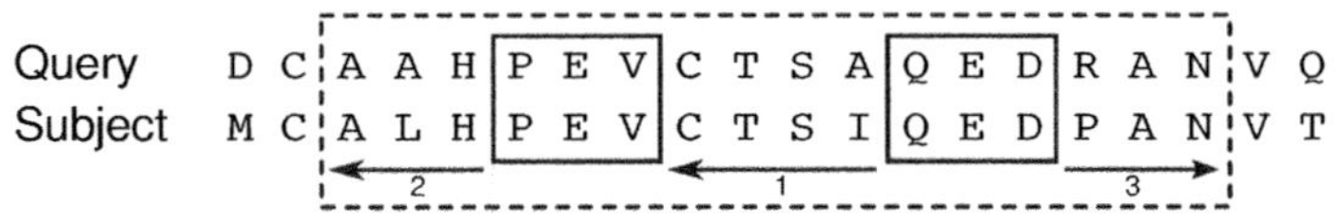

```
Query    D C │A A H │P E V│C T S A│Q E D│R A N│V Q
Subject  M C │A L H │P E V│C T S I│Q E D│P A N│V T
             └──2──→      ←──1──        ──3──→
```

Fig. 1. Example for the ungapped extension of a two-hit in the NCBI BLAST implementation. The solid rectangles mark the two-hit, the dashed an extension. Arrows indicate the direction and the attached numbers the order of the extensions.

the original Needleman-Wunsch algorithm, the traceback to complete the final alignment starts at the matrix cell with the calculated maximum score and not at the lower right corner. Additionally, runtime is being reduced by using the X-drop mechanism again, i.e. omitting the calculation of matrix cell values where the score declines below a certain cut-off distance from the so far calculated maximum cell value.

3 RIVYERA Architecture

The massively parallel FPGA-based hardware platform RIVYERA was first introduced as COPACOBANA 5000 for applications in the fields of bioinformatics [13,17].

For the application presented here, the specific RIVYERA S3-5000 rev 1 is used, developed and distributed by SciEngines GmbH [3]. It consists of two basic elements, the in-built multiple FPGA-based supercomputer and a standard server grade mainboard, running a Linux operating system on an Intel Core i7-930 processor with 12GB of RAM. The FPGA computer is equipped with up to 128 user configurable Xilinx Spartan3-5000 FPGAs, distributed over 16 FPGA cards, each containing eight user FPGAs. Additionally, a DRAM module with a capacity of 32MB is attached to each user FPGA.

The FPGAs are connected by a systolic-like bus system, i.e. each FPGA on an FPGA card is connected with two neighbors forming a ring including a communication controller. The FPGA card slots are connected as a ring as well, providing the connection of the communication controllers on each FPGA card. At least one communication interface of the FPGA cards is connected via PCIe to the host mainboard.

For application development, the RIVYERA provides an API for software development and hardware design, i.e. an API controlling the data transfer between the host software and the FPGAs including broadcast facilities, and an API for the user defined hardware configuration of the FPGAs controlling the data transfer to other FPGAs and the host as well as a communication interface for the attached DRAM.

A picture of the RIVYERA S3-5000 and the overview of its hardware structure is shown in Fig. 2. RIVYERA is packed in a standard rack mountable $3U$ housing and powered by two redundant 650W supplies.

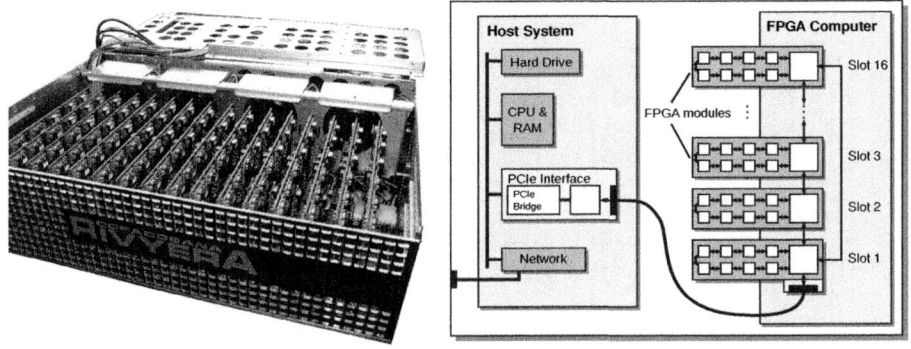

Fig. 2. RIVYERA S3-5000 picture and hardware structure

4 Application Structure and Implementation

As in our previous implementation, we concentrate on BLASTp for protein queries and databases. Our main modifications focus on the hardware part while the software part is left nearly untouched. Still, the software is responsible for the preprocessing of the query sequences to identify all k-words. This list of k-words is submitted to the hardware pipelines as well as the database and the query sequences. The most compute intensive tasks to identify hits from the k-words and generate HSPs as candidates for an alignment, is processed by the hardware pipelines on the Xilinx Spartan3-5000 FPGAs of the RIVYERA. The final human-readable alignments are generated by the software part again. In contrast to our previous implementation, we reduced the number of final alignment candidates by more than 80% and hence, widened the bottleneck for the post-processing step significantly. Additionally, we accelerated our hardware pipeline by replacing the original two-hit method. The details of the particular components can be found in the following Sect. 4.1 while a description of the software part follows in Sect. 4.2.

4.1 Modified BLASTp Hardware Pipeline

In contrast to our previous implementation we changed the structure of our hardware pipeline significantly. The first stage stays the *HitFinder* component which analyzes the neighborhood, i.e. searching for k-words in the submitted subject sequence. For the second stage we replaced the four *TwoHitMethod* components of quadratic runtime complexity with only one of linear complexity. Hence, a splitting into sub-pipelines, as there has been in the previous implementation, is not necessary anymore. The new component, referred to as *TwoHitFinder* in the following, follows the HitFinder directly. As before, the identified two-hits are stored in a *TwoHitFIFO* for further processing in the *UngappedExtender*. For the ungapped extension process, this component accesses a *QueryBuffer* and a *SubjectBuffer*, holding the actual query and subject sequence respectively.

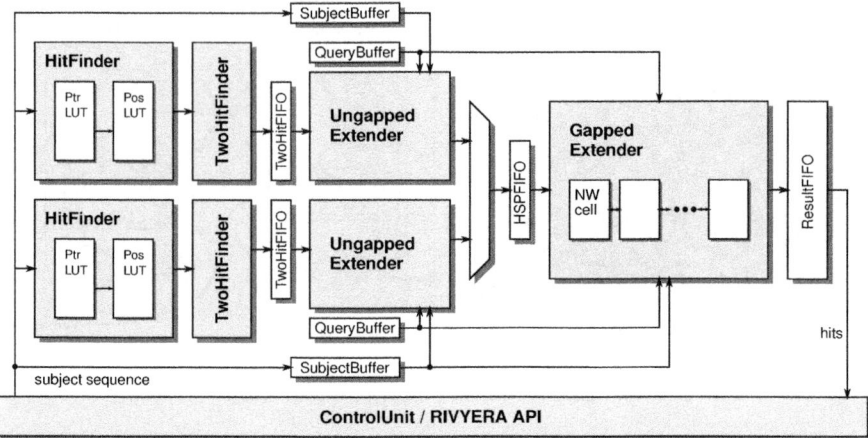

Fig. 3. Structure of two BLASTp hardware pipelines sharing one GappedExtender component

Instead of processing the resulting high-scoring pairs (HSPs) on the host-system for a gapped extension, we inserted a new gapped extension filter, the *GappedExtender*, as the last stage in our pipeline. A *HSPFIFO* buffers the HSPs for this purpose. Again, the gapped extension process requires access to the query and subject.

As it turned out, the GappedExtender is the most resource occupying component, but it is utilized the most infrequently as well. Thus, for a full utilization of a Spartan3-5000 FPGA, we doubled all pipeline components but the HSP-FIFO and the GappedExtender. Therefore, we effectively get two independent BLASTp pipelines, sharing the same GappedExtender component. The final results are stored in a large *ResultFIFO*, implemented in the attached DRAM. An overall structure is shown in Fig. 3.

HitFinder. The HitFinder searches for occurences of k-words in the subject sequence. Each k-mer of the subject is simply looked up in a hashtable, which is precomputed and initialized by the host software (s. Sect. 4.2). It is organized in two lookup tables. The first (*PtrLUT*) is directly addressed by the k-mer and provides pointers to a variable memory space implemented in the second lookup table (*PtrLUT*). If occurences of a k-mer exist in the query sequence, the return values are valid positions and will be routed to the TwoHitFinder component, otherwise the k-mer is ignored for further processing.

TwoHitFinder. In contrast to the previous implementation, where every possible pair of hits was actively tested to hold equation (1) in Sect. 2, we focus on a new strategy to reduce the runtime complexity of this component from clearly quadratic to linear. The idea is very similar to the one published by Jacob et al. for Mercury BLASTp [8].

First, we provide an array of length $l = 1024$. This corresponds to our maximum query length plus the parameter A for the bounds of (1). This array stores at position p the most recent subject position s_0 to the corresponding query position q_0. The position p is calculated from the following equation:

$$p = (s_0 - q_0) \bmod 1024 \qquad (2)$$

Before inserting a new position, the content of the array cell is read. If this cell contains a valid subject position s_1, it holds $s_0 > s_1$ and:

$$s_0 - q_0 = s_1 - q_1 \quad (\bmod\ 1024) \qquad (3)$$
$$\Leftrightarrow \quad s_0 - s_1 = q_0 - q_1 \quad (\bmod\ 1024) \qquad (4)$$

Assuming $s_0 - s_1 < A$, it follows from (4):

$$s_0 - s_1 = \begin{cases} q_0 - q_1 & \text{if } q_0 \geq q_1 \\ 1024 - (q_1 - q_0) & \text{if } q_0 < q_1 \end{cases} \qquad (5)$$

The second case (assuming $q_0 < q_1$) results in:

$$A > 1024 - (q_1 - q_0) \qquad (6)$$
$$\Leftrightarrow \quad q_1 - q_0 > 1024 - A \qquad (7)$$

This is in direct contradiction to the bounds of the query length which is $l = 1024 - A$. Hence, if $s_0 - s_1 < A$ it directly follows $s_0 - s_1 = q_0 - q_1$, and if (1) holds, i.e. $k \leq s_0 - s_1$, this result can be reported as two-hit.

UngappedExtender. The UngappedExtender conforms with the implementation of the original NCBI BLASTp. First, the extension is directed left, starting with the right hit of the hit pair. In every clock cycle the score of a pair of residues from the query and the subject sequence is calculated from a scoring matrix, implemented in dual-ported ROM, and summed up to the total score. The X-drop mechanism is implemented by checking in every clock cycle that the new calculated score does not drop a predefined cut-off distance below the so far calculated maximum score.

The extension continues directed right from the right hit of the hit pair, but only if the gap between the two hits has been crossed. The two positions where both extensions reached their highest scores now form a high-scoring pair (HSP), which is reported if the score exceeds another predefined threshold value.

GappedExtender. The GappedExtender component basically performs a modified Needleman-Wunsch alignment [12] with a banded matrix and a HSP at its center. We keep our implementation close to the one in NCBI BLASTp. The Needleman-Wunsch alignment takes a fixed width of the matrix band ($\omega = 64$), and uses the X-drop mechanism, as for the ungapped extension, to stop the extension process. Thus, in contrast to Mercury BLASTp, the length of our matrix band stays variable.

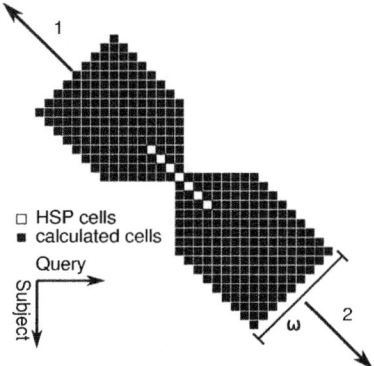

Fig. 4. Principle of the gapped extension of a high-scoring pair (HSP). White cells in the middle indicate the HSP, black cells the calculated cells in the Needleman-Wunsch alignment. Arrows with attached numbers indicate the direction and the order of the extension. The extension uses the X-drop mechanism to stop.

The alignment starts at the center of the HSP and first, extends backward, using the reverse sequences for Needleman-Wunsch. Afterwards, a forward directed alignment, again starting from the HSPs center, is performed in the same way (s. an example in Fig. 4). If the sum of both alignment scores exceeds some report threshold parameter, the original HSP is being reported to the host software. This way, we save valuable software runtime by filtering nearly every HSP in advance, which would be omitted by the host software after the software-based alignment anyway.

The subcomponents of the GappedExtender component basically consist of ω processing elements connected in a chain. Each processing element represents a cell of the Needleman-Wunsch matrix whose score is to be calculated next, in the following referred to as *NWcell*. The part of the query sequence, which is to be analyzed, is inserted from the one end of the chain, while the corresponding part of the subject sequence is inserted from the other end. With every clock cycle one residue of either the query or the subject is inserted. The same way, this information is passed from one cell to its left or right neighbor respectively.

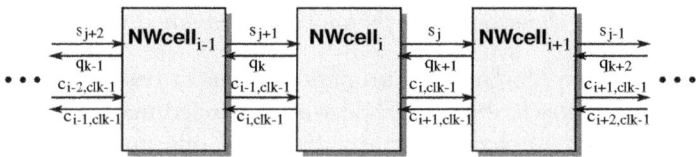

Fig. 5. Structure of the NW cell chain implemented in the GappedExtender component. s_j is the subject symbol at position j, q_k is the query symbol at position k, $c_{i,clk}$ is the current score of cell i.

The score of each matrix cell is dependent on the scores of its upper, left and upper-left neighbors (s. [12]). Since we insert residues from both ends of the NWcell chain, we calculate all scores of the matrix cells within an anti-diagonal of the matrix band of width ω in one clock cycle. Due to the alternating insertion of residues, the current anti-diagonal "moves" alternating rightward and downward in each clock cycle. Hence, for the calculation of an NWcell's score, each NWcell requires access to the scores of *both* neighbors in the chain, calculated in the previous clock cycle. The structure of an NW cell chain is depicted in Fig. 5.

4.2 Software Control

We completely integrated the software control of our hardware implementation into the software core of NCBI BLASTp version 2.2.25+. Thus, the necessary preprocessing of the query sequences to generate the neighborhood of the k-mers is done by this software as well as the postprocessing, including the generation of the final alignments resulting from the reported high-scoring pairs. This way the hardware core becomes completely transparent for the user.

The neighborhood information as well as the plain-text query and parameters is stored on each FPGA on the RIVYERA. The database information is passed to the RIVYERA via broadcast, while all results are fetched from each FPGA individually and passed directly to the postprocessing part of the software.

Since our design only supports query lengths of about $l = 1024 - A$ (s. Sect. 4.1, default: $A = 40$), longer queries have to be split. Luckily, the NCBI software performs likewise. Hence, besides adaption to our supported query length, no further adjustments had to be applied on the software part.

5 Performance Evaluation

The BLASTp pipeline implementation described in Sect. 4.1 is written in the VHDL programming language and targets a Xilinx Spartan3-5000 FPGA. We used Xilinx ISE 13.2 for synthesis and generation of the configuration file. The software part is written in C++ and integrated into the core of NCBI BLASTp version 2.2.25+. We are able to utilize a Spartan3-5000 with two pipelines sharing the same GappedExtender component. Disregarding the GappedExtender, block RAM utilization was the limitative factor for the device utilization. We chose the width of the matrix band in the GappedExtender to be $\omega = 64$, as in Mercury BLASTp. The total device utilization, including the RIVYERA API, is 25,499 slices (76%), 24,848 slice flipflops (37%), 32,654 LUTs (49%) and 93 BRAMs (89%). The base clock frequency of our implementation is 50MHz, although all pipeline components, excluding the GappedExtender, are clocked at 100MHz.

The reference system was a high-performance PC equipped with two Intel Xeon E5520 CPUs, each containing 4 cores (8 threads) running at 2.26GHz, 48GB DDR3-RAM, and 64bit Linux OS. We compared NCBI BLASTp version 2.2.25+ with default parameters, BLOSUM62 scoring matrix and a varying number of threads (BLASTp option "-num_threads") to our implementation utilizing a fully equipped RIVYERA machine (128 FPGAs), and several

Table 1. BLASTp runtimes of three randomly reduced query sets against part one of the NCBI *RefSeq* database. The 2x Xeon E5520 reference system runs NCBI BLASTp v. 2.2.25+. The marked (*) runtimes are estimations calculated from published runtimes adapted to the changed database and query set.

Query set	RIVYERA (n FPGAs)					2x Xeon E5520		Mercury	CUDA
	128	64	32	16	8	16 thr.	8 thr.	BLASTp	BLASTp
A. thaliana	353s	648s	1106s	1934s	3531s	8301s	9995s	3780s*	7780s*
P. trichoc.	482s	808s	1323s	2309s	4210s	10226s	12506s	5161s*	9615s*
H. sapiens	561s	987s	1723s	2817s	4409s	9464s	11602s	6007s*	8026s*

partly equipped configurations (64, 32, and 16 FPGAs) downto one single FPGA card (8 FPGAs). We chose three different query sets (Proteomes of *Arabidopsis thaliana*, *Populus trichocarpa*, and human (*Homo sapiens*) from *SUPERFAMILY* database [4]), randomly reduced to 2,335, 3,151, and 1,990 sequences respectively, such that each set contains about 1 million residues. Each query set was aligned against the first part of the NCBI *RefSeq* BLAST database, release 50 [2] containing 2,996,372 sequences ($\approx$ 1 billion residues).

We compared the runtimes of our design to Mercury BLASTp and CUDA-BLASTp v2.0 as well. Unfortunately, CUDA-BLASTp apparently does not support large databases (our test system, equipped with an nVidia GeForce GTX480 GPU, refused to work with databases larger than 1/1000 of our benchmark database). It would have been possible to use the same queries and databases as in [10] for our system, but as the published results were only measured against a single query sequence, a comparison would be impossible since our design relies on the concurrent processing of several queries. Thus, we adapted and interpolated their best result linearly according to the changed size of the database and query sets and the runtime relations of our three query sets. The same applies for Mercury BLASTp, since we did not have access to the required hardware as well, and the results published in [8] refer to already outdated databases and query sets we were unable to find. Hence, due to extreme dependencies on the quality of the query, these results are only to be seen as a rough estimation. All results are stated in Table 1. Figure 6 illustrates the speedups of RIVYERA with a different number of utilized FPGAs versus our test system. The energy consumption of RIVYERA and the reference PC was measured with a customary power measurement device.

5.1 Discussion

According to the runtime analysis in Table 1, RIVYERA clearly outperforms the reference system. In its fastest configuration, a fully equipped RIVYERA with 128 FPGAs, a speedup factor of about 376 is reached against one thread on the fully utilized reference system (i.e. 23.5 against the complete reference system or 47 against one Xeon CPU). Hence, the runtime performance of one single FPGA conforms to about three CPU threads.

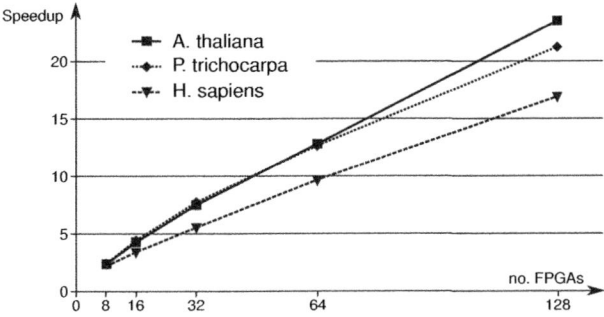

Fig. 6. Speedups of RIVYERA S3-5000 with different number of utilized FPGAs vs. two Xeon E5520 (16 threads)

The measured energy consumption of a fully-equipped RIVYERA is only 590W. Regarding the energy consumption of 290W by the reference system, we save up to 92.3% compared to a PC cluster with the same performance.

In contrast to our previous implementation, we are able to process up to 256 queries at once now, which is 33% less than before. However, with an increasing degree of parallelization, i.e. equipping RIVYERA from one up to 16 FPGA cards, our speedup grows approximately linear (s. Fig. 6). This indicates, that we succesfully removed the former bottleneck formed by the postprocessing step on the host system.

Regarding quality analysis, we chose a small query subset (109 sequences, 28,483 residues) and compared all alignments from our implementation to those from NCBI BLASTp. We counted 21,918 hits from our implementation while NCBI found 22,167 hits for our test query set. A detailed analysis showed that 63 hits (0.29%) are additional results not found by NCBI, and 312 hits (1.41%) from the NCBI results are not found by our implementation. Another 24 hits (0.11%) in both sets were differing only in their alignment positions for the same query and subject sequence. This shows that our alignment quality is almost the same as for the NCBI software. However, since BLAST is heuristic, small discrepancies in the alignments do not necessarily imply a difference in quality. Hence, we decided to take this potential tradeoff for the benefit of runtime performance.

6 Conclusion

BLAST remains one of the most important available tools to perform biological database searches. With our FPGA-based implementation of BLASTp on RIVYERA valuable time can be saved. The massive parallelization benefits from processing large query sets, since we are able to process up to 256 queries concurrently. We clearly outperform other hardware accelerators, such as Mercury BLASTp [8] or GPUs running CUDA-BLASTp [10], as well as an eight core (16 threads) high-performance PC with a speedup factor of up to 376 against one thread. In addition, we keep almost the same alignment quality as in NCBI BLASTp, and RIVYERA saves up to 92.3% of the energy required by the PC reference system.

In the future, we are addressing an again four times runtime improvement of our design by taking advantage of the latest available Spartan6-based RIVYERA S6-LX150.

References

1. NCBI BLAST, http://blast.ncbi.nlm.nih.gov/Blast.cgi
2. NCBI RefSeq database, http://www.ncbi.nlm.nih.gov/RefSeq/
3. SciEngines GmbH, http://www.sciengines.com
4. Superfamily HMM library and genome assignments server, http://supfam.cs.bris.ac.uk/SUPERFAMILY/
5. Altschul, S.F., Gish, W., Miller, W., Myers, E.W., Lipman, D.J.: Basic Local Alignment Search Tool. Journal of Molecular Biology 215(3), 403–410 (1990)
6. Altschul, S.F., Madden, T.L., Schäffer, A.A., Zhang, J., Zhang, Z., Miller, W., Lipman, D.J.: Gapped BLAST and PSI-BLAST: a new generation of protein database search programs. Nucleic Acids Research 25, 3389–3402 (1997)
7. Guo, X., Wang, H., Devabhaktuni, V.: Design of a FPGA-Based Parallel Architecture for BLAST Algorithm with Multi-hits Detection. In: Proceedings of ITNG 2011, pp. 689–694. IEEE Computer Society (2011)
8. Jacob, A., Lancaster, J., Buhler, J., Harris, B., Chamberlain, R.D.: Mercury BLASTp: Accelerating Protein Sequence Alignment. ACM Transactions on Reconfigurable Technology and Systems 1, 9:1–9:44 (2008)
9. Kasap, S., Benkrid, K., Liu, Y.: Design and Implementation of an FPGA-based Core for Gapped BLAST Sequence Alignment with the Two-Hit Method. Engineering Letters 16, 443–452 (2008)
10. Liu, W., Schmidt, B., Müller-Wittig, W.: CUDA-BLASTP: Accelerating BLASTP on CUDA-Enabled Graphics Hardware. IEEE/ACM Transactions on Computational Biology and Bioinformatics 8, 1678–1684 (2011)
11. Mahram, A., Herbordt, M.C.: Fast and Accurate BLASTP: Acceleration with Multiphase FPGA-Based Prefiltering. In: Proceedings of ICS 2010, pp. 73–28 (2010)
12. Needleman, S.B., Wunsch, C.D.: A general method applicable to the search for similarities in the amino acid sequence of two proteins. Journal of Molecular Biology 48(3), 443–453 (1970)
13. Schimmler, M., Wienbrandt, L., Gneysu, T., Bissel, J.: COPACOBANA: A Massively Parallel FPGA-Based Computer Architecture. In: Schmidt, B. (ed.) Bioinformatics – High Performance Parallel Computer Architectures, pp. 223–262. CRC Press (2010)
14. Smith, T.F., Waterman, M.S.: Identification of common molecular subsequences. Journal of Molecular Biology 147, 195–197 (1981)
15. Sotiriades, E., Dollas, A.: A General Reconfigurable Architecture for the BLAST Algorithm. VLSI Signal Processing 48, 198–208 (2007)
16. Wienbrandt, L., Baumgart, S., Bissel, J., Schatz, F., Schimmler, M.: Massively parallel FPGA-based implementation of BLASTp with the two-hit method. In: ICCS 2011. Procedia Computer Science, vol. 1, pp. 1967–1976 (2011)
17. Wienbrandt, L., Baumgart, S., Bissel, J., Yeo, C.M.Y., Schimmler, M.: Using the reconfigurable massively parallel architecture COPACOBANA 5000 for applications in bioinformatics. In: ICCS 2010. Procedia Computer Science, vol. 1, pp. 1027–1034 (2010)

A Polynomial Time Solution for Protein Chain Pair Simplification under the Discrete Fréchet Distance

Tim Wylie and Binhai Zhu

Department of Computer Science, Montana State University, Bozeman, MT 59717-3880, USA
{timothy.wylie,bhz}@cs.montana.edu

Abstract. The comparison and simplification of polygonal chains is an important and active topic in many areas of research. In the study of protein structure alignment and comparison, a lot of work has been done using RMSD as the distance measure. This method has certain drawbacks, and thus recently, the discrete Fréchet distance was applied to the problem of protein (backbone) structure alignment and comparison with promising results. Another important area within protein structure research is visualization, due to the number of nodes along each backbone. Protein chain backbones can have as many as 500~600 α-carbon atoms which constitute the vertices in the comparison. Even with an excellent alignment, the similarity of two polygonal chains can be very difficult to see visually unless the two chains are nearly identical. To address this issue, the chain pair simplification problem (CPS-3F) was proposed in 2008 to simultaneously simplify both chains with respect to each other under the discrete Fréchet distance. It is unknown whether CPS-3F is **NP**-complete, and so heuristic methods have been developed. Here, we first define a version of CPS-3F, denoted CPS-3F$^+$, and prove that it is polynomially solvable by presenting a dynamic programming solution. Then we compare the CPS-3F$^+$ solutions with previous empirical results, and further demonstrate some of the benefits of the simplified comparison. Finally, we discuss future work and implications along with a web-based software implementation, named FPACT (The Fréchet-based Protein Alignment & Comparison Tool), allowing users to align, simplify, and compare protein backbone chains using methods based on the discrete Fréchet distance.

1 Introduction

The comparison and simplification of polygonal chains have been well studied in several fields including computer vision, bioinformatics, computational geometry, and parametric curve approximations [1,4,21]. Within structural biology, polygonal chain similarity is one of the central problems of protein research. In general, it is believed that a protein's structure implies its function, and thus to compare the functionality of proteins their structures must be compared [15]. This is known to be true for certain situations, especially with homologous traits between proteins, and the empirical evidence between proteins in general is in agreement [15,14]. The structure is defined by the α-carbon atoms of the residues (amino acids) along the backbone of each chain. These atoms represent the vertices that constitute our 3D polygonal chains.

Since the structure of the protein is intrinsically related to its function, there have been many software systems designed for protein structure alignment and comparison in the last couple of decades. A few of the more well-known systems are SSAP [20], DALI [12,11], CATH [16], CE [18], SCOP [8], MAMMOTH [17], ProteinDBS [19] and 3D-BLAST [24]. None of these systems use the discrete Fréchet distance, and the majority of the work previously done on protein global structure alignment and protein local structure alignment use the RMSD (Root Mean Square Distance) distance measure. Given two m-vectors $V_1 = \langle u_1, u_2, ..., u_m \rangle$ and $V_2 = \langle v_1, v_2, ..., v_m \rangle$, RMSD is defined as:

$$RMSD(V_1, V_2) = \sqrt{\frac{\sum_i (u_i - v_i)^2}{m}}.$$

This gives an average pair-wise distance along the two vectors which provides some insight into the similarity of the two chains, but the reliance on m shows one of the major drawbacks of using RMSD. The comparison hinges on the necessity that the two vectors be the same length and that the vertices at a given index in each chain be pairwise similar. If we modified the chains at all we could receive very different RMSD values. This suggests that a measure independent of the number of vertices or a pair-wise alignment would be a better indication on the similarity of the two chains.

To achieve a more accurate measure of similarity between two protein structures, Jiang et al. proposed using the discrete Fréchet distance for the protein backbone similarity comparison [13]. The two main problems they addressed were the alignment of the two chains, and then the comparison itself. They showed that the optimal alignment problem, as defined in [13], between two 3D chains under the discrete Fréchet distance takes $O(n^7 m^7 \log(n+m))$ time to solve [13]. Due to the high time complexity they proposed a heuristic method not dependent on the discrete Fréchet distance. Recently, we revisited the optimal alignment problem by proposing a possibly PTAS heuristic algorithm in which all translations and rotations were based on the current discrete Fréchet distance of the two chains [22]. We also showed that this was at worst a 2-approximation algorithm for the optimal alignment problem. The new algorithm provided better alignment results than the previous method for all empirical evaluations.

When comparing polygonal structures, alignment is just one of the issues surrounding the research. Given that protein backbones can have as many as 500~600 vertices (α-carbon atoms) in each chain, even with an optimal alignment, visualizing the similarity of the two chains is difficult unless those chains are nearly identical. To address this issue the chain pair simplification (CPS-3F) problem was proposed by Bereg et al. in 2008 [6]. They were unable to show whether CPS-3F is **NP**-complete or not. They proved that even the 2D Hausdorff version of the problem, which simplified the chains via the Hausdorff distance (CPS-2H), was **NP**-complete [6]. This led them to postulate that under the discrete Fréchet distance the problem was likely to be **NP**-complete. Thus, in our previous work [22] we used a heuristic $O(n)$ time algorithm to simplify pairs of chains. Here we show that a restricted version of CPS-3F, denoted CPS-3F$^+$, is polynomially solvable. This restriction is beneficial because as we simplify the proteins, we want to visually compare the two backbone chains without distorting their lengths.

As previously mentioned, the majority of software systems for aligning, comparing, and simplifying protein backbones use the RMSD measure. Thus, we have created an on-line software program called FPACT (The Fréchet-based Protein Alignment &

Comparison Tool) which uses the alignment and comparison algorithms (including the optimal CPS-3F$^+$) based on the discrete Fréchet distance [23]. This allows anyone to use their own PDB (Protein Data Bank) files to align, simplify, and compare the backbone structures of the proteins. This is discussed in more detail later.

The paper is organized as follows. In Section 2 we discuss the discrete Fréchet distance, CPS-3F$^+$, and the related background. In Section 3 we present a polynomial time solution for CPS-3F$^+$, cover the time complexity, and then give a pseudocode algorithm to implement the solution. In Section 4 we present some comparison results to our previous heuristic method for CPS-3F. Finally, we outline some implications and future work as well as discussing an available software implementation in Section 5, and then we conclude the paper with some open problems for future work.

2 Background

The Fréchet distance was first defined by Maurice Fréchet in 1906 as a measure of similarity between two parametric curves [10]. Subsequently, it has become a standard measure used between parametric curves used in many areas. In the early 90s, the Fréchet distance was applied to polygonal curves by Alt and Godau [2,3]. Then in 1994 Eiter and Mannila defined the discrete Fréchet distance as an approximation of the Fréchet distance to be used between two polygonal chains using only the nodes along the chains for the measurements [9]. They also referred to this discrete form as the coupling distance which is used synonymously. Furthermore, they proved the discrete version can be computed in $O(mn)$ time where m, n are the number of vertices in each polygonal chains.

The discrete Fréchet distance has since been applied in several fields of research, but recently, one of the prominent applications has been in aligning and comparing the similarity of protein backbones [5,25,13]. In this comparison between backbones each vertex in the polygonal chains represent an α-carbon atom, which gives the comparison based on the atoms a clear biological meaning. Thus, a comparison between two backbones using the discrete Fréchet distance is more meaningful than one using the continuous version. For clarity, we next cover the definition of the discrete Fréchet distance.

Given two paths, we define their discrete Fréchet distance as follows. (We use the graph-theoretic term "paths" instead of the geometric term "polygonal chains" here because our definition makes no assumption that the underlying space of points is geometric.) We use $d(a,b)$ to represent the Euclidean distance between two 3D points a and b, but certainly it can be replaced with some other distance measure, depending on applications.

Definition 1. *Given a path* $P = \langle p_1, \ldots, p_n \rangle$ *of n vertices, a **t-walk** along P is a partitioning of P along the path into t disjoint non-empty subpaths $\{P_i\}_{i=1..t}$ such that $P_i = \langle p_{n_{i-1}+1}, \ldots, p_{n_i} \rangle$ and $0 = n_0 < n_1 < \cdots < n_t = n$.*

Given two paths $A = \langle a_1, \ldots, a_m \rangle$ *and* $B = \langle b_1, \ldots, b_n \rangle$, *a **paired walk** along A and B is a t-walk $\{A_i\}_{i=1..t}$ along A and a t-walk $\{B_i\}_{i=1..t}$ along B for some t, such that, for $1 \leq i \leq t$, either $|A_i| = 1$ or $|B_i| = 1$ (that is, either A_i or B_i contains exactly one vertex). The **cost** of a paired walk $W = \{(A_i, B_i)\}$ along two paths A and B is*

$$d_{\mathcal{F}}^W(A, B) = \max_i \max_{(a,b) \in A_i \times B_i} d(a,b).$$

The **discrete Fréchet distance** between two paths A and B is

$$d_{\mathcal{F}}(A, B) = \min_W d_{\mathcal{F}}^W(A, B).$$

*The paired walk that achieves the discrete Fréchet distance between two paths A and B is called the **Fréchet alignment** of A and B.*

Definition 2. *The **moving cost** of a paired walk $W = \{(A_i, B_i)\}$ is*

$$m_c^W(A_i, B_i) = \begin{cases} |A_i|, & \text{if } |B_i| = 1 \\ |B_i|, & \text{if } |A_i| = 1 \end{cases}$$

Then the moving cost of a paired walk W between A and B is

$$m_c^W(A, B) = \sum_{i=1}^{t} m_c^W(A_i, B_i).$$

The Fréchet distance is typically explained as the relationship between a person and a dog connected by a leash walking along the two curves and trying to keep the leash as short as possible. Consider the scenario in which a person walks along A and a dog along B. Intuitively, the definition of the paired walk is based on three cases:

1. $|B_i| > |A_i| = 1$: the person stays and the dog hops forward;
2. $|A_i| > |B_i| = 1$: the person hops forward and the dog stays;
3. $|A_i| = |B_i| = 1$: both the person and the dog hop forward.

The moving cost for A and B can be thought of as the sum of the number of times a move (1, 2, or 3) is made where each choice counts as one move along the walk. With enough nodes the discrete Fréchet distance can closely approximate the continuous version, and with a standard dynamic programming approach, it is not hard to obtain the following theorem.

Theorem 1. *[9] The discrete Fréchet distance between two paths with m and n vertices respectively can be computed in $O(mn)$ time.*

In 2008, the chain pair simplification problem in three dimensions under the discrete Fréchet distance was defined in order to allow better visualization of the the two chains. The problem not only allows you to see the two chains in a simplified form, but it also keeps the characteristic similarities that exist between the chains. Although the problem does not necessarily need to be limited to only 3D space, we state the original decision problem as it was defined relating to protein backbone chains.

The CPS-3F problem is defined as follows:

Instance: Given a pair of 3D chains A and B, with lengths $O(m), O(n)$ respectively, an integer $K > 0$, and three real numbers $\delta_1, \delta_2, \delta_3 > 0$.

Problem: Does there exist a pair of chains A', B' each of at most K vertices such that the vertices of A', B' are from A, B respectively, and $d_\mathcal{F}(A, A') \leq \delta_1, d_\mathcal{F}(B, B') \leq \delta_2, d_\mathcal{F}(A', B') \leq \delta_3$?

The CPS-3F$^+$ problem is now defined as follows:

Instance: Given 3D chains A and B in 3D, with lengths $O(m), O(n)$ respectively, an integer $K' > 0$, and $\delta_1, \delta_2, \delta_3 \in \mathbb{R}^+$.
Problem: Does there exist a pair of chains A', B' where the vertices are from A, B respectively, such that for some paired walk W between A', B', $m_c^W(A', B') \leq K'$, and $d_\mathcal{F}(A, A') \leq \delta_1, d_\mathcal{F}(B, B') \leq \delta_2, d_\mathcal{F}(A', B') \leq \delta_3$?

3 Polynomial Time Solution for CPS-3F$^+$

3.1 CPS-3F$^+$ ∈ P

In this section we present a polynomial time solution for CPS-3F$^+$. Several versions of the single chain simplification problem were addressed and shown to be polynomially solvable by Bereg et al. [6]. However, CPS-2H (where the Hausdorff distance is used for $d(A, A')$ and $d(B, B')$) was shown to be **NP**-complete and thus it is believed that the Fréchet version might be as well. The solution presented here proves that under the discrete Fréchet distance, the restricted version of the chain pair simplification problem (CPS-3F$^+$) is polynomially solvable when the dimension is fixed. The algorithm returns the optimal K', specified in the definition of the decision problem, which is equal to

$$m_c(A', B') = \min_W m_c^W(A', B'),$$

among all feasible W. We now define several necessary terms and data structures.

Given two polygonal chains $A = \langle a_1, a_2, ..., a_m \rangle$, $B = \langle b_1, b_2, ..., b_n \rangle$, and constraints $\delta_1, \delta_2, \delta_3 \in \mathbb{R}^+$, we can design a dynamic programming algorithm to find the optimal K'. First we let $\mathcal{D} = \{(a_i, b_j) | a_i \in A, b_j \in B \text{ and } d(a_i, b_j) \leq \delta_3\}$. This is the set of all pairs of nodes between the two chains which are at a distance of at most δ_3 from each other. Then we can define a matrix $\mathcal{C}$ of size $m \times n$ which in any cell $\mathcal{C}_{i,j}$ contains the minimum number, K', of pairs $(a_k, b_l) \in \mathcal{D}$ which given δ_1, δ_2, and δ_3 simplify A and B via CPS-3F$^+$ from (a_1, b_1) up to (a_i, b_j).

In order to maintain $\mathcal{C}$, we need another data structure $\mathcal{R}$ and some other helpful definitions. We define $S_X(x_i, \delta)$ as the maximal continuous subchain containing x_i on the polygonal chain X such that all the vertices on this subchain are contained in the sphere centered at x_i and with radius δ. Now let $r_{i,j}$ be the rectangle on $\mathcal{C}$ defined as $\langle \min(S_A(a_i, \delta_1)), \max(S_A(a_i, \delta_1)), \min(S_B(b_j, \delta_2)), \max(S_B(b_j, \delta_2)) \rangle$ such that $(a_i, b_j) \in \mathcal{D}$. Here, min and max refer to the minimum or maximum indexed element within $S_X(x_i, \delta)$. For every pair in $\mathcal{D}$, we envision the corresponding rectangles as being overlayed on $\mathcal{C}$. A rectangle $r_{i,j}$ covers all the cells of $\mathcal{C}$ that are analogous to the vertices in $S_A(a_i, \delta_1) \cup S_B(b_j, \delta_2)$ as shown in Figure 1.

For convenience we also define the set of all rectangles that a cell in $\mathcal{C}$ belongs to: $\mathcal{Q}_{k,l} = \{r_{i,j} | a_k \in A, b_l \in B \text{ and } \min(S_A(a_i, \delta_1)) \leq a_k \leq \max(S_A(a_i, \delta_1)) \text{ and } \min(S_B(b_j, \delta_2)) \leq b_l \leq \max(S_B(b_j, \delta_2))\}$.

Now we define $\mathcal{R}$ as a matrix of sets where the matrix is of size m by n, and $\mathcal{R}$ provides information needed to fill out $\mathcal{C}$ by storing a list of rectangles for each cell. $\mathcal{R}_{i,j}$ contains a set of rectangles (dynamic array) which pertain to the number of coverings (rectangles) still viable at any (i,j) relating to the number already calculated for $\mathcal{C}_{i,j}$. These are computed by the recurrence below.

Initial Conditions: $\mathcal{Q}_{1,1} \neq \emptyset$, $\mathcal{R}_{1,1} = \mathcal{Q}_{1,1}$, and $\mathcal{C}_{1,1} = 1$.

$$\mathcal{C}_{i,j} = \min_{(k,l) \in \{(i-1,j),(i,j-1),(i-1,j-1)\}} \begin{cases} \mathcal{C}_{k,l}, & \text{if } \mathcal{Q}_{i,j} \cap \mathcal{R}_{k,l} \neq \emptyset \\ \mathcal{C}_{k,l} + 1, & \text{if } \mathcal{Q}_{i,j} \cap \mathcal{R}_{k,l} = \emptyset, \mathcal{Q}_{k,l} \neq \emptyset, \mathcal{R}_{k,l} \neq \emptyset \\ NULL, & \text{if } \mathcal{Q}_{i,j} = \emptyset \end{cases}$$

$$\mathcal{R}_{i,j} = \bigcup_{(k,l) \in \{(i-1,j),(i,j-1),(i-1,j-1)\}} \begin{cases} \mathcal{R}_{k,l} \cap \mathcal{Q}_{i,j}, & \text{if } \mathcal{C}_{i,j} = \mathcal{C}_{k,l}, \mathcal{R}_{k,l} \cap \mathcal{Q}_{i,j} \neq \emptyset \\ \mathcal{Q}_{i,j}, & \text{if } \mathcal{C}_{i,j} = \mathcal{C}_{k,l} + 1, \mathcal{R}_{k,l} \neq \emptyset, \\ & \mathcal{R}_{k,l} \cap \mathcal{Q}_{i,j} = \emptyset \end{cases}$$

The idea is to find the minimum covered xy-monotone increasing path from (a_1, b_1) to (a_m, b_n) which corresponds to $\mathcal{C}_{1,1}$ to $\mathcal{C}_{m,n}$. This is the minimum path by basic dynamic programming with all feasible options explored. If we visited a cell that was not covered, that would mean one of the nodes is not covered by a pair in $\mathcal{D}$. By finding this covered walk, one guarantees that every column and every row is covered by at least one rectangle which means all of the nodes of A and B are covered.

The increasing xy-monotone path is necessary in the recurrence due to the definition of the discrete Fréchet distance. Without the requirement of a monotonically increasing path this would be using the weak discrete Fréchet distance.

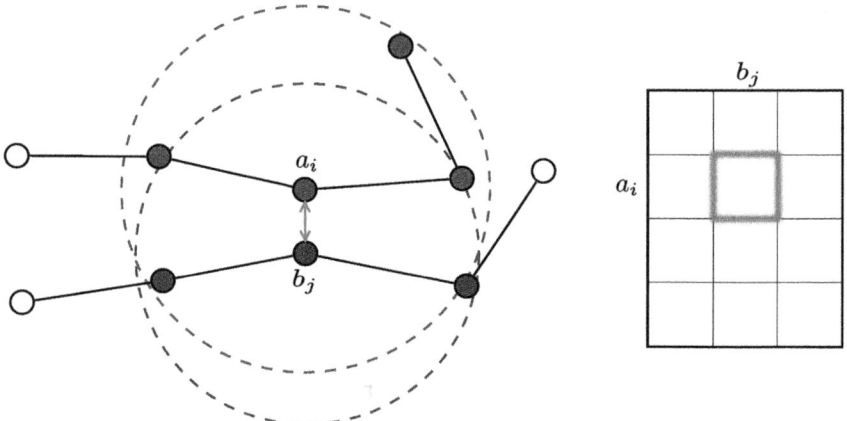

Fig. 1. The rectangle $r_{i,j}$ constructed from subchains of A, B where $d(a_i, b_j) \leq \delta_3$. Here $S_A(a_i, \delta_1)$ contains the vertices a_{i-1} to a_{i+2}, and $S_B(b_j, \delta_2)$ contains the vertices b_{j-1} to b_{j+1}. Thus, $r_{i,j}$ is defined by the min and max node indices in each subchain.

We first characterize the optimal substructure of CPS-3F$^+$ as an optimization problem given our definitions, and then show this yields the optimal solution for K and thus decides CPS-3F$^+$.

Theorem 2. *Optimal substructure of CPS-3F$^+$:*
Let $A = \langle a_1, \ldots, a_m \rangle$ and $B = \langle b_1, \ldots, b_n \rangle$ be two polygonal chains, $\delta_1, \delta_2, \delta_3 \in \mathbb{R}^+$, and let $Z_i = \langle z_1, \ldots, z_i \rangle$ such that every z_j is a rectangle, be any CPS-3F$^+$ solution.

1. *If (a_k, b_l) is covered by z_i where $(k, l) \in \{(m\text{-}1, n), (m, n\text{-}1), (m\text{-}1, n\text{-}1)\}$, then Z_i is a CPS-3F$^+$ solution for A_k, B_l.*
2. *If (a_k, b_l) is covered by z_{i-1} where $(k, l) \in \{(m\text{-}1, n), (m, n\text{-}1), (m\text{-}1, n\text{-}1)\}$, then Z_{i-1} is a CPS-3F$^+$ solution for A_k, B_l.*
3. *If (a_k, b_l) is not covered by z_i or z_{i-1} where $(k, l) \in \{(m\text{-}1, n), (m, n\text{-}1), (m\text{-}1, n\text{-}1)\}$, then $\nexists$ a CPS-3F$^+$ solution for A_k, B_l.*

Proof. **(1)** If z_i covers (a_k, b_l) where $(k, l) \in \{(m\text{-}1, n), (m, n\text{-}1), (m\text{-}1, n\text{-}1)\}$, then $z_i \in \mathcal{Q}_{m,n} \cap \mathcal{R}_{k,l}$, and $C_{k,l} = |Z_i|$. Suppose $C_{k,l} \neq |Z_i|$, then for (a_k, b_l) there are two possibilities: either $C_{k,l} > |Z_i|$ or $C_{k,l} < |Z_i|$. $C_{k,l} > |Z_i|$ implies the solution required another rectangle at the previous step, but since the recurrence is monotonically increasing this is impossible. If $C_{k,l} < |Z_i|$ then given the addition of z_i for (m, n) implies $z_i \notin \mathcal{Q}_{m,n} \cap \mathcal{R}_{k,l}$, but that contradicts our assumption.

(2) If z_{i-1} covers (a_k, b_l) where $(k, l) \in \{(m\text{-}1, n), (m, n\text{-}1), (m\text{-}1, n\text{-}1)\}$, but not (a_m, b_n) then $z_i \notin \mathcal{Q}_{m,n} \cap \mathcal{R}_{k,l}$, and $C_{k,l} = |Z_{i-1}| = |Z_i|\text{-}1$. If we suppose $C_{k,l} \neq |Z_{i-1}|$, then either $C_{k,l} > |Z_{i-1}|$ or $C_{k,l} < |Z_{i-1}|$. If $C_{k,l} > |Z_{i-1}|$, then $C_{k,l} = |Z_{i-1}|+1 = |Z_i|$, and since Z_i is an optimal solution we have added another rectangle that must cover (a_k, b_l) and (a_m, b_n) in order to be a solution. However, this means $\exists z_i \in \mathcal{Q}_{m,n} \cap \mathcal{R}_{k,l}$ which contradicts our assumption. Suppose $C_{k,l} < |Z_{i-1}|$, then to cover (a_m, b_n) we must add another rectangle, but that contradicts Z_i being an optimal solution since we have covered A_m, B_n with $|Z_i|\text{-}1$ rectangles.

(3) Since $\nexists$ any rectangles z_i and z_{i-1} that cover both (a_m, b_n) and (a_k, b_l) where $(k, l) \in \{(m\text{-}1, n), (m, n\text{-}1), (m\text{-}1, n\text{-}1)\}$, then $\mathcal{Q}_{k,l} = \emptyset$ or $\mathcal{Q}_{m,n} = \emptyset$. By definition, if no rectangles cover the cells, then there is no solution for A_k, B_l or A_m, B_n. □

Theorem 3. *Constrained chain pair simplification, under the discrete Fréchet distance, is polynomially solvable, i.e. $CPS\text{-}3F^+ \in \mathbf{P}$.*

Proof. Since we have shown that CPS-3F$^+$ has an optimal substructure, given A, B, K', δ_1, δ_2, and δ_3, we can find an optimal K'' from our algorithm. Then we decide CPS-3F$^+$ by comparing whether $K' \leq K''$. □

Corollary 1. *Constrained chain pair simplification gives a factor 2-approximation to the chain pair simplification problem under the discrete Fréchet distance, i.e., CPS-3F$^+$ provides a 2-approximation of CPS-3F.*

3.2 Time Complexity

The time complexity is largely dependent on δ_1, δ_2, and δ_3 because they define the size and the number of rectangles. We allow δ_1, δ_2, and δ_3 to be absorbed in the complexity because their values do not guarantee a specific number of nodes covered, however, with small values this is a constant number and the algorithm runs in $O(mn)$ time. The number of rectangles, $Q_{i,j}$, grows as $\mathcal{D}$ increases–which is largely dependent on δ_3. The worst case is when δ_1, δ_2, and δ_3 are set larger than the lengths of the chains causing every $Q_{i,j}$ to contain all possible rectangles. Thus, building $\mathcal{Q}$ naïvely before running the algorithm requires $O(m^2n^2)$ time. The optimal solution for CPS-3F$^+$ would be $K' = 1$, but would require $O(m^2n^2)$ time to run as well.

Filtering steps could be added to watch for large δ values. With filtering, the time at any step could be kept to a constant (or logarithmic) value and the time complexity would be $O(mn)$ or have an extra logarithmic factor dependent on δ_1, δ_2, and δ_3.

3.3 Dynamic Programming Algorithm

By directly applying the recurrences the optimal solution can be found. Since the recurrence only requires the previous row and column for any decision the function can be implemented as a simpler iterative algorithm. Similar to the edit distance problem between two strings, only two rows at a time are needed, so the amount of memory can be drastically reduced by only storing those two rows of $\mathcal{C}$ and $\mathcal{R}$ that the algorithm is currently using. This approach will give the optimal K' value, but in order to retrieve the actual K' pairs, $\mathcal{C}$ and $\mathcal{R}$ must be stored in full.

Algorithm 1 assumes that $\mathcal{C}$ is an $m \times n$ matrix initialized to NULL, and that $\mathcal{R}$ is an $m \times n$ dynamic 3D jagged array where every element, $\mathcal{R}_{i,j}$, is its own array. We assume that $\mathcal{Q}$ has been calculated before this function is called. For simplicity, to dynamically append to the end of an array or create a new row we use a generic function 'add'.

Algorithm 2 defines a recursive function that finds the rectangles through which the optimal path exists. The 'Level' variable begins as the optimal K' value ($\mathcal{C}_{m,n}$) when the function is originally called. 'i','j' are originally set to m, n respectively, and 'CurrRect' is set to NULL. The method assumes a jagged array '$Path$' of size K' that must exist to store the rectangles of the path. In an actual implementation with each of these stored rectangles you would want to add child or parent pointers to make it easier to follow the path. An alternative to having pointers is to add an 'if' statement to check if $Path[1]$ has a value and then exit the function if it does. This means at least one optimal path has been found (there could be multiple).

4 Comparison of Results

We present some results comparing our previous heuristic method SIMPLIFY [22] and the optimal solution (Algorithm 1) for CPS-3F$^+$ in this section. The algorithm has been implemented in both Python and C#. Using the same format as our previous results, we set $\delta_1 = \delta_2$ for simplicity and to ensure chains A', B' will have the same reduced length. δ_3 is set to the minimum integer that will work for the two chains given δ_1, δ_2. The comparison tables in both cases are using the protein backbone 1o7j.a (protein A) and comparing it with seven other chains from the Protein DataBank: 1hfj.c, 1qd1.b, 1toh, 4eca.c, 1d9q.d, 4eca.b, 4eca.d. These seven chains were reported to be similar to 1o7j.a by the ProteinDBS software [19] (this took a few seconds searching the whole PDB,

Algorithm 1. FIND-CPS-3F$^+$ ▷ Compute Optimal K Iteratively

```
 1: procedure FIND-CPS-3F⁺
 2:     for i ← 1, m do                                              ▷ Loop over A
 3:         for j ← 1, n do                                          ▷ Loop over B
 4:             if i = 1 and j = 1 then                              ▷ Initial Conditions
 5:                 C_{1,1} ← 1
 6:                 R_{1,1} ← Q_{1,1}
 7:             else
 8:                 Values, Rectangles ← Array                       ▷ Arrays to hold possible values
 9:                 for each (k, l) ∈ {(i-1, j), (i, j-1), (i-1, j-1)} do
10:                     if k > 0 and l > 0 then                      ▷ For the edge cases
11:                         for each rect ∈ R_{k,l} do
12:                             if (i, j) ∈ rect then                ▷ If the cell (i,j) is covered by rect
13:                                 Values.add( C_{k,l} )
14:                                 Rectangles.add( rect )
15:                             else                                 ▷ No rectangles in common
16:                                 Values.add( C_{k,l} + 1 )
17:                                 Rectangles.add( NULL )
18:
19:                 C_{i,j} ← min( Values )                          ▷ The min value (K')
20:                 for r ← 1, | Values | do                         ▷ Populate R_{i,j}
21:                     if Values[ r ] = C_{i,j} then
22:                         if Rectangles[ r ] = NULL then
23:                             R_{i,j}.add( Q_{i,j} \ R_{i,j} )      ▷ Add all possible rectangles
24:                         else
25:                             R_{i,j}.add( Rectangles[ r ] )       ▷ Add the single rectangle
26:     return C_{m,n}                                               ▷ Return the optimal K
```

Algorithm 2. GET-CPS-3F$^+$-PATH ▷ Return an optimal path of rectangles

```
 1: procedure GET-CPS-3F⁺-PATH( i, j, Level, CurrRect )
 2:     if CurrRect = NULL then                                      ▷ Starting conditions
 3:         for each rect ∈ R_{i,j} do
 4:             Path[ Level ].add( rect )
 5:             GET-CPS-3F⁺-PATH( i, j, Level, rect )
 6:     for each (k, l) ∈ {(i-1, j), (i, j-1), (i-1, j-1)} do
 7:         if k < 0 or l < 0 or C_{k,l} = NULL then                 ▷ Do not do anything
 8:         else if C_{k,l} = C_{i,j} then                           ▷ Moving inside a rectangle
 9:             Get-CPS-3F⁺-Path( k, l, Level, CurrRect )
10:         else if C_{k,l} = C_{i,j} − 1 then                       ▷ Moving into a new rectangle
11:             for each rect ∈ R_{k,l} do
12:                 if rect ∉ Path[ Level ] then
13:                     if rect ∉ Path[ Level - 1 ] then             ▷ Add rectangle to previous level
14:                         Path[ Level - 1 ].add( rect )
15:                     GET-CPS-3F⁺-PATH( k, l, Level - 1, rect )
```

which contained over 30,000 protein backbones at that time). Previously, [13] used a heuristic algorithm based on the discrete Fréchet distance and showed that three of the seven chains were not actually similar to 107j.a, and ProteinDBS has subsequently updated their page to reflect this. 107j.a has 325 nodes along the backbone and all but one of the seven other chains do as well.

For the CPS-3F$^+$ algorithm, all chains are assumed to be aligned, and we use the alignments from our previous algorithm ALIGN [22]. In Table 1 we fixed $\delta_1 = \delta_2 = 4$ since the distance between two α-carbon atoms in the backbone is approximately ≈ 3.7 to 3.8 (angstroms). This value ensures that we will be simplifying the chains a minimal amount. We can see that we get an approximate reduced length of $1/3$ which is what we would expect (since this distance will only use the neighboring nodes). The optimal algorithm allows for δ_3 to be much smaller than the heuristic because it can simplify the chains with a value often less than $d_\mathcal{F}(A,B)$, and hence $d_\mathcal{F}(A',B')$ is a lower value.

Table 1. Comparison of Algorithm SIMPLIFY[22] and FIND-CPS-3F$^+$ with 107j.a (Chain A), and $\delta_1 = \delta_2 = 4$

Protein Chain(B)	δ_1	δ_2	δ_3 [22]	δ_3	Length	Reduced Length [22]	Reduced Length	$d_\mathcal{F}(A,B)$	$d_\mathcal{F}(A',B')$ [22]	$d_\mathcal{F}(A',B')$
1hfj.c	4	4	1	1	325	109	109	0.95	0.95	0.95
1qd1.b	4	4	50	21	325	109	150	22.65	24.96	20.70
1toh	4	4	60	21	325	110	178	22.06	23.39	20.54
4eca.c	4	4	17	6	325	109	111	5.55	7.96	5.97
1d9q.d	4	4	43	20	297	109	166	20.87	23.68	19.86
4eca.b	4	4	17	5	325	109	111	5.64	7.51	4.89
4eca.d	4	4	18	5	325	109	113	5.71	7.82	4.94

In Table 2 we vary δ_1, δ_2 for different amounts of simplification and again set δ_3 to the minimum integer value that works. We keep $\delta_1 = \delta_2$ for simplicity and to keep a similar reduced size for both chains. Here we have a more dramatic difference in δ_3 and in $d_\mathcal{F}(A',B')$ because of the greater simplification possibilities between A, A' and B, B' since δ_1, δ_2 are much larger. This demonstrates how CPS-3F$^+$ is able to simplify the two chains simultaneously while highlighting the similarities between the chains.

Table 2. Comparison of Algorithm SIMPLIFY[22] and FIND-CPS-3F$^+$ with 107j.a (Chain A) with $\delta_1=\delta_2$, and different δ_3

Protein Chain(B)	δ_1	δ_2	δ_3 [22]	δ_3	Length	Reduced Length [22]	Reduced Length	$d_\mathcal{F}(A,B)$	$d_\mathcal{F}(A',B')$ [22]	$d_\mathcal{F}(A',B')$
1hfj.c	12	12	4	1	325	26	26	0.95	3.77	0.95
1qd1.b	15	15	33	12	325	17	24	22.65	22.64	11.94
1toh	16	16	44	13	325	17	22	22.06	27.24	12.80
4eca.c	12	12	12	3	325	28	27	5.55	11.73	2.90
1d9q.d	15	15	34	13	297	21	26	20.87	23.65	12.99
4eca.b	12	12	13	3	325	28	26	5.64	12.57	2.94
4eca.d	12	12	14	3	325	28	32	5.71	13.65	2.99

We can see that the optimal results far exceed the heuristic approximation. If we look at 4eca.d, the difference between the heuristic (13.65) and the optimal (2.99) is dramatic. The optimal δ_3 is 3 to 4 times smaller in general, and the discrete Fréchet distance between A' and B' is smaller than the original distance between A, B.

5 Concluding Remarks

In this paper we show that this restricted version of the chain pair simplification problem under the discrete Fréchet distance (CPS-3F$^+$) is polynomially solvable. We then present algorithms to find K', the minimum moving cost between chains A' and B', via CPS-3F$^+$, and a backtracking method to return the vertices of the simplified chains. Along these lines the FPACT software is now available to use these algorithms as well as some of the past methods based on the discrete Fréchet distance [23].

The web-based implementation runs within the Silverlight framework, and can be used in any browser supporting the Silverlight or Moonlight runtime. The software is available to the public for general use thus providing the ability to compare protein backbones with the discrete Fréchet distance [23].

There are still several issues that need to be addressed to fully utilize the benefits of the discrete Fréchet distance in comparing polygonal chains, and specifically protein backbones:

(1) Without any restrictions, is the chain pair simplification problem under the discrete Fréchet distance (CPS-3F) **NP**-complete?

(2) The ALIGN [22] running time still needs to be improved since all comparisons, including CPS-3F$^+$, rely on the two polygonal chains being aligned. Can the alignment be further simplified?

(3) More generally, the question of whether it is theoretically possible to design a practical PTAS (global structure-structure) alignment algorithm based on the discrete Fréchet distance needs to be answered.

(4) For protein backbone structures, can we exploit the physical properties of the chains in order to speed up alignment, comparison, or simplification. e.g. the fixed distance between each node (α-carbon atom), and the minimum distance two atoms can be in relation to each other (they can not touch).

(5) There are several possible strategies to keep CPS-3F$^+$ running in $O(mn)$ time including filtering redundant rectangles. Can these be identified and integrated easily into our algorithm?

Acknowledgments. This research is partially supported by NSF under grant DMS-0901034 and by NSF of China under project 60928006.

References

1. Alt, H., Behrends, B., Blömer, J.: Approximate matching of polygonal shapes (extended abstract). In: Proceedings of the 7th Annual Symposium on Computational Geometry (SoCG 1991), pp. 186–193 (1991)
2. Alt, H., Godau, M.: Measuring the resemblance of polygonal curves. In: Proceedings of the 8th Annual Symposium on Computational Geometry (SoCG 1992), pp. 102–109 (1992)

3. Alt, H., Godau, M.: Computing the Fréchet distance between two polygonal curves. Internat. J. Comput. Geom. Appl. 5, 75–91 (1995)
4. Alt, H., Knauer, C., Wenk, C.: Matching Polygonal Curves with Respect to the Fréchet Distance. In: Ferreira, A., Reichel, H. (eds.) STACS 2001. LNCS, vol. 2010, pp. 63–74. Springer, Heidelberg (2001)
5. Aronov, B., Har-Peled, S., Knauer, C., Wang, Y., Wenk, C.: Fréchet Distance for Curves, Revisited. In: Azar, Y., Erlebach, T. (eds.) ESA 2006. LNCS, vol. 4168, pp. 52–63. Springer, Heidelberg (2006)
6. Bereg, S., Jiang, M., Wang, W., Yang, B., Zhu, B.: Simplifying 3D Polygonal Chains Under the Discrete Fréchet Distance. In: Laber, E.S., Bornstein, C., Nogueira, L.T., Faria, L. (eds.) LATIN 2008. LNCS, vol. 4957, pp. 630–641. Springer, Heidelberg (2008)
7. Cole, R.: Slowing down sorting networks to obtain faster sorting algorithms. J. ACM 34, 200–208 (1987)
8. Conte, L., Ailey, B., Hubbard, T., Brenner, S., Murzin, A., Chothia, C.: SCOP: a structural classification of protein database. Nucleic Acids Research 28, 257–259 (2000)
9. Eiter, T., Mannila, H.: Computing discrete Fréchet distance. Tech. Report CD-TR 94/64, Information Systems Department, Technical University of Vienna (1994)
10. Fréchet, M.: Sur quelques points du calcul fonctionnel. Rendiconti del Circolo Mathematico di Palermo 22, 1–74 (1906)
11. Holm, L., Park, J.: DaliLite workbench for protein structure comparison. Bioinformatics 16, 566–567 (2000)
12. Holm, L., Sander, C.: Protein structure comparison by alignment of distance matrices. J. Mol. Biol. 233, 123–138 (1993)
13. Jiang, M., Xu, Y., Zhu, B.: Protein structure-structure alignment with discrete Fréchet distance. J. of Bioinformatics and Computational Biology 6, 51–64 (2008)
14. Mauzy, C., Hermodson, M.: Structural homology between rbs repressor and ribose binding protein implies functional similarity. Protein Science 1, 843–849 (1992)
15. Needleman, S., Wunsch, C.: A general method applicable to the search for similarities in the amino acid sequence of two proteins. J. Mol. Biol. 48, 443–453 (1970)
16. Orengo, C., Michie, A., Jones, S., Jones, D., Swindles, M., Thornton, J.: CATH—a hierarchic classification of protein domain structures. Structure, 5, 1093–1108 (1997)
17. Oritz, A., Strauss, C., Olmea, O.: MAMMOTH (matching molecular models obtained from theory): an automated method for model comparison. Protein Science 11, 2606–2621 (2002)
18. Shindyalov, I., Bourne, P.: Protein structure alignment by incremental combinatorial extension (CE) of the optimal path. Protein Engineering 11, 739–747 (1998)
19. Shyu, C.-R., Chi, P.-H., Scott, G., Xu, D.: ProteinDBS: a real-time retrieval system for protein structure comparison. Nucleic Acids Research 32, W572–W575 (2004)
20. Taylor, W., Orengo, C.: Protein structure alignment. J. Mol. Biol. 208, 1–22 (1989)
21. Wenk, C.: Shape Matching in Higher Dimensions. PhD thesis, Freie Universitaet Berlin (2002)
22. Wylie, T., Luo, J., Zhu, B.: A Practical Solution for Aligning and Simplifying Pairs of Protein Backbones under the Discrete Fréchet Distance. In: Murgante, B., Gervasi, O., Iglesias, A., Taniar, D., Apduhan, B.O. (eds.) ICCSA 2011, Part III. LNCS, vol. 6784, pp. 74–83. Springer, Heidelberg (2011)
23. Wylie, T.: FPACT: The Fréchet-based Protein Alignment & Comparison Tool (2012), http://www.cs.montana.edu/~timothy.wylie/frechet
24. Yang, J.-M., Tung, C.-H.: Protein structure database search and evolutionary classification. Nucleic Acids Research 34, 3646–3659 (2006)
25. Zhu, B.: Protein local structure alignment under the discrete Fréchet distance. J. Computational Biology 14(10), 1343–1351 (2007)

Designing RNA Secondary Structures in Coding Regions

Rukhsana Yeasmin and Steven Skiena

Department of Computer Science
Stony Brook University, Stony Brook, NY 11794-4400, USA
{ryeasmin,skiena}@cs.stonybrook.edu

Abstract. Sophisticated regulatory structures appear within highly-constrained coding regions of genes. We study the extent to which such structures can be constructed. Can such stem loops appear essentially anywhere within coding regions, or is the potential restricted to rare amino acid sequences? While predicting secondary structures of an RNA sequence is an extensively studied problem in computational biology, the inverse problem, designing sequences based on a known structure is also important. We work on a particular version of the inverse RNA folding problem, where our goal is to achieve a targeted energy level. For a particular RNA structure, we design sequences with the maximum and minimum folding energy while maintaining desired codon distribution. Our major contributions in work is to optimize RNA secondary structure under codon constraints via fast estimation of folding energies following local modification.

Keywords: Synthetic biology, RNA structure prediction, inverse RNA folding.

1 Introduction

The secondary structures formed by RNA molecules are critical to understanding their molecular interactions and biological functions. Computationally predicting the secondary structure of an RNA sequence is a classical problem in computational biology, and has been extensively studied [1,2,3,4]. Indeed programs such as Mfold (http://mfold.rna.albany.edu) and the Vienna RNA package (http://rna.tbi.univie.ac.at) are widely used tools throughout molecular biology. The inverse problem, that of designing RNAs which fold into specific desired structures, is also important. RNA structures (such as transfer RNAs) are critical to a host of biological processes, motivating the need to design sequences to achieve desired shapes and functions.

We consider a particular version of the inverse RNA folding problem for gene coding sequences, where we seek to achieve a targeted energy level as opposed to a particular structure/shape. Cohen and Skiena [5,6] have previously developed effective algorithms for designing optimal RNA sequences which code for specified amino acid sequences while maximizing or (as desired) minimizing the folding energy of the sequence. However, the gene sequences produced by these algorithms tend to use extremely skewed distributions of codons, because C-G bonds are roughly twice as stable as A-U bonds. The optimized genes thus exhibit codon usage distributions very different from that of the host organism, typically resulting in very poor expression.

In this paper, we study the algorithmic design of RNA sequences which code for a specific amino acid sequence using a desired distribution of codons – maximizing or minimizing the folding energy of resulting RNA. Our work is motivated by designing genes to modulate gene expression. Recent studies [7,8] have shown that the amount of secondary structure on the 5' end of the coding sequence plays a critical role in maximizing protein production from a given gene. Typically synthetic genes are designed to match targeted codon distributions, but these studies suggest that RNA secondary structure should be an important part of the design considerations. Another study suggests that folding energy is an important factor in determining translation efficiency [9]. Secondary structures also play important roles as signals in viral replication, and hence designs minimizing the size of these structures have proven important in our experience. Our group routinely designs and synthesizes virus-length coding sequences for a few thousand dollars each [10,11,12].

Our major contributions in this paper include:

- *Optimizing RNA Secondary Structures under Codon Constraints* – We present what we believe to be the first algorithms for modulating (either minimizing or maximizing) the RNA folding energy of a gene while respecting codon constraints. As described above, the demand for such tools is destined to grow as large-scale synthesis costs decline and turnaround times improve. Indeed, with our collaborators we are planning to synthesize high/low secondary structure variants of particular genes to study their effect on translation and replication.
- *Improvements in Unconstrained Secondary Structure Optimization* – We demonstrate that our algorithms produce structures with equal or less (greater for minimizing structures) energy on the codon distributions employed by the previous best inverse design algorithms. In particular, we have employed our optimization algorithms on codon distributions resulting from designs produced by the Cohen-Skiena (CS) algorithms. As validated by Mfold, we show that our algorithms design genes with better energy than those produced by [5]. This is particularly impressive as the Cohen-Skiena minimum-energy algorithm guarantees an optimal solution, albeit under a simpler energy function. These results validate the quality of our designs on wildtype codon distributions where direct comparisons for optimality are unavailable.
- *Fast Estimation of Folding Energies Following Local Modification* – The high $O(n^3)$ running time of traditional RNA folding algorithms limits the number of iterations possible in search-based optimization strategies like ours. We have investigated the tradeoff between the accurate but slow computations of Mfold in quickly recalculating the energy change resulting from small local changes in a given RNA sequence. We find that we generally can reduce the number of calls of this expensive operation (and hence the running time of our algorithms) by a factor of five with little degradation in accuracy.

2 Preliminaries

Predicting RNA secondary structure is an important problem in computational biology. Several groups implemented algorithms for accurate energy determination of the folded RNA structure. Michael Zuker's Mfold/ UNAfold (http://mfold.rna.albany.edu) and Ivo Hofacker's Vienna RNA package named RNAfold (http://rna.tbi.univie.ac.at) are among the most popular.

Mfold [13] uses a nearest neighbor energy rule to determine the structure. The program implements a dynamic programming (DP) technique where they maintain a DP table to store the calculated substructure energies and the optimal structure is obtained by backtracking. Dynamic programming based methods can correctly predict about 73% of known base pairs on domain of fewer than 700 nucleotides [14]. To calculate the structure energy the entire structure is divided in parts consisting of stacked pairs, hairpins, bulges, internal loops and multi loops. There can also be single stranded bases. Total structure energy is the sum of all substructure energies. By default energy values are calculated at 37 °C. For a given RNA sequence S, Mfold program predicts the non-crossing, minimal energy structure P for S in $O(n^3)$ time and $O(n^2)$ space.

RNAfold [4] uses similar dynamic programming techniques to calculate the minimum free energy structure. The parameters used in the program are described in [14]. The Vienna RNA package uses three kinds of dynamic programming algorithms for structure prediction: the minimum free energy algorithm of Zuker and Stiegler [15] which yields a single optimal structure, the partition function algorithm of McCaskill [16] which calculates base pair probabilities in thermodynamic ensemble, and the suboptimal folding algorithm of Wuchty et.al [17] which generates all suboptimal structures within a given energy range of optimal energy. For secondary structure comparison, the package uses string alignment or tree-editing [18] methods to measure distance or dissimilarities. Finally they use inverse folding algorithm to design sequences with predefined structures, where they search for sequences folding into a predefined structure. In case of unsuccessful searches, a structure distance to the target structure is provided.

Inverse RNA Folding was first introduced in [4,19]. RNAinverse in the Vienna RNA package [4] was developed to perform inverse RNA folding. Later an extended *Inverse RNA Folding* problem was studied by Dromi, Avihoo and Barash [20], where they added several non-structural constraints to the output such as thermodynamic stability and mutational robustness. Dahiyat and Mayo [21] worked on the *Inverse Protein Folding* problem where given a folded structure, we seek the primary sequence that folds into it. When they worked on it, designing the three dimensional structure from the sequence alone seemed difficult. However the inverse problem would be more tractable as one could over-engineer the system to favor the desired folding pattern.

INFO-RNA [22] is an web server for *Inverse RNA Folding* maintaining sequence constraints. Here, they apply dynamic programming algorithm to find the initial RNA sequence that satisfies given secondary structure. It is not guaranteed to fold to the target structure as it might have another minimum free energy (mfe) structure. Thus the sequence is further processed by performing stochastic local searches to minimize the structure distance between the mfe structure of the obtained sequence and the given target structure. RNAexinv [23] is another software that performs extended *Inverse RNA folding* by considering not only the desired structure while generating the sequence but also other favorable attributes (i.e. thermodynamic stability and mutational robustness).

Stochastic context-free grammars (SCFGs) are alternative probabilistic methodologies for modeling RNA structure [24,25]. Specific grammar rules are used to induce a joint probability distribution over all possible RNA structures and sequences. Parameters of SCFG models specify probability distributions over possible transformations that may be applied to a nonterminal symbol. These parameters do not have direct physical interpretations, they are learned from collections of RNA sequences with known secondary structures, no external laboratory experiments are needed [26].

CONTRAfold [27] is another secondary structure prediction tool which is based on a flexible probabilistic model called a conditional log-linear model (CLLM). Like SCFGs, CLLMs use the computationally driven parameter learning. However, unlike SCFGs they also have the generality to represent complex scoring schemes, such as those used in energy based predictions i.e. Mfold. CONTRAfold thus closes the gap between probabilistic and thermodynamic models.

Recently studies are being performed on improved parameter sets. Andronescu et al. [28,29] applied Constraint-Generation and Boltzman-likelihood methods for better parameter estimation. Using these parameters they obtained much better RNA structure prediction models. Zakov et al. [30] further refined previous models by examining more types of structural elements and a larger sequential context for these elements. Their study showed that use of more detailed models with rich parameter sets improves prediction quality.

3 Heuristics for Inverse RNA Folding

We start with an initial wild type structure maintaining constraints imposed by the codon constraint table. The initial RNA sequence is formed by arbitrarily positioning one of the several possible codons for each amino acid. Now our goal is to find the RNA sequences with maximum and minimum energy that code for the given amino acid sequence using this codon distribution. According to Zuker and Stiegler [15], dinucleotide composition is a primary contributor to folding free energy. Our main goal was to maximize (minimize) the number of bonds to get the most (least) stable or the max (min) structure. Here by most (least) stable, we mean the structure with the minimum (maximum) energy.

3.1 Structure Maximization

To maximize structure (minimize energy) we give the wild type sequence to Mfold as input to find the initial bonded structure. Next we build a free codon table that keeps track of all the codons that does not form bond with any other codon, i.e. all the nucleotides of that codon are free. Hence we can replace the codons that are already bonded with any other free synonymous codon if the replacement improves the bond energy or we can replace two free codons without breaking any existing bond.

Now, we start with a random position of the structure and check for all the codons of the RNA sequence whether swapping it with any other codon from the free codon table (which codes for the same amino acid as this one) improves the bond energy. If there is an improvement in the bond structure, we swap this codon with that free codon. We also maintain a neighbor list for each codon. Initially we allowed options, where after a swap we do not allow neighbor swap. That means, before swapping a codon we first check whether any of its neighbors has already been swapped or not. The check is performed to keep other parts of the structure relatively stable while changing one part. Later we allowed neighbors to swap and figured out allowing neighbor swap gives more freedom to find the structure with minimum energy. After checking the whole sequence, we apply the new sequence again to the Mfold program. This way we continue until we enter a position when no more improvement is possible. In this process sometimes we encounter the same sequence repeatedly. At this position we are stuck in a local minima. To leave the local minima, we perform some arbitrary change to the structure without violating the constraints. We penalize the codons that have been changed in the previous step so that they cannot change their position in the next step. Moreover, if there is no improvement in the structure energy for a specific number of iterations, we perform some random codon swaps to get out of local minima.

Algorithm 1. RNAfoldMaximizeStructure(Input: Amino Acid sequence, Codon constraints; Output: Stable structure S_1)

Construct initial wild type RNA sequence from the given amino acid sequence maintaining codon constraints
Call Mfold to find the current best structure S from the initial RNA sequence
while there is an improvement in the structure S **do**
    if no improvement in the structure energy for a specific period of time **then**
        Perform random changes to the structure
    else
        if the structure has previously occurred **then**
            Penalize the codons that have been changed in the previous step
        end if
        Find free codons for current RNA structure S
        Change the whole structure by swapping two codons that correspond to same amino acid whenever possible if the swap improves bond energy
    end if
    Call Mfold to find the best bonding $S = S_1$ for the modified structure
end while

In general we run 400 iterations and the complexity in each iteration is dominated by the $O(n^3)$ running time of Mfold.

3.2 Structure Minimization

To find the minimum structure (maximum energy), we employed two different strategies:

- *Region break:* – Here we find the strongest bonded parts (a fixed percentage of the total number of codons) of the current folded RNA sequence. Then we perform an internal swap of the codons corresponding to same amino acid in these regions. For each codon, we find out the codon with maximum mismatch (based on three letters of the codon) with current codon corresponding to same amino acid and perform the swap. We do this for all codons in those most bonded regions.
- *Random swap:* – This process is quite similar to that we did for maximizing structure. However, here we swap two codons if the swap reduces the structure stability or maximizes energy.

We tried each of these strategies separately and also in combination. After each iteration we call Mfold for the modified structure. The combined strategy performs better than individual ones. Fig. 1 shows the comparison plot of the distribution of energies for all of these strategies. From the figure, we see though region break strategy achieves almost same best result as the combined strategy, the frequency of getting good results is smaller.

According to Doshi et al. [31], Mfold RNA secondary structure prediction accuracy degrades as the contact distance between base-pairs increases. One potential reason could be Mfold assumes much more long range base pairs than it occurs in general. However, in our algorithm we are maximizing (or minimizing) bonds in small local regions, so we generally avoid creating long range stem loops.

Algorithm 2. RNAfoldMinimizeStructure(Input: Amino Acid sequence, Codon constraints; Output: Stable structure S_1)

Construct initial wild type RNA sequence from the given amino acid sequence
Call Mfold to find the current best structure S from the initial RNA sequence
while there is an improvement in the structure S **do**
    if the structure has previously occurred **then**
        Penalize the codons that have been changed in the previous step
    end if
    Apply random swap, or region break, or the combined strategy to change the current structure
    Call Mfold to find the best bonding $S = S_1$ for the modified structure
end while

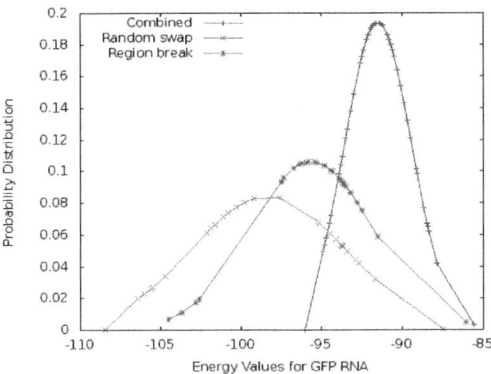

Fig. 1. Comparison plot of Min Structure Energy for Region break, Random swap and Combined strategy for GFP RNA of yeast. Most of the energy values for the distribution of the Combined strategy (rightmost) are higher than other two strategies.

3.3 Algorithmic Variants

We experimented with several different algorithmic variants. To find the minimum and maximum energy sequences we considered two different variations:

- *Random walk:* – Always take the current changed structure even if it is worse than the previous one.
- *Gradient descent approach:* – Before taking a bad move wait for a few (5) iterations. If after the specified number of iterations still get a worse structure, then take the best of all these bad moves as a sequence to move forward. If any of these six moves were better than the previous move but it was a duplicate one, in that case also take the next good move rather than taking the duplicate one.

The *gradient descent* approach performs better than the *random walk* approach, possibly because the *random walk* approach has greater chance of allowing bad moves and trapping into local minima or maxima.

As mentioned before, for minimizing structures we used two strategies: *region break* and *random swap*. In *region break* we had to consider several different criteria. We tried with several different region sizes which was specified by window sizes (let W) and also varied the number of windows to consider at the same time. Here we used a parameter Z to specify that we will break $Z\%$ of the entire RNA sequence. Later we counted the number of windows it will take to add up to $Z\%$ of the entire RNA length. Another issue is whether while breaking regions we should allow codons to swap only in regions internal to those windows or allow external swaps too. Initially we allow only internal swaps, if we encounter a duplicate structure found before we use external swaps for the next step. Fig. 2 shows an example of *region break* strategy.

Amino Acid Sequence:

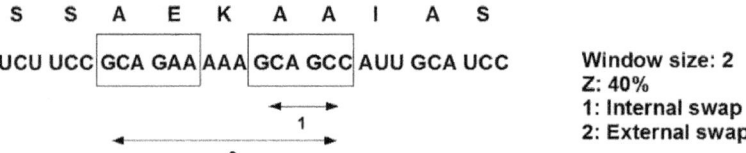

Fig. 2. Region breaking strategy: Here, 1 shows internal swap of two codons that code for amino acid A (allow swap only in codons internal to window), 2 shows external swap of two codons corresponding to A between two windows

Table 1. Summary of Maximized and Minimized Sequence Energies

Sequence	Length	Initial dG	dG(Max)	dG(Min)
Banana Virus1	405	-77.830	-142.43	-62.690
Pumpkin Virus	405	-91.570	-139.28	-83.780
Xenopus tropicalis hemoglobin	429	-111.19	-167.13	-98.520
Banana Virus2	531	-137.09	-189.56	-123.16
StCroix River virus	675	-191.50	-271.58	-160.57
GFP Jelly fish	720	-134.21	-243.56	-112.37
Citirus variegation Virus	849	-236.89	-333.42	-189.97
Honeysucklin	1035	-288.25	-434.46	-268.76
Potato Virus	1077	-299.97	-423.91	-249.43
Vibrio sp. Ex25	1260	-367.07	-500.60	-326.23
Polio Virus	2643	-666.68	-1003.2	-604.90

In *random swap* the first consideration is whether we should always start with a fixed starting point or a random one. In case of fixed starting point, we always check codons for swap from the beginning of the RNA sequence. In random starting point, all codons of the sequence starting from a randomly generated position are checked for swap in a circular manner. We found the performance of random starting point is better than fixed starting point as it allows more variations thus allowing a larger search space. The next issue is whether we should swap codons only with free codons or allow arbitrary swaps. For minimum energy structure, allowing swaps only with free codons seems reasonable as swapping two bonded codons might cause reduction in the stability of the bonded parts of the structure. For maximum energy structure it could be a reasonable one, however its performance was not good. There is a possibility that, arbitrary swap might give good result after searching for long. Here one issue is the speed of merging to a good solution. Allowing free codon swap gives good result even within a reasonably short time. Again in this case we waited five steps to get a better move before accepting a bad one.

Later for minimizing structure we combined two strategies to allow more variations while searching for the best structure. The performance of the combined one was quite better than any single one for minimizing structure as shown before in Fig. 1. Here the question is whether we use two strategies in an interleaving manner or continue with one until we encounter a duplicate. There is no significant variation in the obtained results for any of these strategies. However, rather than changing strategy every next move, sticking with the current strategy until a duplicate structure is encountered seems reasonable as changing strategy every next move might undo the good moves of the previous step.

For structure maximization we used only the *random swap* one as breaking regions arbitrarily to find the most stable structure does not seem reasonable.

4 Experimental Results

We performed experiments for the *Green Fluorescent Protein* (GFP) RNA of jelly fish, yeast and also for polio virus RNA with different initial wild type structures. We also checked for several other RNA sequences. We see from the experimental results that our algorithm always gives a more stable structure than the initial wildtype sequence. Table 1 shows results for some of these sequences.

Fig. 3 shows sample min and max structure output along with initial wildtype sequence for GFP RNA of jelly fish. Initial wildtype sequence energy was -134.21 $kcal/mol$. Our min program output energy is -112.37 $kcal/mol$ and max program output energy is -243.56 $kcal/mol$. From this point on by GFP RNA we will mean GFP RNA of jelly fish.

It is impossible to judge the quality of a heuristic without knowledge of the correct answer. The CS program produces a provably minimum energy sequence for a model close to that of Mfold, but provides no constraint on codon usage. We evaluate our heuristics starting from a random design using the same codon

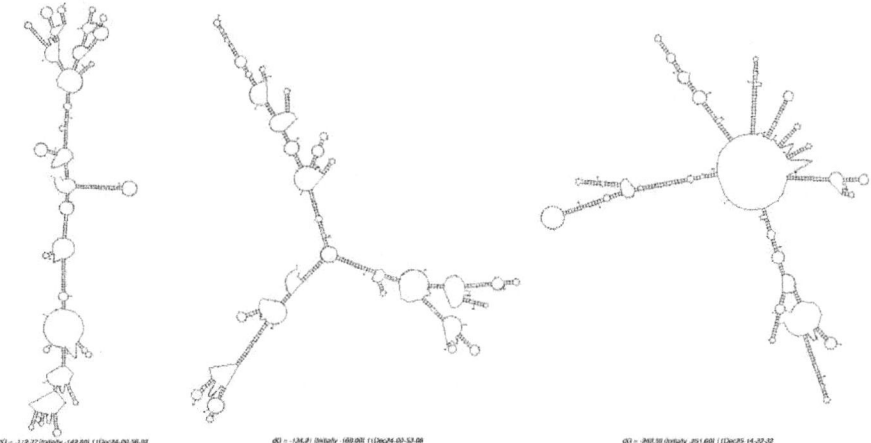

Fig. 3. GFP RNA (jelly fish) Min, wildtype and Max Folded Structures with wildtype (middle) Energy of -134.21 $kcal/mol$, Min Structure (left) Energy of -112.37 $kcal/mol$ and Max Structure (right) Energy of -243.56 $kcal/mol$

distribution as optimized in the CS sequence. Thus if our program is doing a good job of optimization, it will produce similar energies to the optimal design. We ran our algorithm for polio virus RNA with the same codon distribution as obtained from Cohen and Skiena [5] for both minimization and maximization program, starting with a wild type sequence based on that codon distribution. Our algorithm found structures with better energy than that from CS algorithm for both minimized and maximized structures. Later we checked for GFP RNA sequence. We plotted the distribution of energy values for both max and min sequence. In Fig. 4, the left plot shows max energy distribution for GFP RNA and the right one is for polio virus RNA based on CS codon distribution. Fig. 5 shows results for the same amino acid sequences maintaining given codon constraint. Fig. 6 and Fig. 7 shows corresponding min structure energy distribution. Here we see, both of our max and min programs perform much better with respect to CS program output for polio virus RNA. For GFP RNA our algorithm always gets much better structure than CS optimized sequence for the minimized structure and does as good as CS optimal one for the max structure.

We performed experiment to compare the change in the structure energy with the change of codon distribution from wildtype to CS. Fig. 8 shows results from that study for both maximized and minimized structures generated randomly. Here at point 0 of X-axis, the structure follows wildtype codon distribution; at 100, it completely follows CS codon distribution. We see from Fig. 8 that the trend of energy plot is from lower to higher for minimized structure and from higher to lower for maximized structure as the percentage of CS codon distribution goes higher as expected. We optimized the structures using our program. As the figure shows, we always get much better structure than the initial random one. However, the gap between initial energy and optimized energy decreases as we move toward CS codon distribution.

4.1 Parameter Optimization

We have several different parameters that we need to optimize.

- For RNA structure minimization, in *region break* strategy while we are evaluating energy to determine the most structured parts of the folded RNA, we are moving a sliding window W around the RNA sequence, calculating energy for that window. These windows may overlap. Let P be the number of codons by which two windows overlap. Now we want to break top $Z\%$ of the entire RNA structure. We tried to determine the best values of W, P, and Z that fasten the RNA structure minimization process. We checked for $W = 10, 15, 20$ with $P = 5, 8, 10$ and for different Z values i.e. $Z = 10, 15, 20, 25$. We figured out, structure minimization process is quite independent of these parameter values. However, $Z = 15$ might be a good choice to break the structure. Fig. 9 shows the plot of the progress of structure minimization process for *region break* strategy with different W and P values for $Z = (10, 15)$ with same initial wild type structure for GFP RNA. For other sequences the effect of these parameters are quite similar.

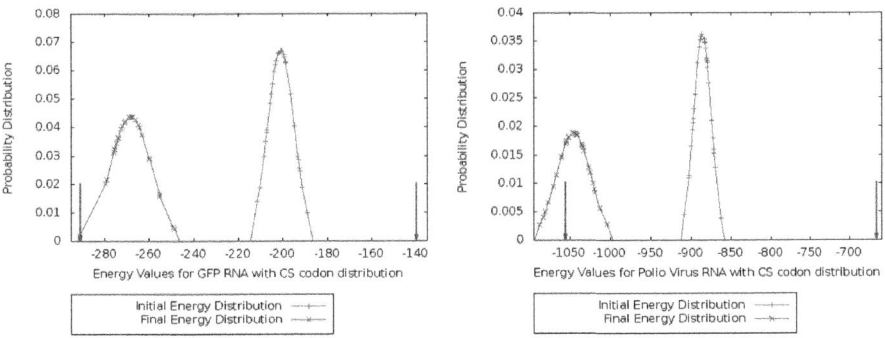

Fig. 4. Max Structure Energy Distribution for GFP RNA and Polio Virus RNA based on CS Codon Distribution; wildtype Energy for $GFP = -139.8$ $kcal/mol$, for $PolioVirus = -666.6$ $kcal/mol$, CS opt Energy for $GFP = -290.7$ $kcal/mol$, for $PolioVirus = -1056$ $kcal/mol$; Max Structure Energy with CS distribution from our program is -291.3 $kcal/mol$ for GFP and -1080.6 $kcal/mol$ for Polio Virus; RED arrow indicates wildtype Energy and BLUE arrow indicates CS opt Energy

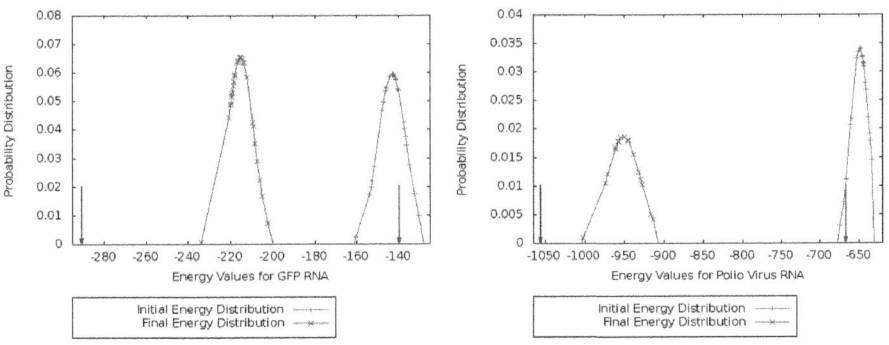

Fig. 5. Max Structure Energy Distribution for GFP RNA and Polio Virus RNA maintaining given Codon Constraints; wildtype Energy for $GFP = -139.8$ $kcal/mol$, for $PolioVirus = -666.6$ $kcal/mol$, CS opt Energy for $GFP = -290.7$ $kcal/mol$, for $PolioVirus = -1056$ $kcal/mol$; Max Structure Energy from our program is -234 $kcal/mol$ for GFP and -1003.2 $kcal/mol$ for Polio Virus; RED arrow indicates wildtype Energy and BLUE arrow indicates CS opt Energy

- Again, for structure maximization rather than calling Mfold every iteration, we reduced the number of Mfold call which is controlled by the parameter X. For $X = i$ we call Mfold every i^{th} iteration. When $X = 1$, Mfold is called every iteration. Here, the reason to reduce the number of Mfold call is to reduce the total running time of the algorithm as the running time is dominated by Mfold call. We tried for $X = 1, 2, 3, 4, 5, 10, 20$. Experimental results show that up to $X = 5$ the algorithm's output remains the same, i.e. even running

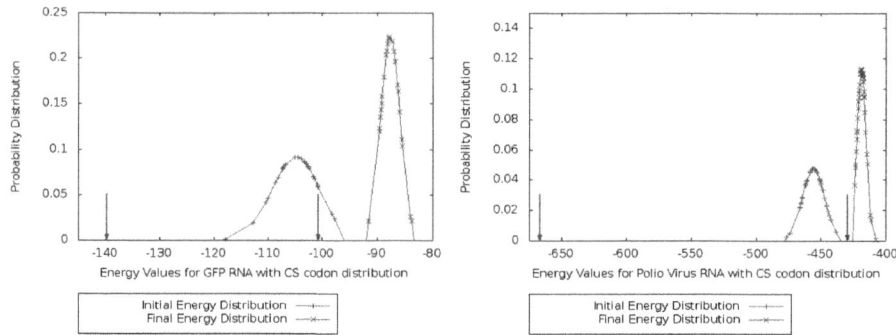

Fig. 6. Min Structure Energy Distribution for GFP RNA and Polio Virus RNA based on CS Codon Distribution; wildtype Energy for $GFP = -139.8$ $kcal/mol$, for $PolioVirus = -666.6$ $kcal/mol$, CS opt Energy for $GFP = -100.7$ $kcal/mol$, for $PolioVirus = -429.8$ $kcal/mol$; Min Structure Energy with CS distribution from our program is -83.84 $kcal/mol$ for GFP and -407.7 $kcal/mol$ for Polio Virus; RED arrow indicates wildtype Energy and BLUE arrow indicates CS opt Energy

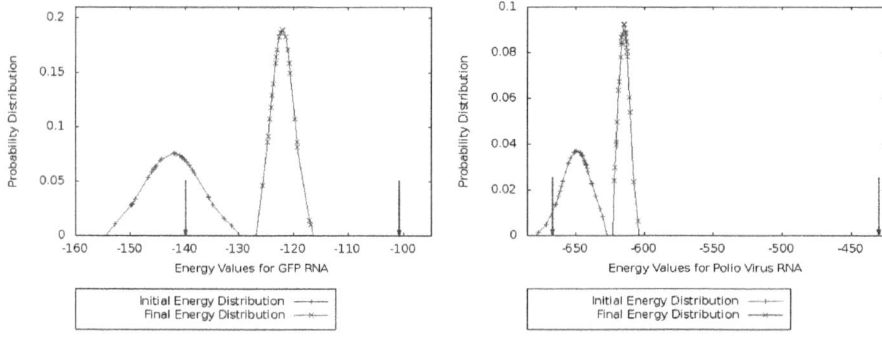

Fig. 7. Min Structure Energy Distribution for GFP RNA and Polio Virus RNA maintaining given Codon Constraints; wildtype Energy for $GFP = -139.8$ $kcal/mol$, for $PolioVirus = -666.6$ $kcal/mol$, CS opt Energy for $GFP = -100.7$ $kcal/mol$, for $PolioVirus = -429.8$ $kcal/mol$; Min Structure Energy from our program is -116.9 $kcal/mol$ for GFP and -604.9 $kcal/mol$ for Polio Virus; RED arrow indicates wildtype Energy and BLUE arrow indicates CS opt Energy

Mfold only every fifth iteration gave us the maximized structure. However, after that (e.g. $X = 10, 20$) algorithm's performance degrades substantially. Fig. 10 shows the comparison of the distribution of energy values for different X values (left) and also the plot of the distribution of energy values for different maximum number of iterations (right). We developed algorithm to determine energy of the current structure based on the information available in the ct file obtained from Mfold program. Once we know the neighbors of each codon, after swaps at each iteration we update the neighbors based

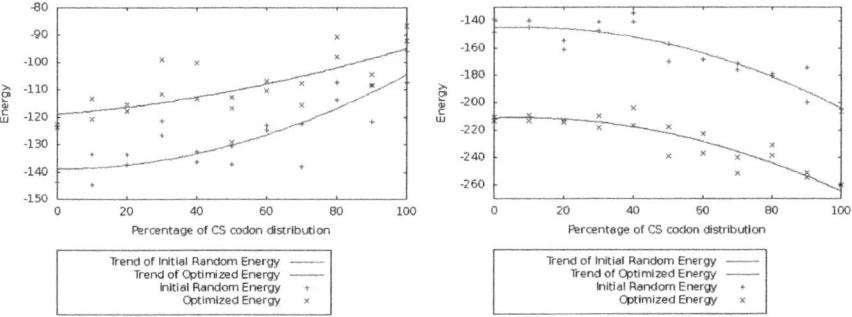

Fig. 8. Plot of Initial and Optimized Energy Curve of GFP RNA structure for different percentage of the wildtype and CS Codon Distribution. In the left figure (minimize structure), the lower curve shows trend of initial energy values; the leftmost point (0% CS codon distribution) maintains wildtype codon distribution, rightmost point maintains CS codon distribution, intermediate points maintain different ratios of codon distributions from wildtype and CS one. The upper plot shows corresponding optimized energies by RNAfoldMinimizeStructure Algorithm. In the right figure (maximize structure), the upper curve is for wildtype structure energy and lower curve indicates the trend of energy values after maximizing structures using RNAfoldMaximizeStructure algorithm.

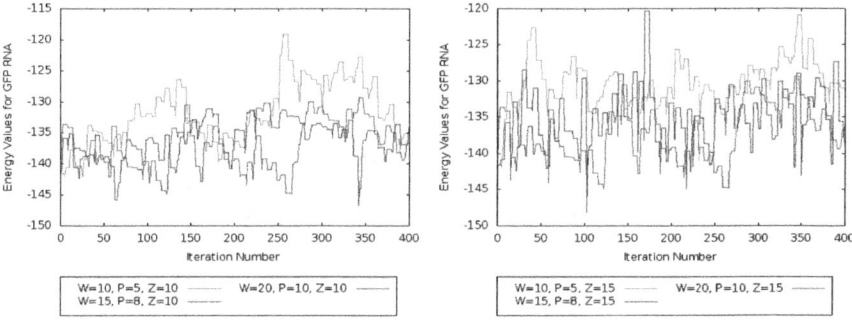

Fig. 9. Comparison plot of *region break* strategy for different $W = 10, 15, 20$ and $P = 5, 8, 10$ with $Z = 10, 15$, where W stands for sliding window size, P is the number of overlapping codons between two windows, Z indicates the percentage of the entire structure that will be broken each step. In the figure we see, there is no significant difference in the energy optimization path that was followed for different parameter values.

on the swap. Next we calculate energy from the updated structure information. As we are updating neighbor information from the changed structure based on local swap decisions without folding the structure, it might not predict the accurate structure energy. However, we figured out, the updated information is good enough to continue changing the structure even without calling Mfold up to five iterations.

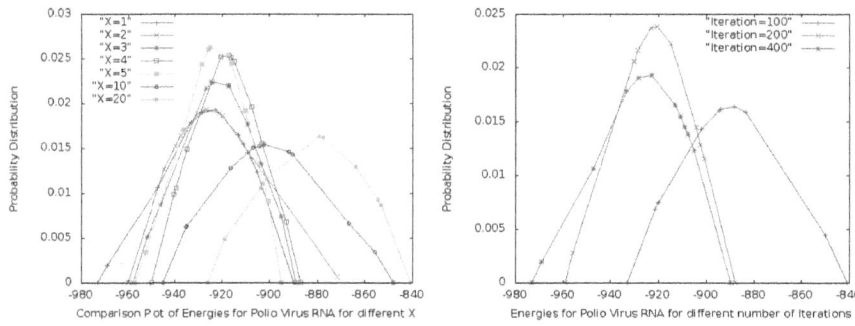

Fig. 10. Comparing results of the distribution of energy values for different X (left) for Polio Virus RNA; $X = 1, 2, 3, 4, 5, 10, 20$, and for different number of maximum iteration limit (right); $I = 100, 200, 400$. From the left figure, up to $X = 5$ (i.e. running Mfold every fifth iteration) performance of the algorithm is close to that for $X = 1$ (i.e. running Mfold every iteration), after that performance degrades. From the right one, energy optimization improves significantly when maximum iteration limit (I) goes from 100 to 200. When moving from $I = 200$ to 400, improvement is not that much.

- We checked whether optimizing the same initial wild type sequence for longer is better than starting from several initial starting positions. Suppose we want to make optimum use of total time T. We want to run the algorithm in S steps with maximum iterations I at each step. Now, if each iteration of the program takes t time, then we can run $\frac{T}{t}$ iterations (say $I = i$) starting from a single step $S = 1$. Alternately, we can run several steps $S = s$ ($s > 1$) with number of iterations $i\prime$ where $i\prime < i$. Here, $i \times t = s \times i\prime \times t$. For polio virus we found the algorithm finds the best result with approximately around 400 iterations. After that there is no improvement in the output energy. Hence, we increased the number of steps S keeping I fixed at 400 to get the optimized result. We see from the right part of Fig. 10, the difference in energy improvement from 100 to 200 iterations is much higher than that from 200 to 400 iterations. Here, we note that the maximum number of iterations to converge to an optimal solution is dependent on the length of the RNA sequence as longer the sequence there are more options for codon swap. In general, for a sequence shorter than polio virus RNA 400 iterations should be sufficient to converge to the optimal solution. For a longer one it might take more iterations to converge.

5 Conclusion

In this paper, we describe programs to find the minimum and maximum energy structures of a given amino acid sequence with codon constraints. We used two algorithmic variants: *random walk* and *gradient descent* approach. Simulated annealing or the Metropolis algorithm [32] could be another approach, where at every step we could take a backward move based on a probability value. Our *gradient descent* approach is similar but we wait for several steps to find a better move before taking a bad one as Mfold is slow.

We checked for several different RNA sequences. In most cases improvement toward minimum energy structure (max structure) is better than toward maximum energy structure (min structure). Finding the least stable structure is harder. For polio virus, our optimized max structure energy is -1003.2 $kcal/mol$ and min structure energy is -604.9 $kcal/mol$, where initial wild type energy was around -666.68 $kcal/mol$. We checked the output from CS max and min program, where they do not follow any codon constraint. The max structure energy for their program was -1056 $kcal/mol$ and min structure energy was -429.8 $kcal/mol$. We ran our program with the same codon distribution as used by CS and got max structure energy -1080.6 $kcal/mol$ and min structure energy -407.7 $kcal/mol$. This indicates our algorithm's performance is better than CS algorithm output, which is a clear indication of obtaining optimized structures.

Acknowledgement. Our work is available at: http://www.algorithm.cs.sunysb.edu/RNAdesign. We specially thank Yanqing Chen for assistance during the study. This work was partially supported by NIH Grant AI075219 and NSF Grants DBI-1060572 and IIS-1017181.

References

1. Zuker, M., Mathews, D.H., Turner, D.H.: Algorithms and thermodynamics for rna secondary structure prediction (1999)
2. Jaeger, J., Turner, D.H., Zuker, M.: Improved predictions of secondary structures for rna. Proc. Natl. Acad. Sci. USA 86, 7706–7710 (1989)
3. Tinoco, I.J., Borer, P., Dengler, B., Levin, M., Uhlenbeck, O., Crothers, D., Bralla, J.: Improved estimation of secondary structure in ribonucleic acids. Nat. New Biol. 246, 40–41 (1973)
4. Hofacker, I.L., Fontana, W., Stadler, P.F., Bonhoeffer, L.S.: Fast folding and comparison of rna secondary structures. Monatshefte fur Chemie 125, 167–188 (1994)
5. Cohen, B., Skiena, S.: Natural selection and algorithmic design of mRNA. J. Computational Biology 10, 419–432 (2003)
6. Cohen, B., Skiena, S.: Optimizing rna secondary structure over all possible encodings of a given protein. In: RECOMB (2000)
7. Kudla, G., Murray, A., Tollervey, D., Plotkin, J.: Coding-sequence determinants of gene expression in escherichia coli. Science 324, 255–258 (2009)
8. Plotkin, J., Kudla, G.: Synonymous but not the same: the causes and consequences of codon bias. Nature Reviews Genetics 12, 32–42 (2011)
9. Tuller, T., Waldman, Y.Y., Kupiec, M., Ruppin, E.: Translation efficiency is determined by both codon bias and folding energy. Natl. Acad. Sci. USA
10. Sitaraman, V., Hearing, P., Ward, C., Gnatenko, D., Wimmer, E., Mueller, S., Skiena, S., Bahou, W.: Computationally designed adeno-associated virus (aav) rep 78 is efficiently maintained within an adenovirus vector. Proc. National Academy of Sciences 108, 14294–14299 (2011)
11. Mueller, S., Coleman, R., Papamichail, D., Ward, C., Nimnual, A., Futcher, B., Skiena, S., Wimmer, E.: Live attenuated influenza vaccines by computer-aided rational design. Nature Biotechnology 28 (2010)

12. Coleman, J., Papamichial, D., Futcher, B., Skiena, S., Mueller, S., Wimmer, E.: Virus attenuation by genome-scale changes in codon-pair bias. Science 320, 1784–1787 (2008)
13. Zuker, M.: Mfold web server for nucleic acid folding and hybridization prediction. Nucleic Acids Res. 31, 3406–3415 (2003)
14. Mathews, D., Sabina, J., Zuker, M., Turner, D.: Expanded sequence dependence of thermodynamic parameters improves prediction of rna secondary structure. J. Mol. Biol. 288, 911–940 (1999)
15. Zuker, M., Stiegler, P.: Optimal computer folding of large rna sequences using thermodynamics and auxiliary information. Nucleic Acids Res., 133–148 (1981)
16. McCaskill, J.: The equilibrium partition function and base pair binding probabilities for rna secondary structure. Biopolymers 29, 1105–1119 (1990)
17. Wuchty, S., Fontana, W., L.H. I., Schuster, P.: Complete suboptimal folding of rna and the stability of secondary structures. Biopolymers 49, 145–165 (1999)
18. Shapiro, B.A., Zhang, K.: Comparing multiple rna secondary structures using tree comparisons. Comput. Appl. Biosci. 6, 309–318 (1990)
19. Hofacker, I.L.: The rules of the evolutionary game for rna: A statistical characterization of the sequence to structure mapping in rna. PhD thesis
20. Dromi, N., Avihoo, A., Barash, D.: Reconstruction of natural rna sequences from rna shape, thermodynamic stability, mutational robustness, and linguistic complexity by evolutionary computation. Biomol. Struct. Dyn. 26, 147–162 (2008)
21. Dehiyat, B.I., Mayo, S.L.: De novo protein design: Fully automated sequence selection. Science 278, 82–87 (1997)
22. Busch, A., Backofen, R.: Info-rna: a server for fast inverse rna folding satisfying sequence constraints. Nucleic Acids Res. (2007)
23. Avihoo, A., Churkin, A., Barash, D.: Rnaexinv: An extended inverse rna folding from shape and physical attributes to sequences. BMC Bioinformatics
24. Durbin, R., Eddy, S., Krogh, A., Mitchison, G.: Biological sequence analysis: Probabilistic models of proteins and nucleic acids. Cambridge University Press (1998)
25. Knudsen, B., Hein, J.: Rna secondary structure prediction using stochastic context-free grammars and evolutionary history. Bioinformatics 15, 446–454 (1999)
26. Dowell, R., Eddy, S.: Evaluation of several lightweight stochastic context-free grammars for rna secondary structure prediction. BMC Bioinformatics 5 (2004)
27. Do, C.B., Woods, D.A., Batzoglou, S.: Contrafold: Rna secondary structure prediction without physics-based models. Bioinformatics 22, 90–98 (2006)
28. Andronescu, M., Condon, A., Hoos, H.H., Mathews, D.H., Murphy, K.P.: Efficient parameter estimation for rna secondary structure prediction. Bioinformatics 23 (2007)
29. Andronescu, M.: Computational approaches for rna energy parameter estimation. PhD thesis, University of British Columbia, Vancouver, Canada
30. Zakov, S., Goldberg, Y., Elhadad, M., Ziv-Ukelson, M.: Rich Parameterization Improves RNA Structure Prediction. In: Bafna, V., Sahinalp, S.C. (eds.) RECOMB 2011. LNCS, vol. 6577, pp. 546–562. Springer, Heidelberg (2011)
31. Doshi, K.J., Cannone, J.J., Cobaugh, C.W., Gutell, R.R.: Evaluation of the suitability of free-energy minimization using nearest-neighbor energy parameters for rna secondary structure prediction. BMC Bioinformatics 5, 105 (2004)
32. Metropolis, N., Rosenbluth, A.W., Rosenbluth, M.N., Teller, A.H., Teller, E.: Equation of state by fast computing machines. The Journal of Chemical Physics 21, 1087–1092 (1953)

Phylogenetic Tree Reconstruction with Protein Linkage

Junjie Yu[1], Henry Chi Ming Leung[1], Siu Ming Yiu[1], Yong Zhang[1],
Francis Y.L. Chin[1], Nathan Hobbs[2], and Amy Y.X. Wang[2]

[1] Department of Computer Science, The University of Hong Kong
{jjyu,cmleung2,smyiu,yzhang,chin}@cs.hku.hk
[2] Institute for Interdisciplinary Information Sciences, Tsinghua University
marscheese@gmail.com, amywang@mail.tsinghua.edu.cn

Abstract. When reconstructing a phylogenetic tree, one common representation for a species is a binary string indicating the existence of some selected genes/proteins. Up until now, all existing methods have assumed the existence of these genes/proteins to be independent. However, in most cases, this assumption is not valid. In this paper, we consider the reconstruction problem by taking into account the dependency of proteins, i.e. protein linkage. We assume that the tree structure and leaf sequences are given, so we need only to find an optimal assignment to the ancestral nodes. We prove that the Phylogenetic Tree Reconstruction with Protein Linkage (*PTRPL*) problem for three different versions of linkage distance is NP-complete. We provide an efficient dynamic programming algorithm to solve the general problem in $O(4^m \cdot n)$ and $O(4^m \cdot (m+n))$ time (compared to the straight-forward $O(4^m \cdot m \cdot n)$ and $O(4^m \cdot m^2 \cdot n)$ time algorithm), depending on the versions of linkage distance used, where n stands for the number of species and m for the number of proteins, i.e. length of binary string. We also argue, by experiments, that trees with higher accuracy can be constructed by using linkage information than by using only hamming distance to measure the differences between the binary strings, thus validating the significance of linkage information.

Keywords: Phylogenetic tree reconstruction, Protein linkage, NP-complete.

1 Introduction

Discovering evolutionary relationships among species is an important problem in bioinformatics. Given a set of species, the phylogenetic tree reconstruction problem is to reconstruct an evolutionary tree where each leaf of the tree represents an input species and each internal node u represents a hypothetical ancestor of the species represented by all leaf nodes in the subtree rooted at u. There are two common representations of the species in this problem: a DNA (or protein) sequence [1] of some selected genes (or proteins) which are believed to be relevant to the evolution process, or a binary string [2] that indicates the existence/absence of those selected genes (or proteins) in the species. Based on these representations, there are three general approaches for the reconstruction of a phylogenetic tree: (1) distance methods, (2) parsimony methods, and (3) likelihood methods.

Distance methods first define an evolutionary distance measure between two DNA sequences (or two binary strings) that represent the two species being compared. A distance matrix that captures the evolutionary distance between each pair of species is used as an input to the phylogentic tree construction problem. The distance method then infers a phylogenetic tree from the distance matrix by grouping closer species together in a subtree. The most well-known distance method is the Neighbor-joining (NJ) algorithm [3], which was later improved in terms of time complexity by the Fast Neighbor Joining (FNJ) algorithm [4]. Other distance methods such as ProfDists [5], Profile Neighbor Joining [6], and Weighbor [7], are efficient algorithms with polynomial time complexities, but they lack robustness. Moreover, the effectiveness of these methods relies very much on the evolutionary distance measure being used.

Maximum parsimony methods aim to find a phylogenetic tree which can explain the given sequences with a minimum number of substitutions. This problem is known to be NP-hard [8]. Heuristic methods, such as the genetic algorithm [9], tabu search [10], simulated annealing [11], and hill climbing [12] were introduced to find near optimal results. If the tree structure is known (usually referred to as the "small parsimony problem"), the problem can be solved in polynomial time using the Fitch-Hartigan algorithm [13] or Sankoff algorithm [14].

Likelihood methods (sometimes called probabilistic methods) are based on evolutionary models, such as the gene evolution model [15], Jukes-Cantor's one-parameter model, or Kimura's two-parameters model [16]. Maximum likelihood methods [17-18] and Bayesian methods [19] are examples of probabilistic methods that find phylogenetic trees by maximizing the likelihood (or posterior probability) of generating the observed sequences. Probabilistic methods and parsimony methods are usually more robust but are more expensive computationally.

There are other methods for which the input is a set of small subtrees, each of which captures the evolutionary relationship of three (called triplets) or four (called quartets) species among the given set of species. The problem is to combine these subtrees into a single phylogenetic tree such that the evolutionary relationship of the species is consistent with the small subtrees. However, this problem has been shown to be difficult. Even when all given quartets are consistent and can be combined to form a phylogenetic tree, finding such a tree is still NP-complete [20]. There are heuristic methods [21-22] and PTAS [23-24] algorithms for the case when all $\binom{n}{4}$ quartets are given. Another issue for this approach is that the availability of the triplets and quartets is usually limited. In this paper, we do not consider this type of input but focus on the representation of using a binary string to indicate the existence of selected genes/proteins.

In distance methods and maximum parsimony methods where binary strings are input, a common assumption is that each bit of the binary string is independent, i.e. the existence of a particular gene/protein in a species is independent of the existence of another gene/protein in the same species. This assumption is obviously an oversimplification because genes/proteins in real life usually work together with other genes/proteins in various ways, e.g. in metabolic pathways [25], protein complexes [26], and signal transduction pathways [27]. Thus, the existence of different genes in a species may be dependent [28], and when reconstructing a phylogenetic tree, this kind of dependence should be considered.

There is strong evidence [28] that some proteins/genes should co-exist, which is referred to as protein linkage. From an evolutionary point of view, these functionally dependent proteins should usually be present (or absent if the function is no longer needed) in the same generation. Based on the principle of parsimony, in addition to finding a phylogenetic tree with the minimum number of insertions/deletions (hamming distance), it is not desirable to have some proteins present and others absent if they belong to the same functional group. To capture this biological property, we introduce a new model for calculating the cost of a phylogenetic tree by incorporating protein linkage information using the parsimony approach. Given a tree topology, we define the *phylogenetic tree reconstruction with protein linkage problem* (the *PTRPL* problem) as finding an assignment of internal nodes which minimize the total number of protein insertions/deletions (hamming distance) and different versions of linkage cost. We show that the *PTRPL* problem under three different versions of linkage costs resolve to be NP-complete.

Although the versions of the problem we consider are NP-complete, we believe that linkage information is important for evolutionary studies. To further this area of research, we provide an efficient dynamic programming algorithm to find an optimal assignment in $O(4^m \cdot n)$ or $O(4^m \cdot (m+n))$ time when the tree topology is given (compared to the straight-forward brute-force $O(4^m \cdot m \cdot n)$ or $O(4^m \cdot m^2 \cdot n)$ time algorithms), where n is the number of species and m is the length of the binary string for each species. We demonstrate the effectiveness of our algorithm on real biological data that contain protein and protein linkage information. We incorporate our algorithm into the Nearest Neighbor Interchange algorithm [33]. For every step of the Nearest Neighbor Interchange, based on the current topology, we compute the minimum cost using our approach. The results show that we are able to reconstruct more accurate phylogenetic trees (with 11% higher in accuracy) than other existing approaches that ignore linkage information. The running time of our algorithm is acceptable for some practical sizes of m and is about 5 – 35 times faster than the brute-force algorithm according to our experimental results.

2 Different Versions of Linkage Distance

The cost of each evolution step depends on the number of protein insertions/deletions (hamming distance) and the degree of protein association (protein linkage). Given two length m binary strings, u and v, that represent the existence of m proteins in two species, the cost between u and v is the sum of the hamming distance and the protein linkage distance (described in the following subsections) between u and v. Given a tree T with n leaf nodes, each labeled with a binary string of length m, and the protein linkage information of these m proteins, the *PTRPL* problem is to assign length-m binary strings to the internal nodes of T that minimize the total cost of the tree T, i.e. the sum of the cost of all edges in T. We define different versions of linkage distance according to various assumptions about evolution and show that the *PTRPL* problem is NP-complete under each of the versions of linkage distance we define.

2.1 Whole Block Linkage Distance

Whole block linkage distance assumes that all proteins should be present or absent in the species as a block (consistent state) during evolution. From this presence or absence assumption, blocks are usually disjoint, i.e. each protein normally belongs to at most one block. The linkage cost will be w if the species evolves from a consistent state to an inconsistent one and zero otherwise. For all $\sigma, \sigma' \in \{0,1\}^m$, and $w > 0$,

$$l_{wb}(\sigma, \sigma') = \begin{cases} w & \sigma \in \{0^m, 1^m\} \text{ and } \sigma' \notin \{0^m, 1^m\} \\ 0 & else \end{cases}$$

The NP-complete proof is by reduction from the NP-complete NOT-ALL-EQUAL 3SAT problem [32], defined as: given a Boolean formula $\phi = C_1 \wedge C_2 \wedge \cdots \wedge C_t = (x_{11} \vee x_{12} \vee x_{13}) \wedge (x_{21} \vee x_{22} \vee x_{23}) \wedge \cdots \wedge (x_{t1} \vee x_{t2} \vee x_{t3})$ over t clauses and k variables, the problem is to find a truth assignment for the variables such that each clause C_i of ϕ has at least one true literal and one false literal, $1 \leq i \leq t$.

The NP-complete proof is to transform ϕ into an instance of the *PTRPL* problem such that ϕ is satisfiable if and only if the *PTRPL* instance has an internal node assignment that makes total distance of the tree at most $(2k+15)t + (2k+2)k$.

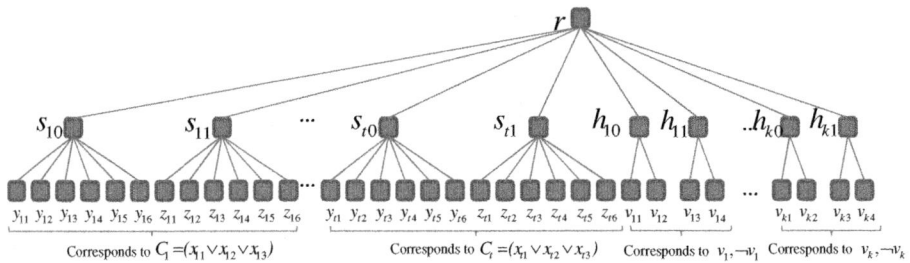

Fig. 1. PTRPL instance from the reduction of NOT-ALL-EQUAL 3SAT problem

The *PTRPL* instance consists of a 3-level tree T (Figure 1) with each node represented by a length-$2k$ binary string, one bit for each variable or its negation, where all $2k$ bits are considered to be one block. Root r has $(2t + 2k)$ children with s_{i0} and s_{i1} corresponding to clause C_i in ϕ, $1 \leq i \leq t$ (satisfaction testing), and h_{j0} and h_{j1} corresponding to variable v_j, $1 \leq j \leq k$ (truth setting). Node s_{i0} (together with s_{i1} representing C_i) has 6 children (leaf nodes) whose binary strings represent 6 different satisfiable assignments (100, 010, 001, 110, 101, 011) on its 3 literals x_{i1}, x_{i2}, x_{i3} at the corresponding bit positions of the $2k$-binary string, and 0 at all the other $(2k-3)$ bits. The binary strings for the children of s_{i1} are similar to that of s_{i0}, except that all the other $(2k - 3)$ bits are 1s (Figure 2). Thus, in order to have the total minimum hamming distance $(2k-3)$ and 18 from s_{i0} and s_{i1} to r and their leaf nodes respectively, $s_{i0}(s_{i1})$ would have 0's (1's) at these $(2k-3)$ bit positions and the same bit assignments as root r at the remaining three bit positions (for clause satisfaction). Similarly, both h_{j0} and h_{j1} each have 2 children that encode the truth assignment for v_j and $\neg v_j$ by having (10, 01) at the corresponding two bit positions of the 2k-binary string, with all other $(2k-2)$ bits assigned 0 or 1 for the children of h_{j0} or h_{j1} respectively. Again, h_{j0} or h_{j1} would have 0 or 1 at the $(2k-2)$ bit positions and the same bit assignments as root r at the two bit positions for the truth assignments of v_j and $\neg v_j$. When the internal nodes are assigned in this way, the total hamming distance is minimum, i.e. $(2k+15)t + (2k+2)k$. In order to have zero linkage distance, none of the internal nodes can be 0^{2k} or 1^{2k}, i.e. both h_{j0} and h_{j1} ensure that either v_j or $\neg v_j$ (but not both) is assigned 1, and s_{i0} (resp. s_{i1}) ensure that not all the literals in clause C_i are assigned 0's (resp. 1's) (the NOT-ALL-EQUAL 3SAT problem). The binary string of root r would contain an encoding of the truth assignment of each variable, and each internal node (with the exception of r) would have the same binary string as one of its children, i.e. the clause C_i is satisfiable iff ϕ is satisfiable.

	Bit representing literal x_{i1}				Bit representing literal x_{i2}				Bit representing literal x_{i3}			
$y_{i,1}$:	0	...	0	1	0	...	0	0	0	...	0	0	0	...	0
$y_{i,2}$:	0	...	0	0	0	...	0	1	0	...	0	0	0	...	0
$y_{i,3}$:	0	...	0	0	0	...	0	0	0	...	0	1	0	...	0
$y_{i,4}$:	0	...	0	1	0	...	0	1	0	...	0	0	0	...	0
$y_{i,5}$:	0	...	0	1	0	...	0	0	0	...	0	1	0	...	0
$y_{i,6}$:	0	...	0	0	0	...	0	1	0	...	0	1	0	...	0
$z_{i,1}$:	1	...	1	1	1	...	1	0	1	...	1	0	1	...	1
$z_{i,2}$:	1	...	1	0	1	...	1	1	1	...	1	0	1	..	1
$z_{i,3}$:	1	...	1	0	1	...	1	0	1	...	1	1	1	...	1
$z_{i,4}$:	1	...	1	1	1	...	1	1	1	...	1	0	1	...	1
$z_{i,5}$:	1	...	1	1	1	...	1	0	1	...	1	1	1	...	1
$z_{i,6}$:	1	...	1	0	1	...	1	1	1	...	1	1	1	...	1

	Bit representing literal v_j				Bit representing literal $\neg v_j$			
$v_{j,1}$:	0	...	0	1	0	...	0	0	0	...	0
$v_{j,2}$:	0	...	0	0	0	...	0	1	0	...	0
$v_{j,3}$:	1	...	1	1	1	...	1	0	1	...	1
$v_{j,4}$:	1	...	1	0	1	...	1	1	1	...	1

Fig. 2. The binary string for leaf nodes

2.2 Partial Block Linkage Distance

The only difference between the whole block and partial block linkage distance is the addition of a cost when a binary string evolves from an inconsistent state to another inconsistent state (as a state transition of this kind does not make any progress in evolution). For all $\sigma, \sigma' \in \{0,1\}^m$, $w > 0$ and $0 < \delta \leq w$:

$$l_{pb}(\sigma,\sigma') = \begin{cases} 0 & \sigma' \in \{\sigma, 0^m, 1^m\} \\ \delta & \sigma' \notin \{\sigma, 0^m, 1^m\} \text{ and } \sigma \notin \{0^m, 1^m\} \\ w & \sigma' \notin \{0^m, 1^m\} \text{ and } \sigma \in \{0^m, 1^m\} \end{cases}$$

The *PTRPL* problem is also NP-complete under partial block linkage distance. However, the construction of the problem instance from ϕ as shown in the NP-complete proof for the whole block linkage distance might not work for partial block linkage distance, because some decreases in linkage distance might compensate for increases in hamming distance and the resulting assignment might not have to achieve a minimum hamming distance. For example, it is possible that the binary string at r adopts one of its children's binary strings, so as to reduce one linkage distance at the expense of its hamming distances with other children. In order to prevent such scenario, one can extend each binary string by $2(k+t)A$ bits with a different section of A 1-bits, corresponding to each child of r: s_{i0}, s_{i1}, h_{j0} and h_{j1}, $1 \leq i \leq t$ and $1 \leq j \leq k$ (and all other bits 0's). We determine the value of A such that hamming distance will dominate the linkage distance by having $A > w(14t + 6k)$ where the total number of tree edges is $14t + 6k$. For example, the last $2(k+t)A$ bits of s_{10}'s children are $1^A 0^A \cdots 0^A$; the last $2(k+t)A$ bits of s_{11}'s children are $0^A 1^A 0^A \cdots 0^A$; $0^A 0^A 1^A 0^A \cdots 0^A$ for s_{20}'s children, etc. In this case, the root would have an assignment of 0's as its last $2(k+t)A$ bits in order to minimize the total hamming distance to all its $2(k+t)$ children. With this assignment of the internal nodes, the total cost (hamming distance plus linkage distance) will be minimum iff ϕ is satisfiable. Formally, the hamming distance would be $(2k+15)t + (2k+2)k + (2t+2k)A$, and the linkage distance would be $[2(k+t) + 10k + 2t]\delta$.

Note that after the extension of the $2(k+t)A$ bits, the minimum hamming distance can still be achieved if the first $2k$-bits of some internal nodes are assigned 0's or 1's (i.e. some clauses are not satisfiable with the same true/false assignment for all 3 literals or some literals have the same truth assignment as their negations). However, assignments of this kind are avoided because the linkage distance between internal nodes to their children would be increased by at least δ.

2.3 Pairwise Linkage Distance

Pairwise linkage distance is defined through a set P of protein pairs, where two proteins in a protein pair are expected to be present/absent at the same time. In contrast to block linkage, we assume that each protein can belong to more than one pairwise linkage. Note that the whole block and partial block linkage distance can be partially modeled by pairwise linkage distance by having all proteins in a block paired to each other.

Let P be a set of protein pairs (p,q), where p and q represent two bit positions in the binary string, the linkage distance $= \sum_{(p,q) \in P} l_p(\sigma_{pq}, \sigma'_{pq})$ where σ_{pq} (σ'_{pq}) = the two bits at positions p and q of binary string $\sigma(\sigma')$, with $w > 0$, and $0 \leq \delta \leq w$,

$$l_p(\sigma_{pq}, \sigma'_{pq}) = \begin{cases} 0 & \sigma'_{pq} \in \{\sigma_{pq}, 00, 11\} \\ \delta & (\sigma_{pq}, \sigma'_{pq}) \in \{(01,10), (10,01)\} \\ w & \sigma_{pq} \in \{00, 11\} \text{ and } \sigma'_{pq} \in \{01, 10\} \end{cases}$$

The NP-complete proof is by reduction from the NP-complete (3, 4)-SAT problem [29-30], which is a special case of the 3SAT problem, in which each variable appears exactly four times in ϕ.

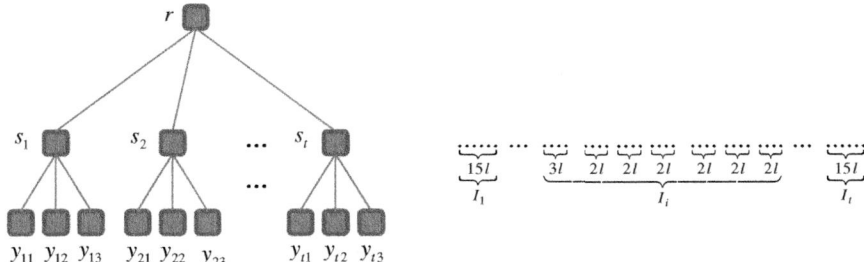

Fig. 3. 3-level tree for reduction   **Fig. 4.** Binary string of a leaf node

Given a formula ϕ with t clauses, we construct a 3-level tree T (Figure 3) for the PTRPL instance where root r has t children s_i, corresponding to clause C_i in ϕ, $1 \leq i \leq t$, and s_i has three children y_{i1}, y_{i2}, y_{i3}, corresponding to literals x_{i1}, x_{i2}, x_{i3} in the clause $C_i = (x_{i1} \lor x_{i2} \lor x_{i3})$.

Each leaf node is represented by a length-$15lt$ binary string that consists of t length-$15l$ sections with the i-th section denoted by I_i, corresponding to clause C_i, $1 \leq i \leq t$ (Figure 4). Note that we determine the value of l such that the linkage distance dominates the hamming distance; formally, $l > [(5t + 3) \cdot t/(4w)]$. The first $3l$ bits in I_i are a dummy section denoted by $I_i(dummy)$, used to balance linkage distance to guarantee strings assigned to s_i same as those of one of its three leaf children when ϕ is satisfiable. The other 6 length-$2l$ literal sections represent the literals or their negations in clause C_i; they are denoted by $I_i(x_{i1})$, $I_i(x_{i2})$, $I_i(x_{i3})$, $I_i(\neg x_{i1})$, $I_i(\neg x_{i2})$, $I_i(\neg x_{i3})$. For leaf node y_{i1} (resp. y_{i2} and y_{i3}), only the length-$3l$ $I_i(dummy)$ section and length-$2l$ $I_i(x_{i1})$ (resp. $I_i(x_{i2})$ and $I_i(x_{i3})$) are assigned 1's; the remaining $(15lt - (3l + 2l))$ bits are assigned 0's. P is defined in such a way so as to enforce satisfaction testing of each clause and consistent truth testing for variables. Set P contains three types of protein pairs: *T-1*, *T-2*, and *T-3*:

1) *T-1* protein linkage: between each bit of $I_i(dummy)$ to each bit of $I_i(x_{i1})$, $I_i(x_{i2})$, $I_i(x_{i3})$, $I_i(\neg x_{i1})$, $I_i(\neg x_{i2})$, $I_i(\neg x_{i3})$; $1 \leq i \leq t$.
2) *T-2* protein linkage: between each pair of bit for sections $I_i(x_{i1})$ and $I_i(x_{i2})$, $I_i(x_{i1})$ and $I_i(x_{i3})$, $I_i(x_{i2})$ and $I_i(x_{i3})$; $1 \leq i \leq t$.
3) *T-3* protein linkage: between each bit in section $I_i(v)$ and each bit in section $I_j(\neg v)$ where literals v or $\neg v$ occur in clause C_i, C_j; $1 \leq i, j \leq t, i \neq j$.

Since each variable appears exactly four times, there are an equal number of *T-1* and *T-2* linkages. When the strings for section I_i of r and s_i are the same and the remaining sections of s_i are all 0's, no *T-1* or *T-2* linkage costs occur between r and s_i. The subsections that represent variable v and its negation $\neg v$ in the two sections of r that correspond to the two different clauses cannot be assigned 1's at the same time; otherwise, the *T-3* linkage distance between r to its children would increase by at least $4wl^2$. Thus, when formula ϕ is satisfiable, s_i will have the same assignment as one of its leaf children and no *T-1*, *T-2* and *T-3* linkage costs will exist between r and s_i. When formula ϕ is not satisfiable, either some clause C_i cannot be satisfied, or some variable and its negation are both assigned 1s at the same time. In either case, the total linkage distance will increase by at least $4wl^2$. Therefore, we can prove that the total cost of the resulting tree of the *PTRPL* instance is at most $(5t+3) \cdot lt + 4(11w + 2\delta)l^2 t$ if and only if the formula in the *(3, 4)-SAT* problem is satisfiable, where t is the number of clauses in the *(3, 4)-SAT* problem, $l > [(5t+3) \cdot t/(4w)]$, $w > 0$ and $0 \leq \delta \leq w$.

3 General Algorithm for the *PTRPL* Problem

In this section, we introduce an algorithm for solving the *PTRPL* problem through the use of dynamic programming.

Similar to Sankoff algorithm [14] which considers each protein independently, we define $C(u, \sigma)$ to be the minimum total cost (including protein linkage distance) of the subtree rooted at node u when node u is assigned string σ, $\sigma \in \{0,1\}^m$, m is the maximum number of proteins being considered, i.e. the length of the binary strings. The value of $min_{\sigma \in \{0,1\}^m} C(r, \sigma)$, where r is the root of the known leaf-labeled phylogenetic tree T, gives the minimum cost of tree T. There are two basic cases for a leaf node u: $C(u, \sigma) = 0$ if u is represented by binary string σ, and $C(u, \sigma) = \infty$ if u is not represented by σ. The value of $C(u, \sigma)$ for all internal nodes can be calculated by the following recursion:

$$C(u, \sigma) = \sum_{v \in children\ of\ u} \min_{\sigma' \in \{0,1\}^m} \{d(\sigma, \sigma') + C(v, \sigma')\},$$

where $d(\sigma, \sigma')$ is sum of hamming distance and linkage distance between σ and σ'. Since there are at most $2^m \cdot 2^m = 4^m$ combinations of σ and σ' to consider, and each $d(\sigma, \sigma')$ takes $O(m)$ time to calculate the hamming distance plus whole/partial block linkage distance, and $O(m^2)$ time to calculate the hamming distance plus pairwise linkage distance, the total takes at most $O(4^m \cdot m \cdot n)$ or $O(4^m \cdot m^2 \cdot n)$ time for a tree with $O(n)$ edges. However, we can optimize the algorithm by precomputing $d(\sigma, \sigma')$ ahead of time for all combinations of σ and σ'. Each $d(\sigma, \sigma')$ can be calculated in $O(1)$ or $O(m)$ time depending on block and pairwise linkage, respectively (by considering each σ' with a particular σ one by one according to their gray code such that two successive binary strings differ by one bit, after exhausting all 2^m different values of σ', the next σ is considered according to its gray code), as each protein can relate to at most one block (resp. *m* pairs) of linkage(s), the values at $d(\sigma,\sigma')$ for all pairs of σ and σ' will take no more than $O(4^m)$ (resp. $O(4^m \cdot m)$) time. As the value of $d(\sigma,\sigma') + C(v,\sigma')$ for each child v of u can be computed independently, we can perform the calculation edge by edge along with the values of $C(u,\sigma)$ for all internal nodes in $O(4^m \cdot n)$ time, where n is the number of leaf nodes. The time complexity for finding $min_{\sigma \in \{0,1\}^m} C(r,\sigma)$ is $O(4^m \cdot n)$ and $O(4^m \cdot (m+n))$ for block and pairwise linkage respectively. Algorithm 1 gives the pseudocode of the algorithm.

Algorithm 1: *PTRPL* problem

Input: Phylogentic tree T with n leaf nodes, each leaf node u labeled with a binary sequence σ_u of length m.

Output: Minimum total cost (hamming distance and linkage distance) of T.

1: Pre-compute $d(\sigma,\sigma')$, for all pair $\sigma,\sigma' \in \{0,1\}^m$

2: Calculating the value of $min_{\sigma \in \{0,1\}^m} C(r,\sigma)$, r is the root of T, by the following recursion:

$$C(u,\sigma) = \begin{cases} 0 & u \text{ is a leaf and } \sigma = \sigma_u \\ \infty & u \text{ is a leaf and } \sigma \neq \sigma_u \\ \sum_{v \in u's \text{ children}} min_{\sigma' \in \{0,1\}^m} \{d(\sigma,\sigma') + C(v,\sigma')\} & u \text{ is not a leaf} \end{cases}$$

---

4 Experiments and Results

In order to evaluate the significance of the use of linkage information in constructing a phylogenetic tree, we downloaded 10 phylogenetic trees from NCBI for experimentation. Each tree contains 10 species: 9 of them bacteria in 3 to 4 different genera and 1 Saccharomyces cerevisiae (yeast) working as an outlier. A total of 10 orthology groups were selected for each tree such that each species is represented by a length-10 binary string with a particular position in the binary string assigned "1" if and only if the species has a protein from the corresponding orthology group. Two orthology groups (positions) are considered as having a pairwise linkage if and only if the two orthology groups are either both present or both absent in at least 8 species. A set of orthology groups are considered to be in the same block if all pairs of orthology groups in the set have pairwise linkage with each other.

Nearest Neighbor Interchange algorithm [33] is applied to reconstruct the phylogenetic tree for the ten species based on different cost functions, including pairwise linkage distance, whole block linkage distance, partial block linkage distance and hamming distance. Given an initial tree topology, the Nearest Neighbor Interchange algorithm greedily updates the tree topology to minimize the total cost based on the corresponding cost function of the tree until no more improvement can occur. Since the initial tree topology has a dramatic effect on the final result, we randomly picked 100 initial trees for each experiment to minimize this effect. Among the 100 final results, the tree with the minimum total cost was considered as the predicted tree. Different values of w and δ are considered, and the resulting assignemnts is not sensitive to the values of w and δ. We use $w = 1$ for the experiments of the whole block linkage distance problem and $w = 1$, $\delta = 0.5$ for the partial block linkage distance and pairwise linkage distance problem. These parameters gave slightly better results than other parameters.

We calculate the number of triplets in each predicted tree that match the NCBI tree. A triplet of three species A, B and C, where species A and B closer to each other than species C in the predicted tree, is considered as a match with the NCBI tree if a) species A and B are closer to each other than species C in the NCBI tree, or b) species A, B and C have the same closest ancestor in the NCBI tree.

Table 1. Percentage of correct triplets for different cost functions

	Pairwise linkage distance	Whole block linkage distance	Partial block linkage distance	Hamming distance
Average number of correct triplets	88.8	82	81.3	75.5
Percentage of correct triplets	74%	68.3%	67.8%	62.9%
Average running time	30 seconds	5 seconds	5 seconds	1 second

From the results shown in Table 1, the trees we predict using both hamming distance and linkage distance information are more accurate than the trees we predict using hamming distance only. Our conclusion is that protein linkage provides information for the reconstruction of a more accurate phylogentic tree. When we compare the trees we construct based on whole block linkage distance and partial block linkage distance, we find that the trees we construct using pairwise linkage distance are more accurate.

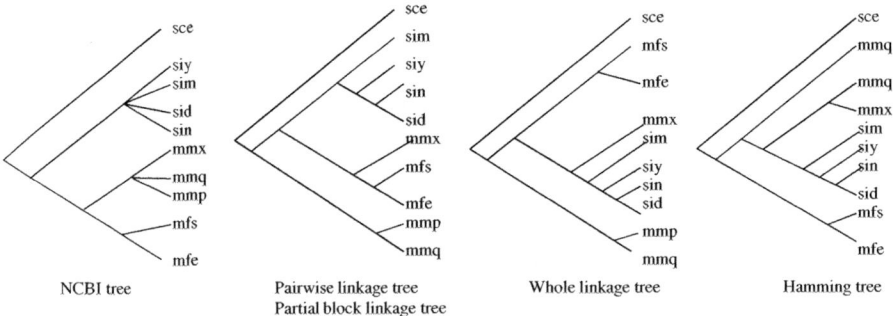

Fig. 5. NCBI tree topology and the predicted tree topologies based on pairwise linkage distance, whole block linkage distance, partial block linkage distance and hamming distance

Figure 5 shows an example of the NCBI tree topology and the predicted tree topologies based on pairwise linkage distance, whole block linkage distance, partial block linkage distance and hamming distance for a set of species (1 Saccharomyces, 2 Methanocaldococcus, 3 Methanococcus and 4 Sulfolobus). The NCBI tree and the predicted tree based on pairwise linkage and partial block linkage distance are similar except that species Methanococcus maripaludis C6 (mmx) in the Methanococcus is misclassified as Methanocaldococcus (70.8% of triplets are correct). The predicted tree based on whole block linkage distance also resembles the NCBI tree except that species mmx in the Methanococcus is misclassfied as Sulfolobus (65.8% of triplets are correct). However, the predicted tree based on hamming distance merges the Methanococcus and Sulfolobus incorrectly and is quite different from the NCBI tree (only 63.3% of triplets are correct).

All experiments were performed on a server machine with eight 2.4 GHZ CPUs and 140G memory. However, only one CPU is used in our experiments and the memory consumption for the program is less than 80G. The running times of the Neighbor Interchange algorithm based solely on hamming distance, hamming distance plus pairwise linkage distance and hamming distance plus block linkage distance on one experiment takes 1 second, 30 seconds and 5 seconds for a single experiment, respectively. Although the algorithm using protein linkage information requires a longer running time, it is acceptable for phylogenetic trees with a small set of species.

5 Conclusion

We have proved the NP-completeness for Phylogenetic Tree Reconstruction with Protein Linkages (*PTRPL*) problem under three definitions of linkage distance. A general algorithm for the *PTRPL* problem is presented for different linkages with time complexity $O(4^m \cdot n)$ and $O(4^m \cdot (m+n))$ for two variations of linkage distance (block linkage and pairwise linkage).

Lastly, we conducted experiments to confirm our hypothesis that the use of linkage information could lead to more accurate phylogenetic trees. For future work, we will try to further evaluate the efficiency and effectiveness of our dynamic programming algorithm. In particular, we shall study the *PTRPL* problem when the phylogenetic tree structure is not given, aim at developing effective heuristics, explore the linkage structures of real world data and develop algorithms for different versions of linkage structures.

Acknowledgements. This work was supported by grant HKU 7116/08E, HKU 719709E, NSFC 11171086.

References

1. Wang, L.-S., Leebens-Mack, J., Wall, P.K., Beckmann, K., Pamphilis, C.W., Warnow, T.: The Impact of Multiple Protein Sequence Alignment on Phylogenetic Estimation. Computational Biology and Bioinformatics 8, 1108–1119 (2011)

2. Zhou, Y., Wang, R., Li, L., Xia, X., Sun, Z.: Inferring Functional Linkages between Proteins from Evolutionary Scenarios. Journal of Molecular Biology 359 (2006)
3. Saitou, N., Nei, M.: The neighbor-joining method: a new method for reconstructing phylogenetic trees. Molecular Biology and Evolution 4, 406–425 (1987)
4. Elias, I., Lagergren, J.: Fast neighbor joining. Theoretical Computer Science (2008)
5. Wolf, M., Ruderisch1, B., Dandekar1, T., Schultz1, J., Müller, T.: ProfDistS (profile-) distance based phylogeny on sequence-structure alignments. Bioinformatics 24 (2008)
6. Muller, T., Rahmann, S., Dandekar, T., Wolf, M.: Accurate and robust phylogeny estimation based on profile distances: a study of the Chlorophyceae (Chlorophyta). BMC (2004)
7. Bruno, W.J., et al.: Weighted Neighbor Joining: A Likelihood-Based Approach to Distance-Based Phylogeny Reconstruction. Molecular Biology and Evolution (2000)
8. Foulds, L.R., Graham, R.L.: The Steiner Problem in Phylogeny is NP-Complete. Advances in Applied Mathematics 3, 43–49 (1982)
9. Ribeiro, C.C., Vianna, D.S.: A hybrid genetic algorithm for the phylogeny problem using path-relinking as a progressive crossover strategy. International Transactions in Operational Research (2009)
10. Lin, Y.-M., Fang, S.-C., Thorne, J.L.: A tabu search algorithm for maximum parsimony phylogeny inference. European Journal of Operational Research 176 (2007)
11. Lin, Y.-M.: Tabu search and genetic algorithm for phylogeny inference (2008)
12. Swofford, D.L.: PAUP*. Phylogenetic Analysis Using Parsimony (*and Other Methods) Version 4 (1998)
13. Hartigan, J.A.: Minimum mutation fits to a given tree. Biometrics 29 (1973)
14. Sankoff, D.: Minimal Mutation Trees of Sequences. SIAM on Applied Mathematics (1975)
15. Arvestad, L., Lagergren, J., Sennblad, B.: The gene evolution model and computing its associated probabilities. J. ACM 56, 1–44 (2009)
16. Kimura, M.: A simple method for estimating evolutionary rates of base substitutions through comparative studies of nucleotide sequences. Journal of Molecular Evolution (1980)
17. Guindon, S., et al.: New Algorithms and Methods to Estimate Maximum-Likelihood Phylogenies: Assessing the Performance of PhyML 3.0. Systematic Biology (2010)
18. Guindon, S., Delsuc, F., Dufayard, J.F., Gascuel, O.: Estimating maximum likelihood phylogenies with PhyML. Methods Mol. Biol. 537, 113–137 (2009)
19. Ronquist, F., Huelsenbeck, J.P.: MrBayes 3: Bayesian phylogenetic inference under mixed models. Bioinformatics 19, 1572–1574 (2003)
20. Steel, M.: The complexity of reconstructing trees from qualitative characters and subtrees. Journal of Classification 9, 91–116 (1992)
21. Cilibrasi, R., Vitany, P.M.B.: A New Quartet Tree Heuristic for Hierarchical Clustering. Presented at the Theory of Evolutionary Algorithms, Dagstuhl, Germany (2006)
22. Schmidt, H.A., Strimmer, K., Vingron, M., Haeseler, A.: TREE-PUZZLE: maximum likelihood phylogenetic analysis using quartets and parallel computing. BMC 18 (2002)
23. Snir, S., Yuster, R.: Reconstructing approximate phylogenetic trees from quartet samples. In: The 21 Annual ACM-SIAM Symposium on Discrete Algorithms, Texas (2010)
24. Tao, J., Kearney, P., Li, M.: Orchestrating quartets: approximation and data correction. In: Proceedings of 39th Annual Symposium on Foundations of Computer Science (1998)
25. G. E. M. L. E., Lupo, P.: Gene-Gene Interactions in the Folate Metabolic Pathway and the Risk of Conotruncal Heart Defects. Journal of Biomedicine and Biotechnology (2010)

26. Pereira-Leal, J., Levy, E.D., Teichmann, S.A.: The origins and evolution of functional modules: lessons from protein complexes. Philosophical Transactions of the Royal Society of London. Series B, Biological Sciences 361, 507–517 (2006)
27. Lu, Y.-C., Yec, W.C., Ohashi, P.S.: LPS/TLR4 signal transduction pathway. Cytokine (2008)
28. Uetz, P., et al.: A comprehensive analysis of protein-protein interactions in Saccharomyces cerevisiae. Nature 403, 623–627 (2000)
29. Craig, T.: A simplified NP-complete satisfiability problem. Discrete Applied Mathematics (1984)
30. Berman, P., Alex, M.K., Scott, E.D.: Computational complexity of some restricted instances of 3-SAT. Discrete Applied Mathematics 155, 649–653 (2007)
31. Doran, R.W.: The Gray Code. Journal of Universal Computer Science 13 (2007)
32. Garey, M.R., Johnson, D.S.: Computers and Intractability: A Guide to the Theory of NP-Completeness. Series of Books in the Mathematical Sciences. W. H. Freeman (1979)
33. Day, W.H.E.: Properties of the nearest neighbor interchange metric for trees of small size. Journal of Theoretical Biology 101, 275–288 (1983)

Protein Structure Prediction and Clustering Using MUFOLD
(Invited Keynote Talk)

Jingfen Zhang and Dong Xu

Department of Computer Science and Christopher S. Bond Life Sciences Center,
University of Missouri, Columbia, MO, 65201, USA
{zhangjingf,xudong}@missouri.edu

Abstract. The 3D structure of protein holds the key in understanding its biological function at the molecular level. Knowledge of protein structure also allows researchers to identify and characterize disease targets and provides a rational approach to drug design. In contrast to high-throughput DNA sequencing, structure determination is relative low-throughput because the experimental methods such as X-ray crystallography and NMR are costly, time-consuming and often technically difficult. Consequently, the gap between the number of known protein sequences and experimentally solved structures has been significantly widening. Computational prediction methods have been a very important alternative approach to protein structure solution, which can often generate useful structure models quickly at little cost.

There have been steady improvements in protein structure prediction during the past two decades. However, current methods are still far from consistently predicting structural models accurately with computing power accessible to common users. To address this challenge, we developed MUFOLD, a hybrid method of using whole and partial template information along with new computational techniques for protein tertiary structure prediction. MUFOLD covers both template-based and *ab initio* predictions using the same framework and aims to achieve high accuracy and fast computing. In MUFOLD, the prediction problem is formulated as a graph-realization problem, and an efficient global optimization approach of multi-dimensional scaling (MDS) is employed to solve it. In this framework, as models are generated efficiently based on predicted distance matrices from template fragment matches, deeper and broader information from PDB can be utilized and the quality of models can be improved with evaluation/refinement processes. MUFOLD speeds up the prediction dramatically over conventional methods. In addition, MUFOLD enhances the predictions consistently by iteratively using the information from generated models. MUFOLD has demonstrated its effectiveness in CASPs and various applications.

Current protein structure prediction methods, including MUFOLD, often generate a large population of candidates (models), and then select near-native models through clustering. Existing structural model clustering methods are time consuming due to pairwise distance calculation between models. To address this issue, we developed a novel method for fast model clustering without losing the clustering accuracy. Instead of the commonly used pairwise RMSD and TM-score values, we propose two new distance measures, Dscore1 and

Dscore2, based on the comparison of the protein distance matrices for describing the difference and the similarity among models, respectively. The analysis indicates that the correlation between Dscore1 and RMSD or between Dscore2 and TM-score is high. Our Dscore1-based clustering achieves a calculation time linearly proportional to the number of models while obtaining almost the same accuracy for near-native model selection in comparison to existing methods with calculation time quadratic to the number of models. By using Dscore2 to select representatives of clusters, we can further improve the quality of the representatives with little increase in computing time. Our method has been implemented in a package named MUFOLD-CL, available at http://mufold.org/clustering.php.

This work has been supported by National Institutes of Health Grant R21/R33-GM078601. Major computer time was provided by the University of Missouri Bioinformatics Consortium.

Keywords: Protein structure prediction, Multi-dimensional scaling, MUFOLD, CASP, Distance matrix, Dscore, Near-native model selection, Protein model clustering.

References

1. Zhang, J., Wang, Q., Barz, B., He, Z., Kosztin, I., Shang, Y., Xu, D.: MUFOLD: A New Solution for Protein 3D Structure Prediction. Proteins: Structure, Function, and Bioinformatics 78, 1137–1152 (2010)
2. Zhang, J., Wang, Q., Vantasin, K., Zhang, J., He, Z., Kosztin, I., Shang, Y., Xu, D.: A multilayer evaluation approach for protein structure prediction and model quality assessment. Proteins: Structure, Function, and Bioinformatics 79(S10), 172–184 (2011)
3. Zhang, J., Xu, D.: Fast Algorithm for Clustering a Large Number of Protein Structural Decoys. In: Proceedings of IEEE International Conference on Bioinformatics & Biomedicine (BIBM 2011), Atlanta, USA, pp. 30–36 (November 2011)

Computational Modeling of Mammalian Promoters
(Invited Keynote Talk)

Michael Q. Zhang[1,2]

[1] Departments of Molecular and Cell Biology & Biomedical Engineering and Center for Systems Biology, The University of Texas at Dallas, USA
[2] Bioinformatics Division, Center for Systems & Synthetic Biology TNLIST, Tsinghua University, China
michael.zhang@utdallas.edu

Gene promoters are important *cis*-regulatory elements that encompass the Transcriptional Start Site (TSS) and their recognition by RNA Polymerase II (RNAPII) is the prerequisite for transcriptional initiation - the first regulation step of gene expression. Understanding the structure and function of promoters is absolutely crucial for understanding gene regulation. Due to its complexity, computational modeling of mammalian gene promoter has been one of the most difficult problems in computational genomics. In the advent of deep sequencing technology, more comprehensive and higher resolution diverse large-scale genetic and epigenetic data have been fueling the development of new predictive models. In this talk, I will review our current understanding of promoter architecture and transcription regulation, summarize recent advances in computational models and outline new challenges.

Author Index

Al-Turaiki, Isra   1

Badr, Ghada   1
Bandyopadhyay, Sanghamitra   237
Bergeron, Anne   201
Bulteau, Laurent   13
Burleigh, J. Gordon   102, 263

Catanzaro, Daniele   24
Chen, Xi   36
Chin, Francis Y.L.   315
Choi, Kwok Pui   153, 165
Christinat, Yann   48
Clarke, Robert   36
Comai, Luca   177
Crosby, Ralph W.   60

Damaschke, Peter   72
DasGupta, Bhaskar   84
Deepak, Akshay   87
Dong, Jianrong   87

Eulenstein, Oliver   102, 115, 263

Fernández-Baca, David   87
Filkov, Vladimir   177

Gibas, Cynthia J.   99
Górecki, Paweł   102, 115
Guo, Fei   127

Halldórsson, Bjarni V.   24
Henry, Isabelle   177
Hobbs, Nathan   315
Hoksza, David   189

Jiang, Minghui   13

Karuturi, R. Krishna Murthy   153
Kirkpatrick, Bonnie   139

Labbé, Martine   24
Leung, Henry Chi Ming   315

Li, Juntao   153
Li, Shuai Cheng   127
Li, Si   165
Lokoč, Jakub   189

Mathkour, Hassan   1
Maulik, Ujjwal   237
Missirian, Victor   177
Molokov, Leonid   72
Moret, Bernard M.E.   48

Nakhleh, Luay   213
Novák, Jiří   189

Ouangraoua, Aïda   201

Park, Hyun Jung   213

Re, Matteo   225
Riggins, Rebecca B.   36

Schimmler, Manfred   275
Schwartz, Russell   250
Sengupta, Debarka   237
Shackney, Stanley   250
Shajahan, Ayesha N.   36
Siebert, Daniel   275
Singh, Ambuj K.   249
Skiena, Steven   299
Skopal, Tomáš   189
Subramanian, Ayshwarya   250
Swenson, Krister M.   201

Valentini, Giorgio   225

Wang, Amy Y.X.   315
Wang, Chen   36
Wang, Lusheng   127
Wehe, André   263
Wienbrandt, Lars   275
Williams, Tiffani L.   60
Wu, Taoyang   165
Wylie, Tim   287

Xu, Dong   328
Xuan, Jianhua   36

Yeasmin, Rukhsana   299
Yiu, Siu Ming   315
Yu, Junjie   315

Zhang, Jingfen   328
Zhang, Louxin   165
Zhang, Michael Q.   330
Zhang, Yong   315
Zhu, Binhai   287

GPSR Compliance

The European Union's (EU) General Product Safety Regulation (GPSR) is a set of rules that requires consumer products to be safe and our obligations to ensure this.

If you have any concerns about our products, you can contact us on ProductSafety@springernature.com

In case Publisher is established outside the EU, the EU authorized representative is:

Springer Nature Customer Service Center GmbH
Europaplatz 3
69115 Heidelberg, Germany

Batch number: 09490872

Printed by Printforce, the Netherlands